EARTH SCIENCE

About the Cover

The photograph was taken from a windy promontory overlooking Peyto Lake and the Mistaya River Valley in the heart of the Canadian Rockies, Banff National Park, Alberta. Even on bright summer days like this one, a chilly wind is virtually constant. Cool, heavy air from nearby Peyto Glacier (not shown in photo) flows down the valley producing an inversion. The "flag tree" in the foreground is mute testimony to the direction and force of the wind.

The Mistaya River Valley and Peyto Lake were formed by a valley glacier that, during the last Ice Age over 7000 years ago, filled the valley with a river of ice reaching the current tree line. Water from Peyto Lake and the Mistaya River flows into the North Saskatchewan River, eventually reaching Hudson Bay.

Peyto Lake's unusual color and opacity are caused by the presence in its waters of a fine, siltlike material called "glacial flour." As Peyto Glacier moves down its valley it grinds the bedrock, producing finely pulverized rock particles that flow into the lake and give it its blue-green color. (For more information on glacial activity read Chapter 6.)

ROBERT J. FOSTER
SAN JOSE STATE UNIVERSITY

EARTH SCIENCE

The Benjamin/Cummings Publishing Company, Inc.
Menlo Park, California • Reading, Massachusetts
London • Amsterdam • Don Mills, Ontario • Sydney

This book was produced by *Ex Libris □Julie Kranhold*
Sponsoring editors *Larry Wilson/Philip Hagopian*
Designers *Dare Porter/Julie Kranhold*
Line illustrations *Ayxa Art*
Color plates and block diagrams *Dennis Tasa, Tasa Graphic Arts, Inc., Minneapolis*
Cover design *Robin Gold*
Cover photograph *James Behnke*

Library of Congress Cataloging in Publication Data
Foster, Robert J. (Robert John), 1929 Apr. 19-
   Earth Science.
   Bibliography: p.
   Includes index.
   1. Earth sciences. I. Title.
QD31.2.F67   550   81-21593
ISBN 0-8053-2660-X   AACR2
ABCDEFGHIJ-DO-898765432

The Benjamin/Cummings Publishing Company, Inc.
2727 Sand Hill Road
Menlo Park, California 94025

PREFACE

Purpose and Focus

It's the only earth we have, so we had better learn to take care of it. This sentence sums up the urgency for an earth science course. It is the reason more and more students are taking earth science courses to fulfill their science requirement. To understand the earth completely, we study its internal makeup and its surface features, its oceans and its atmosphere, its planetary system and its place in the universe. These aspects of the earth are interconnected, so to learn how the earth works, we focus on broader issues than do separate courses in geology, oceanography, meteorology, and astronomy.

Earth Science is classical in coverage but modern in content. It comprehensively presents the essential facts and details upon which a sound and accurate earth science is based. Although chapters and parts are independent of each other, plate tectonics is an important theme throughout the chapters on geology and oceanography and much of the astronomy (see Plate 6 in Chapter 1). This theme allows a degree of unity among the earth sciences that was not possible until recently. The enormity of time in geology and astronomy provides another degree of continuity. These and several other major ideas are displayed in the extraordinary color plates in Chapter 1, which introduce the student to the relationships among the earth sciences.

Audience

Earth Science was written for liberal arts nonscience students and assumes no science background or mathematical abilities. Simple algebra is used in a few places to clarify rather than obscure relationships. Every effort has been made to make this book appealing and useful to the students.

Organization—Flexible Usage

The overriding concept in the organization of the book has been to make it flexible so that it will meet the needs of students and instructors in a wide variety of courses. The book is divided into four parts—geology, oceanography, meteorology, and astronomy. Each of these parts can stand alone, so they can be used in any sequence. The chapters within each part can also stand alone and be used in other sequences. Thus, this book is adaptable to earth science courses with a wide variety in content and length.

Chapter 1 outlines the scope of the book through a graphic presentation. The chapter focuses on the chief concepts of the earth sciences and presents them to the student in sixteen easy-to-understand color plates. These illustrations paint a broad picture of large-scale processes, which are detailed later in the text. The plates can be especially valuable if chapters are omitted in some course outlines.

Learning Aids

Earth Science offers many useful learning and study aids. By employing these aids you will use the book to its best advantage. The following features are available:

1. *Chapter outlines* to preview the key contents of each chapter.
2. *Bold-faced terms* to highlight all important definitions and concepts within the text.
3. *Graphics* to illustrate the text material (over 500 photos, block diagrams, and line drawings, including 16 color plates and 8 pages of color photographs).
4. *Metric (English) units* used throughout.
5. *Chapter end materials* to review chapter content:
   • Summary of chapters
   • Key terms (listed in the same sequence as they were introduced)
   • Review questions
   • Suggested readings (featuring recent short articles in popular magazines)
6. *Glossary* for reference to all terms and definitions.
7. *Appendixes* for reference to pertinent background material:
   • Unit systems
   • Mineral identification
   • Topographic maps
   • Relative humidity and dew point tables
   • Star maps

Acknowledgments

Many individuals have reviewed, commented on, and participated in the development of this book. Each part was critiqued by experts in that field, and, while not all criticisms were taken, the reviewing process was indispensable to accuracy, coverage, and consistency. I am grateful to these colleagues, who indeed care about the quality of contemporary earth science teaching: Lindgren Lin Chyi, The University of Akron; Kenith Exum, Pensacola Junior College, Florida; Marvin L. Ivey, St. Petersburg Junior College, Florida; Eugene Jaworski, Eastern Michigan University; Paul Kirst, Miami Dade Community College, South Campus; David Marczely, Southern Connecticut State University; Charles Stuart, University of Texas, El Paso; Lynn Thompson, Ricks College, Idaho; Tonie A. Toney, Miami Dade Community College, South Campus; Louis Unfer, Southeast Missouri State University; and Ivan Watkins, St. Cloud State University, Minnesota.

The production of a text is complex and involves many creative talents. Julie Kranhold coordinated the production team and organized and directed the design, editing, art, and scheduling. William Waller edited the manuscript. Dennis Tasa illustrated the three-dimensional graphics and created the special color plates in Chapter 1. Ayxa Art developed the two-dimensional line art. Joan Foster created the photo program for the book, including the dramatic chapter openers, and administered the myriad of details of photo collection. She also developed Chapter 1 and most of the art program. Dare Porter designed the book with an open and inviting format. As with most complex endeavors, the production of this book encountered numerous obstacles. I am grateful to this team who not only faced these obstacles, but often provided solutions that were original and creative and always in the spirit of achieving the highest quality.

Larry Wilson was a source of advice, help, and encouragement from the very beginning of this project. The organization and many of the features of this book were developed in discussions with him. Chapter 1 owes much to his vision and encouragement.

Besides her enormous work on the illustrations, Joan Foster supported and influenced the writing of this book in ways that can never be enumerated fully, including typing, editing, proofing, and indexing. She was a critic and companion from its inception to its publication, and this book would not have been completed without her dedication to it.

Robert J. Foster

BRIEF CONTENTS

DETAILED CONTENTS

EARTH SCIENCE

CHAPTER 1

The earth viewed from the moon. (Photo from NASA.)

EARTH SCIENCES: AN ILLUSTRATED OVERVIEW

The main purpose of this chapter is to show how the many aspects of the earth are all interrelated. Graphics rather than text are used to achieve this goal.

Cosmic events created the earth and its life billions of years ago, and our continued existence depends on events in distant parts of the universe. To understand the earth, we must not only study the earth's rocks, oceans, and atmosphere, but also such diverse subjects as magnetic fields, other planets, and stars and their radiation. The traditional divisions of earth science are *geology,* the study of the solid rock earth; *oceanography,* the study of the ocean basins, their water, and its movement; *meteorology,* the study of the atmosphere; and *astronomy,* the study of the universe. The relationships among the earth sciences are the subject of this introduction.

We examine the physical relationships between the earth and the solar system. We then study the time scale of events that led to the formation of the earth. With this basic perspective on distances and spans of time, we can focus on the physical processes of the earth.

In the next section we look at the major geologic processes. The earth is a dynamic planet; and so are its processes—ranging from formation of our magnetic field, the causes of earthquakes and volcanoes, movement of ocean floors and continents, origin of rocks, erosion, and the development of life.

Next, we study the oceans and the atmosphere. Here we learn how the earth's motions relate to oceanic and atmospheric movements. These currents, in turn, produce our major weather systems. Even the celestial bodies influence long-range climatic conditions on earth.

Finally, we examine what we know of the beginnings of the universe and what we postulate for its future. We study the big bang theory of the formation of the universe, and then focus on the processes leading to the birth and evolution of stars and galaxies.

We are all children of the stars. The earth and all its living beings were formed from elements released from the deaths of ancient stars. The earth is not alone in space; its existence and destiny depend on processes in the solar system and beyond.

OUR PLACE IN THE UNIVERSE

We are all familiar with the concepts of *distance* and *time.* Yet when we use these terms to describe events of the universe, we find it impossible to conceive of the distances and time spans required for even the most insignificant of these events to take place—for example, the formation of a small planet. Plate 1 illustrates the immensity of the universe relative to our own solar system. Plate 2 provides a time scale for the main sequence of events that led to the formation of the earth. The universe—its birth, its evolution, its future, and our place in it—continues to be a source of mystery and fascination.

THE EARTH AND ITS DYNAMIC PROCESSES

Our latest scientific thinking suggests that the earth was formed nearly 5000 million (5 billion) years ago from the condensation of a large gaseous cloud that also formed the sun and the rest of our solar system. Unlike our neighbor moon, which is cold and dead, the earth is still in the process of cooling and changing, inside and on the surface. Plates 3–9 describe the earth's structure and the dynamic processes that continue to build, shape, deform, and rebuild it. We discover:

- The size and composition of the earth (Plate 3).
- How the earth's magnetic field protects us from lethal radiation (Plate 4).
- How the earth's processes relate to one another in a unified manner (Plate 5).
- How the earth's crust is composed of large moving plates that cause earthquakes and volcanoes (Plate 6).
- How the three major types of rock (igneous, sedimentary, and metamorphic) are formed (Plate 7).
- How rocks are eroded and landscapes evolve (Plate 8).
- How life has evolved through geologic time (Plate 9).

THE OCEANS AND THE ATMOSPHERE

As we saw in Plates 3–9, the earth's crust is alive and changing even though we hardly notice most of these long-term processes. Much more noticeable and significant to us are the changing seasons, which are controlled by the earth's movement around the sun, as depicted in Plate 10. As the earth moves about the sun, it also spins on its axis, causing large-scale swirling currents in both our oceans and our atmosphere, as illustrated in Plates 11 and 12. These currents, combined with differences in temperature and density within the oceans and the atmosphere, are the fundamental driving forces that produce our weather systems. Plate 13 describes typical weather systems. Finally, even our place in the universe has some influence over the long-term climatic conditions on earth, as illustrated in Plate 14. All of these dynamic processes are interrelated, occur simultaneously, and are fundamental to an understanding of the earth.

THE UNIVERSE: ITS PAST AND FUTURE

This final section of plates brings us full cycle. Having developed in Plate 1 the notion of where we stand in the scale of cosmic events of the universe, we return in this section to another view of the big bang and the processes leading to the birth and evolution of stars, galaxies, dust clouds (nebulae), and novae. Plate 15 describes the essential features of these cosmic processes, and Plate 16 leaps ahead to the future, depicting the eventual old age and death of our sun. Fortunately, time in this case is on our side as such an event will not occur for another 5000 million (5 billion) years!

Plate 1
THE EARTH'S PLACE IN THE UNIVERSE

Each of the parts of this figure is drawn to a different scale because of the immense distances. A light year is the distance a light or radio wave can travel in one year. It is about 10 million, million kilometers (6 million, million miles). A light year also represents time. The image we receive on earth today may be of an object as it actually was millions of years ago—the time the image took to travel the immense distance to our eyes.

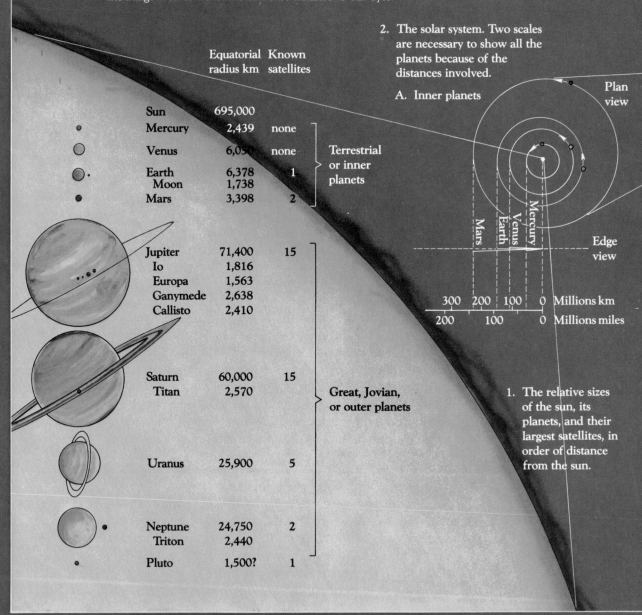

2. The solar system. Two scales are necessary to show all the planets because of the distances involved.

A. Inner planets

Plan view

Edge view

300 200 100 0 Millions km
200 100 0 Millions miles

Mars Venus Mercury Earth

	Equatorial radius km	Known satellites	
Sun	695,000		
Mercury	2,439	none	Terrestrial or inner planets
Venus	6,050	none	
Earth	6,378	1	
Moon	1,738		
Mars	3,398	2	
Jupiter	71,400	15	
Io	1,816		
Europa	1,563		
Ganymede	2,638		
Callisto	2,410		
Saturn	60,000	15	Great, Jovian, or outer planets
Titan	2,570		
Uranus	25,900	5	
Neptune	24,750	2	
Triton	2,440		
Pluto	1,500?	1	

1. The relative sizes of the sun, its planets, and their largest satellites, in order of distance from the sun.

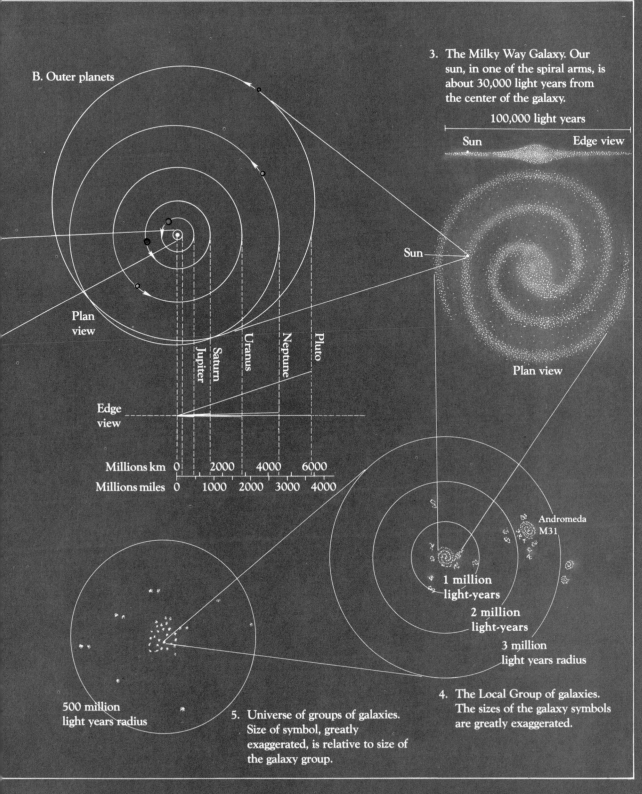

B. Outer planets

Plan view

Plan view

Edge view

Jupiter
Saturn
Uranus
Neptune
Pluto

Millions km 0   2000   4000   6000
Millions miles 0   1000   2000   3000   4000

3. The Milky Way Galaxy. Our sun, in one of the spiral arms, is about 30,000 light years from the center of the galaxy.

100,000 light years

Sun   Edge view

Sun

Plan view

Andromeda M31

1 million light-years

2 million light-years

3 million light years radius

500 million light years radius

4. The Local Group of galaxies. The sizes of the galaxy symbols are greatly exaggerated.

5. Universe of groups of galaxies. Size of symbol, greatly exaggerated, is relative to size of the galaxy group.

5

Plate 2
THE EARTH'S PLACE IN TIME

In order to show time from the formation of the universe to the present day in a single drawing, each column has a very different scale, and the relationship of each to the next is shown.

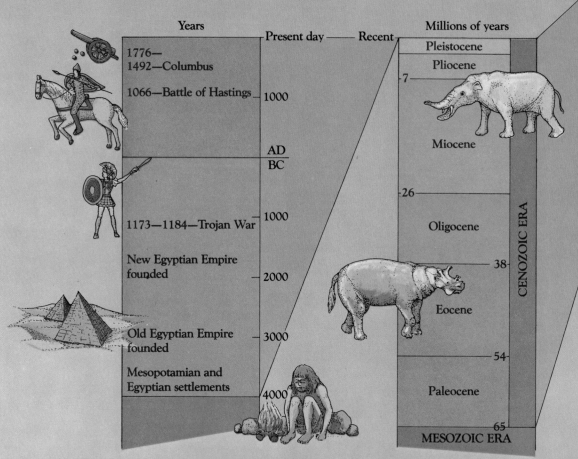

The concept of time is an important element of the study of the earth and its systems. We study many important phenomena, such as weather systems, in terms of hours or days—terms to which we relate easily. Others we study in terms of seconds or fractions of seconds, harder to work with but acceptable to us because we do, after all, experience such times. Others, such as seasons and climates, are in terms of years well within our comprehension. But from the perspective of our own lifetimes it is much harder to perceive time from the standard of recorded human history. The geologic time scale—the history of the earth itself—is even harder to comprehend, yet we still need to conceive the vastness of time back to the formation of the universe. Remember that when we speak of an object 100 million light years distant from us, the radiation that we receive from that object left it 100 million years ago and has been traveling at the speed of light toward us ever since. Our image of that object now is as it was 100,000,000 years ago.

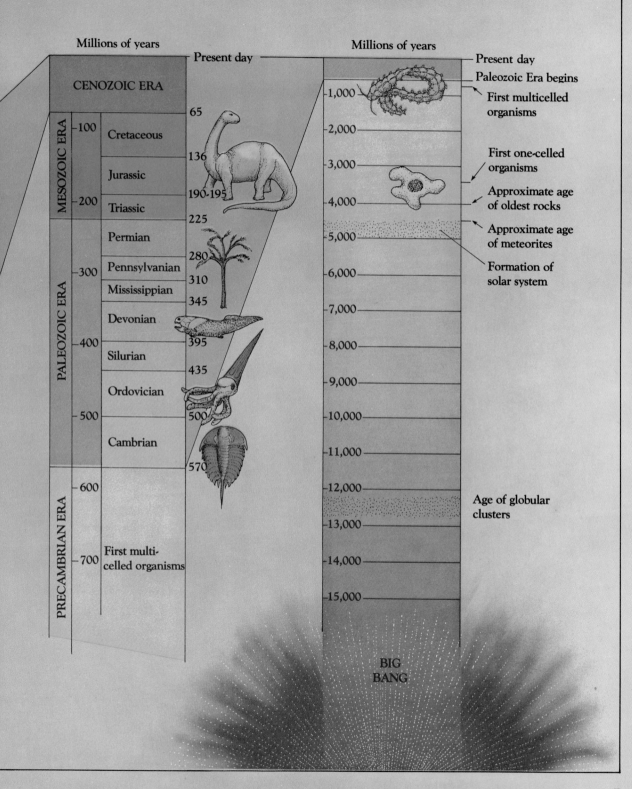

Millions of years

Present day

CENOZOIC ERA

MESOZOIC ERA

— 100   Cretaceous   65

— 200   Jurassic   136

         Triassic   190-195
                            225

PALEOZOIC ERA

         Permian   225

— 300   Pennsylvanian   280

         Mississippian   310
                            345

         Devonian

— 400   Silurian   395

         Ordovician   435

— 500   Cambrian   500

                            570

PRECAMBRIAN ERA

— 600

— 700   First multi-
         celled organisms

Millions of years

Present day
Paleozoic Era begins
First multicelled
organisms

— 1,000

— 2,000

— 3,000   First one-celled
                 organisms

— 4,000   Approximate age
                 of oldest rocks

— 5,000   Approximate age
                 of meteorites

— 6,000   Formation of
                 solar system

— 7,000

— 8,000

— 9,000

— 10,000

— 11,000

— 12,000

— 13,000   Age of globular
                 clusters

— 14,000

— 15,000

BIG
BANG

Plate 3
THE EARTH—A BRIEF DESCRIPTION

These drawings give a brief summary of the size, shape, and layered structure of the earth.

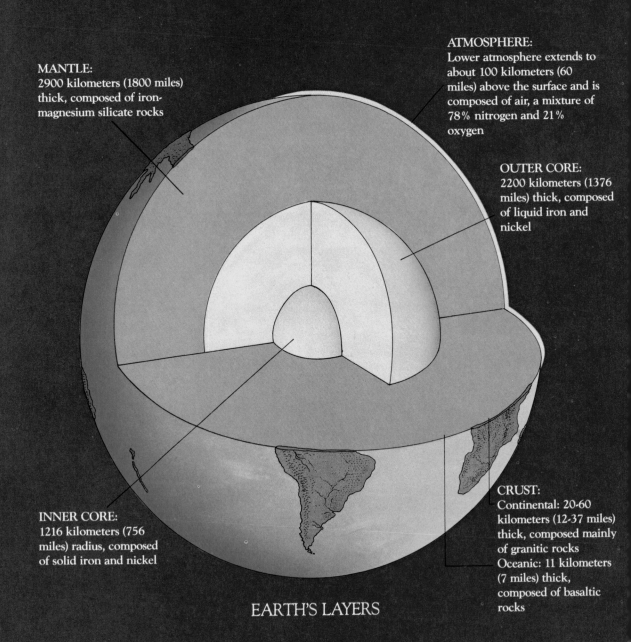

MANTLE:
2900 kilometers (1800 miles) thick, composed of iron-magnesium silicate rocks

ATMOSPHERE:
Lower atmosphere extends to about 100 kilometers (60 miles) above the surface and is composed of air, a mixture of 78% nitrogen and 21% oxygen

OUTER CORE:
2200 kilometers (1376 miles) thick, composed of liquid iron and nickel

INNER CORE:
1216 kilometers (756 miles) radius, composed of solid iron and nickel

CRUST:
Continental: 20-60 kilometers (12-37 miles) thick, composed mainly of granitic rocks
Oceanic: 11 kilometers (7 miles) thick, composed of basaltic rocks

EARTH'S LAYERS

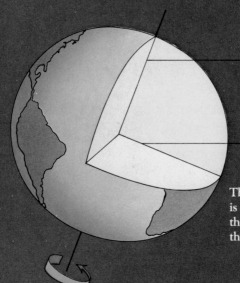

Polar radius:
6357 kilometers (3951 miles)

Equatorial radius:
6378 kilometers (3964 miles)

The difference is 21 kilometers (13 miles), so the earth is very nearly a sphere—not the exaggerated shape in the drawing. The equatorial thickening is caused by the earth's daily spin.

EARTH'S SHAPE AND SIZE

CONTINENTS:
29% area
average elevation
840 meters
(2755 feet)

+840 m

Sea level

OCEANS:
71% area
average elevation
−3802 meters
(−12,465 feet)

Average elevation of earth:
−2442 meters (−8005 feet)

−2242 m

−3802 m

Reduced to the size of a billiard ball, the earth would be much smoother than is a billiard ball.

Deepest:
Marianas
trench
−11,041 meters
(−36,201 feet)

Pacific Ocean

ASIA

Highest:
Mount
Everest
8854 meters
(29,028 feet)

EARTH'S SURFACE ELEVATIONS

Plate 4
THE EARTH'S MAGNETIC FIELD

The solar wind (particles emitted by the sun) interacts with the earth's magnetic field, distorting its shape and forming the Van Allen radiation belts. The magnetic field is compressed on the sunlit side and extends much farther on the dark side of the earth. Low-energy particles are deflected away from the earth by the magnetic field. Higher-energy particles in the solar wind and from cosmic rays (similar particles from distant stars) penetrate the magnetic field but are trapped in the two doughnut-shaped Van Allen radiation belts.

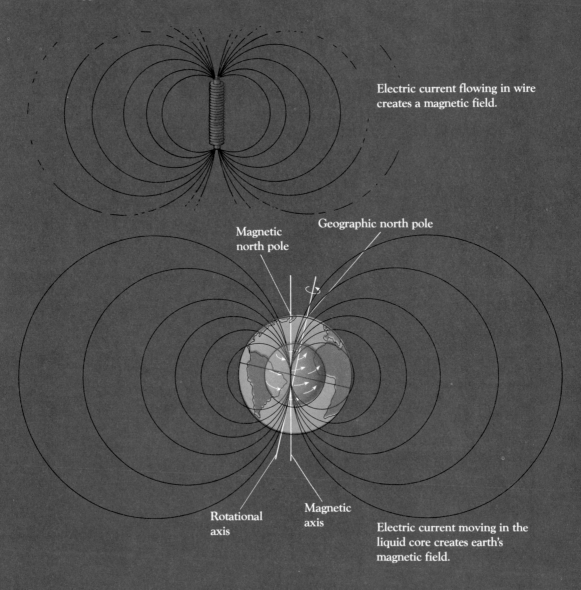

Electric current flowing in wire creates a magnetic field.

Magnetic north pole

Geographic north pole

Rotational axis

Magnetic axis

Electric current moving in the liquid core creates earth's magnetic field.

The high-energy particles in the radiation belts move back and forth between the north and south poles. At times these particles move into the upper atmosphere where they interact with the air molecules, causing the aurora, or northern and southern lights.

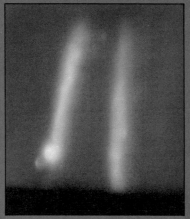

Solar wind

Van Allen
radiation belts

Interactions between solar wind (charged particles from the sun) and the earth's magnetic field cause belts of high-energy particles called radiation belts.

Plate 5
THE RELATIONSHIP OF THE EARTH'S PROCESSES

The earth's surface is the site of constant conflict. Its internal processes build up the surface, and erosion wears it down.

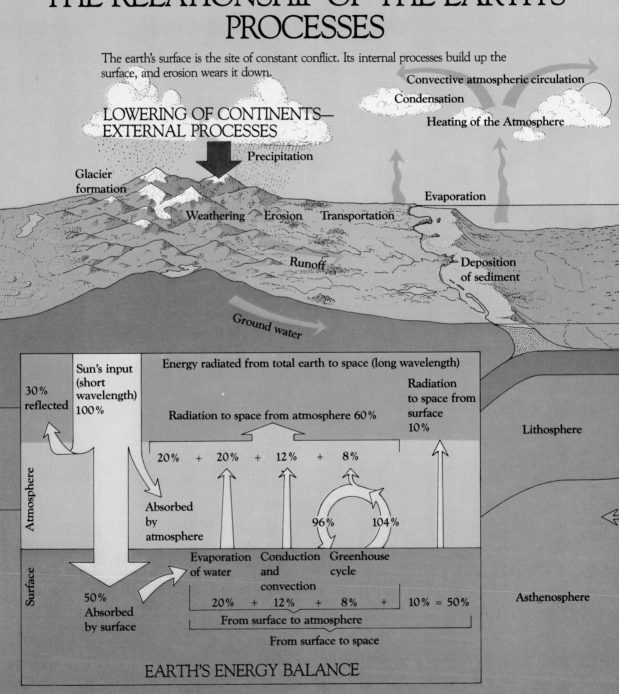

Convective atmospheric circulation

Condensation

Heating of the Atmosphere

LOWERING OF CONTINENTS—
EXTERNAL PROCESSES

Precipitation

Glacier formation

Evaporation

Weathering   Erosion   Transportation

Runoff

Deposition of sediment

Ground water

Energy radiated from total earth to space (long wavelength)

Sun's input (short wavelength) 100%

30% reflected

Radiation to space from atmosphere 60%

Radiation to space from surface 10%

Lithosphere

20% + 20% + 12% + 8%

Absorbed by atmosphere

96%   104%

Atmosphere

Evaporation of water   Conduction and convection   Greenhouse cycle

Asthenosphere

Surface

50% Absorbed by surface

20% + 12% + 8% + 10% = 50%

From surface to atmosphere

From surface to space

EARTH'S ENERGY BALANCE

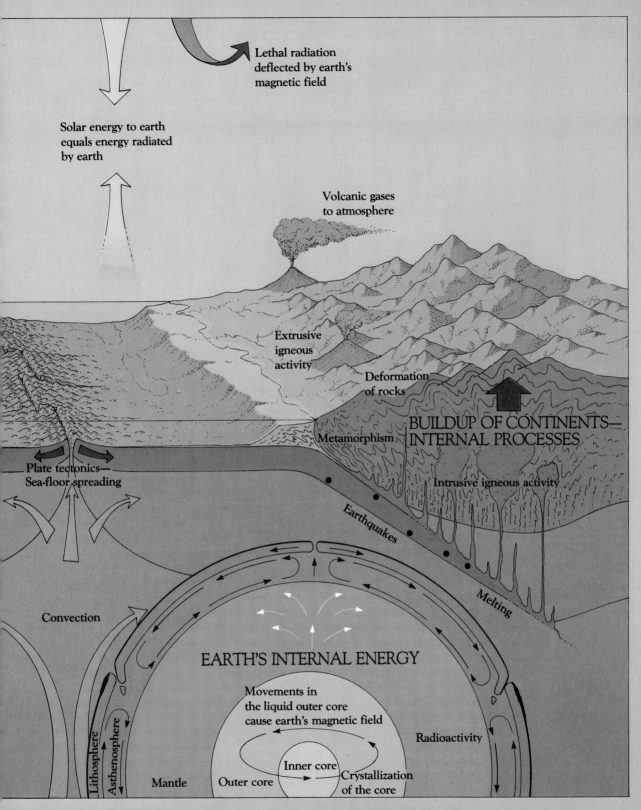

Lethal radiation
deflected by earth's
magnetic field

Solar energy to earth
equals energy radiated
by earth

Volcanic gases
to atmosphere

Extrusive
igneous
activity

Deformation
of rocks

BUILDUP OF CONTINENTS—
INTERNAL PROCESSES

Metamorphism

Plate tectonics—
Sea-floor spreading

Intrusive igneous activity

Earthquakes

Melting

Convection

EARTH'S INTERNAL ENERGY

Movements in
the liquid outer core
cause earth's magnetic field

Radioactivity

Lithosphere

Asthenosphere

Inner core

Mantle

Outer core

Crystallization
of the core

13

Plate 6
PLATE TECTONICS

The earth's surface is a mosaic of moving plates that are formed at mid-ocean ridges and consumed at volcanic arcs. Their movements and collision cause earthquakes, rock deformation, and volcanoes.

The earth's surface is divided into thin plates that move on the surface.

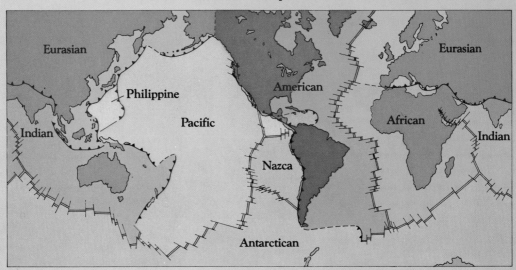

Ocean floors are formed at mid-ocean ridges by partial melting of the mantle. The newly created ocean crust moves away from the ridges. The earth's internal energy is the cause of plate tectonics.

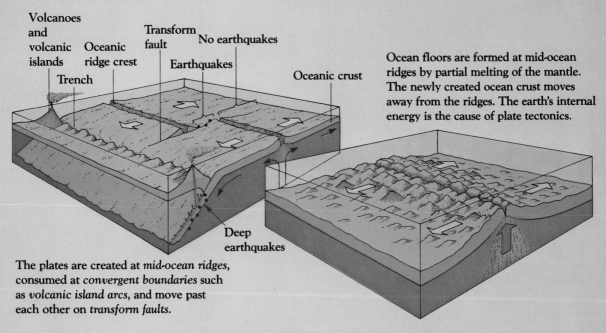

The plates are created at *mid-ocean ridges*, consumed at *convergent boundaries* such as *volcanic island arcs*, and move past each other on *transform faults*.

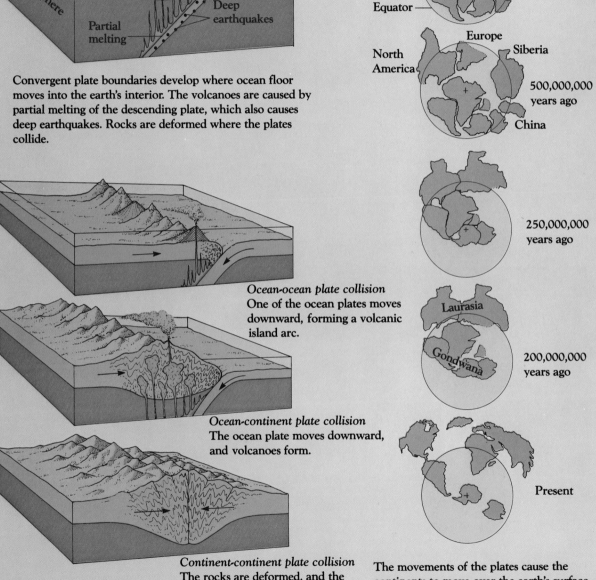

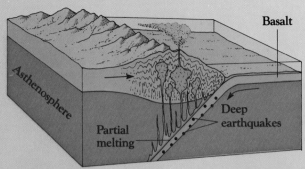

Basalt

Asthenosphere

Partial melting

Deep earthquakes

Convergent plate boundaries develop where ocean floor moves into the earth's interior. The volcanoes are caused by partial melting of the descending plate, which also causes deep earthquakes. Rocks are deformed where the plates collide.

South pole

Equator

600,000,000 years ago

North America

Europe

Siberia

500,000,000 years ago

China

Ocean-ocean plate collision
One of the ocean plates moves downward, forming a volcanic island arc.

250,000,000 years ago

Laurasia

Gondwana

200,000,000 years ago

Ocean-continent plate collision
The ocean plate moves downward, and volcanoes form.

Continent-continent plate collision
The rocks are deformed, and the continents are thickened.

Plate collisions cause rocks to be deformed.

Present

The movements of the plates cause the continents to move over the earth's surface. These maps show the movements of the continents over the last 600,000,000 years.

Plate 7
THE ROCK CYCLE

The rock cycle illustrates the interaction between the earth's internal processes and the atmosphere.

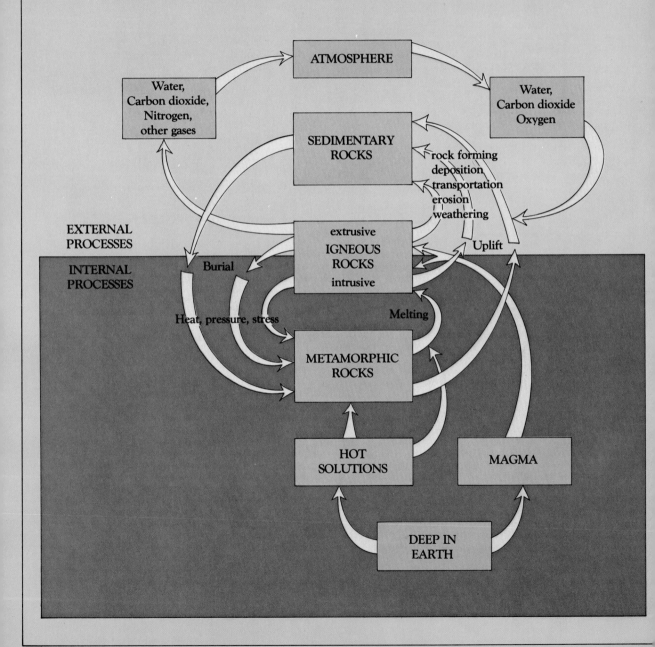

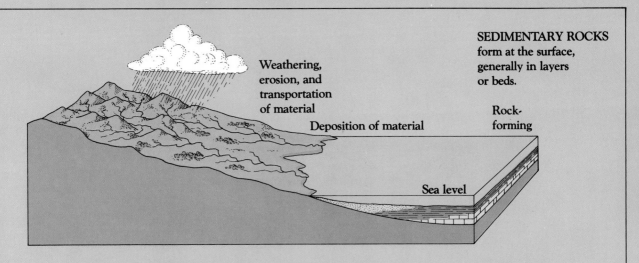

SEDIMENTARY ROCKS form at the surface, generally in layers or beds.

Weathering, erosion, and transportation of material

Deposition of material

Rock-forming

Sea level

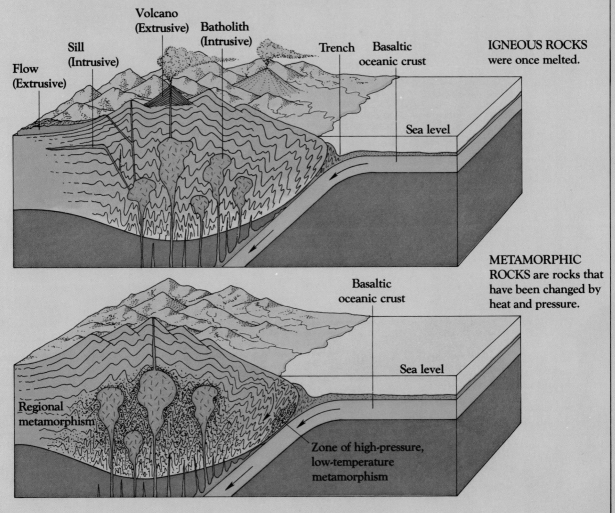

Flow (Extrusive)

Sill (Intrusive)

Volcano (Extrusive)

Batholith (Intrusive)

Trench

Basaltic oceanic crust

IGNEOUS ROCKS were once melted.

Sea level

METAMORPHIC ROCKS are rocks that have been changed by heat and pressure.

Basaltic oceanic crust

Sea level

Regional metamorphism

Zone of high-pressure, low-temperature metamorphism

Plate 8
THE HYDROLOGIC CYCLE—EROSION

The hydrologic cycle is the movement of water on the surface and in the atmosphere.
Evaporation and condensation are very important processes in the hydrologic cycle
and in the movement of heat from the sun. The hydrologic cycle is, therefore, very
important in weather and in erosion.

Precipitation

Snow

Rain

Runoff

EROSION BY RUNNING WATER

Youth

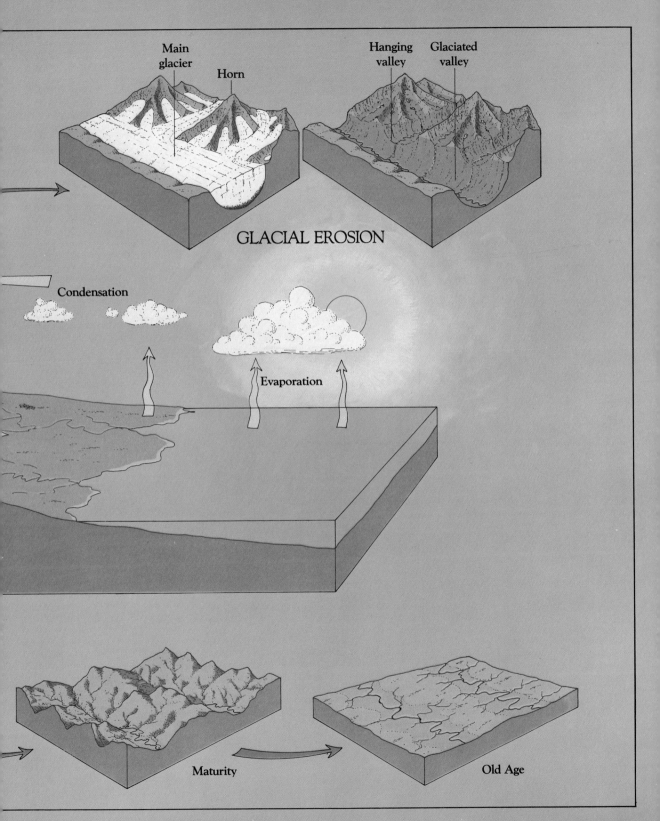

Main glacier

Horn

Hanging valley

Glaciated valley

GLACIAL EROSION

Condensation

Evaporation

Maturity

Old Age

Plate 9
THE EVOLUTION OF LIFE ON EARTH

Here we present the geologic time scale and the development of life. Although the oldest organisms yet found are about 3500 million (3.5 billion) years old, abundant life began only about 600 million years ago.

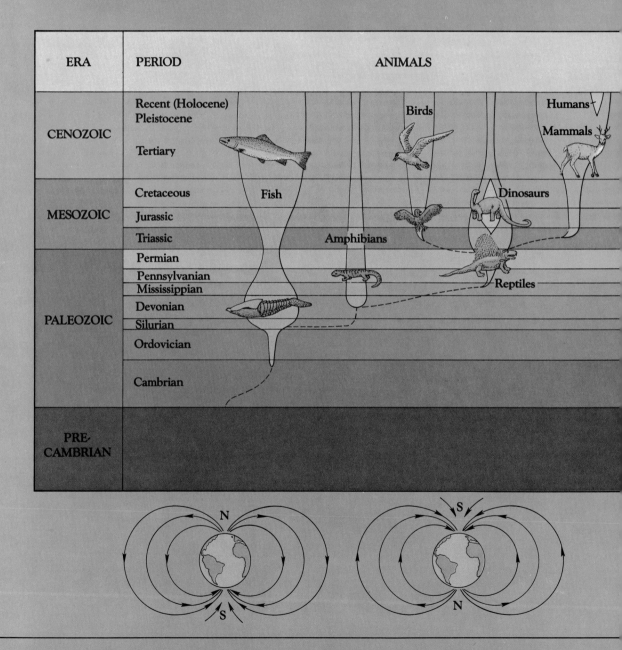

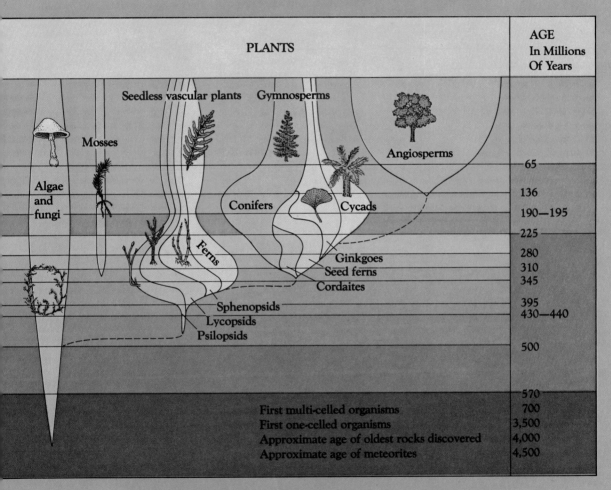

PLANTS	AGE In Millions Of Years
	65
	136
	190—195
	225
	280
	310
	345
	395
	430—440
	500
	570
First multi-celled organisms	700
First one-celled organisms	3,500
Approximate age of oldest rocks discovered	4,000
Approximate age of meteorites	4,500

Algae and fungi

Mosses

Seedless vascular plants

Gymnosperms

Angiosperms

Conifers

Cycads

Ginkgoes
Seed ferns
Cordaites

Ferns

Sphenopsids
Lycopsids
Psilopsids

The stars may also be a factor in the evolution of life. Cosmic rays are a form of radiation and are lethal in large amounts. Smaller doses may cause mutations or cancer. Thus cosmic rays are similar to radioactivity.

The earth's magnetic field changes direction from time to time, probably because the electrical currents in the core change direction. At the time of reversal, the earth's magnetic field passes through zero. At such times, the Van Allen radiation belts are destroyed and cosmic rays reach the earth's surface.

This may account for some of the changes in life in the geologic past, but extinctions, such as the demise of the dinosaurs, do not appear to coincide with magnetic reversals.

Plate 10
THE EARTH'S MOTIONS

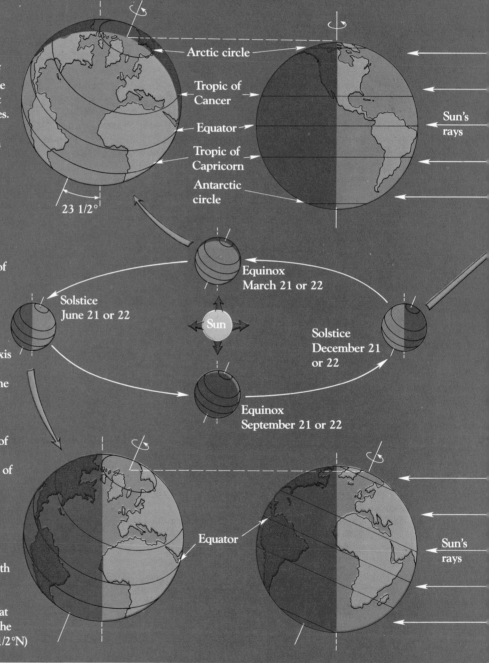

At equinox every place receives 12 hours of sunlight and 12 hours of darkness because the line separating day and night passes through both poles. Sun is directly overhead (no shadows) at noon on the equator.

Arctic circle

Tropic of Cancer

Equator

Tropic of Capricorn

Antarctic circle

Sun's rays

23 1/2°

The earth's yearly orbit of the sun. The earth's spin axis or geographic axis points in the same direction as the earth travels around the sun. The earth's geographic axis is tilted 23 1/2° from the vertical to the plane of the earth's orbit.

Equinox March 21 or 22

Solstice June 21 or 22

Sun

Solstice December 21 or 22

Equinox September 21 or 22

June 21 or 22. First day of summer in the northern hemisphere and first day of winter in the southern hemisphere.

24 hours of sun north of the Arctic circle (66 1/2°N)

24 hours of darkness south of the Antarctic circle (66 1/2°S)

Sun is directly overhead at noon (no shadows) at the Tropic of Cancer (23 1/2°N)

Equator

Sun's rays

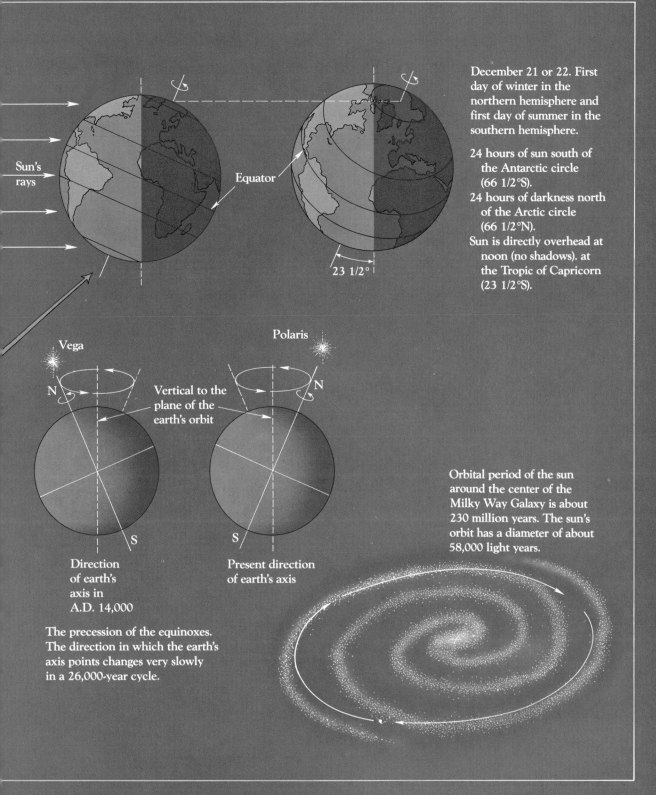

Sun's
rays

Equator

December 21 or 22. First
day of winter in the
northern hemisphere and
first day of summer in the
southern hemisphere.

24 hours of sun south of
the Antarctic circle
(66 1/2°S).
24 hours of darkness north
of the Arctic circle
(66 1/2°N).
Sun is directly overhead at
noon (no shadows). at
the Tropic of Capricorn
(23 1/2°S).

23 1/2°

Vega

Polaris

N

Vertical to the
plane of the
earth's orbit

N

S

S

Direction
of earth's
axis in
A.D. 14,000

Present direction
of earth's axis

Orbital period of the sun
around the center of the
Milky Way Galaxy is about
230 million years. The sun's
orbit has a diameter of about
58,000 light years.

The precession of the equinoxes.
The direction in which the earth's
axis points changes very slowly
in a 26,000-year cycle.

Plate 11
THE CIRCULATION OF
THE OCEANS

The winds cause waves and surface currents; temperature differences cause deep currents.

Warm currents   Cold currents

Surface currents are caused by winds. The positions of the continents also
influence surface currents.

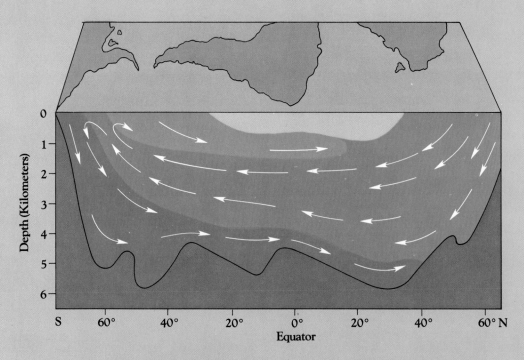

Deep ocean currents are caused by convection. Heavy, cool water near Antarctica sinks, causing this cool water to flow northward.

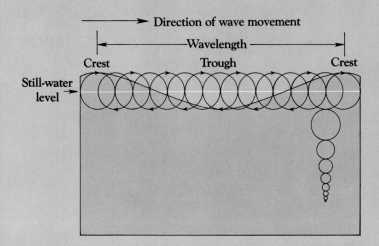

Waves are generated by wind. The water particles move in circles with slight forward motion as the wave form advances. The diameters of the circles die out with depth.

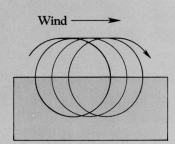

Plate 12
THE CIRCULATION OF
THE ATMOSPHERE

Convection, caused by uneven heating of the earth by the sun, and the earth's rotation cause the circulation of the atmosphere.

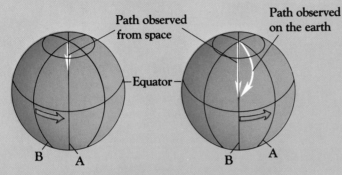

The Coriolis effect is the apparent deflection of a moving object because of the earth's rotation beneath it as it moves. An observer in space would see its path as a straight line, but an observer on the rotating earth would perceive a curved path.

The Coriolis effect is maximum at the poles and zero at the equator. Deflection is to the right in the northern hemisphere and to the left in the southern hemisphere.

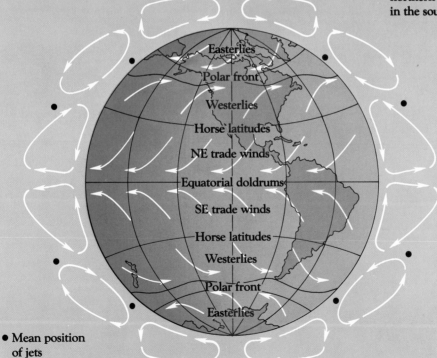

The earth's rotation deflects the convective winds, causing easterly and westerly winds.

● Mean position of jets

Convection is the type of circulation in the atmosphere and the oceans. The fluid is heated at the bottom. The warm fluid expands and rises because it is less dense than the surrounding fluid. Cool fluid from the top flows down to take the place of the rising warm fluid.

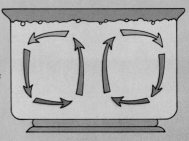

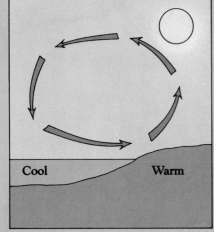

Cool   Warm

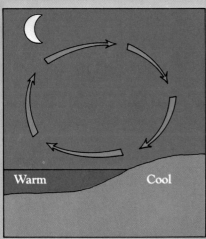

Warm   Cool

An example of convective wind is the on-shore breeze that develops on sunny afternoons at the seashore, reversing at night to an off-shore breeze as the land loses its heat by radiation and becomes cooler than the water.

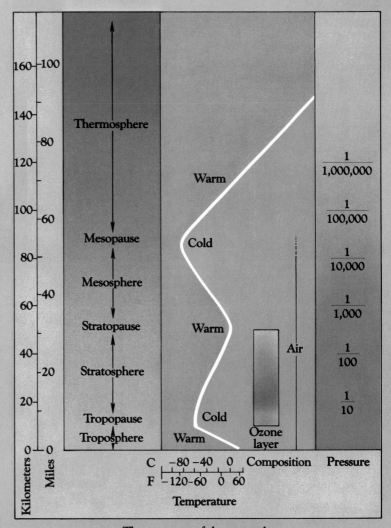

The structure of the atmosphere

Plate 13
WEATHER SYSTEMS

Satellite images give us a look at our weather from a new perspective. The more familiar surface maps are constructed from these images and from data recorded at many stations.

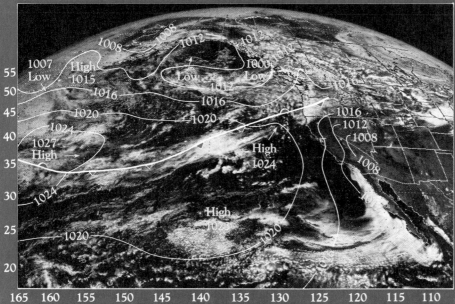

JUNE 9, 1981
A high pressure area off the California coast dominates the weather. Some light showers in the northwestern states and extreme northern part of California. Typical summer pattern of patchy night and morning low clouds along the California coast, becoming sunny or hazy in the afternoon. Southwestern states sunny.

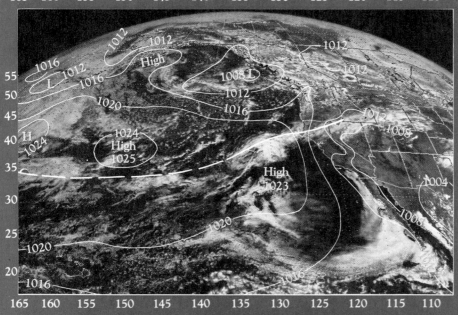

JUNE 10, 1981
Little change from the previous day. Extensive cloudiness and some showers in northern California and inland through Wyoming as a weak system passed through. Low cloudiness in mornings on California coast, mostly cleared in afternoon. Southwest sunny; windy in western Nevada and Arizona.

The simplified maps given here show low- and high-pressure areas defined by isobars (lines between points of equal pressure, labelled in millibars, mb), cold fronts (blue triangles), and warm fronts (red half-circles). Where both symbols are used, the front is stationary where the symbols are on opposite sides of the line, and occluded where they are on the same side. These images, taken at the same time of day on four consecutive days, and the corresponding weather plotting charts from which the maps were derived are courtesy of the Satellite Field Service Station, Redwood City, California.

JUNE 11, 1981
There is a swirl of a weak low-pressure system off central California coast. High clouds spread over northern California. Night and morning clouds continue along southern California coast. Southwestern states sunny, locally windy in Nevada and western Arizona.

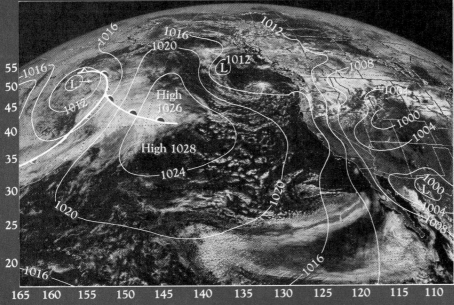

JUNE 12, 1981
Strong high pressure about 1000 miles west of San Francisco combined with low pressure over the western states to cause strong onshore flow and winds in mountains and deserts. Low clouds have cleared along the southern California coast. There are some showers over the northwestern states from the low pressure system.

Plate 14
THE STARS' EFFECT ON THE EARTH'S CLIMATE

Climatic changes may be, in part, of astronomical origin.

As the sun orbits the center of the Milky Way galaxy, it may pass through dust clouds or nebulae. These clouds may reduce the amount of the sun's radiation that reaches the earth and so cool the earth. This could be a cause of glaciation.

The geologic record shows that the earth has experienced glacial° conditions several times in the past. We have only just emerged from the latest glacial age, and earlier periods of glaciation occurred about 275 and 600 million years ago.

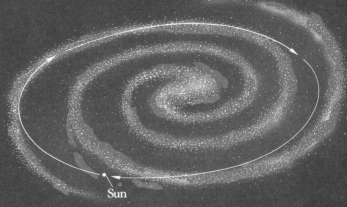

Sun

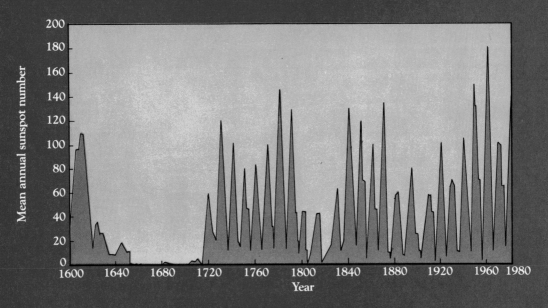

The sunspot cycle, with its 11 and 22 year periods, exerts subtle influences on the earth in such diverse fields as climate and radio communications. We do not yet understand the process, but between the years 1640 and 1715 there were very few sunspots, and this was also the period called the "Little Ice Age" because glaciers advanced during that time.

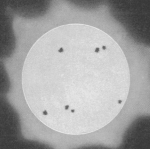

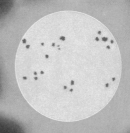

Most climatic changes are caused by movements of the continents caused by plate tectonics. Such movements affect the circulation of atmosphere and ocean.

Variations in the earth's orbit, especially in the tilt of the earth's axis, have been suggested as a cause of glaciation.

Plate 15
THE EARTH'S ANCESTORS

The earth is a very tiny part of the universe, but it could not exist if the universe had not evolved the way it did.

1. The BIG BANG. The universe formed about $16,000 \pm 4000$ million (16 ± 4 billion) years ago in a giant explosion. During this event, most of the matter in the universe was transformed into the light elements hydrogen and helium and only a very small amount of the heavier elements.

2. Stars and galaxies form from hydrogen and helium. A few million years after the big bang, gravitational attraction became stronger than the expansion caused by the big bang, and the matter contracted, forming groups of stars.

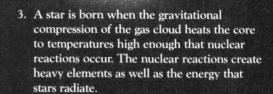

3. A star is born when the gravitational compression of the gas cloud heats the core to temperatures high enough that nuclear reactions occur. The nuclear reactions create heavy elements as well as the energy that stars radiate.

6. The solar system, including the earth, formed as part of the formation of our sun. The elements that form the earth were made in the core of an earlier star. The elements of which we and everything else on the earth are made came from earlier stars that exploded as novae and supernovae. We are indeed children of the stars.

5. Nova and supernova explosions create the dust clouds or *nebulae* from which new stars form. The nebula contains not only hydrogen and helium but the heavy elements created in the former star. Eventually gravitational attraction will cause the nebula to condense, forming a new star.

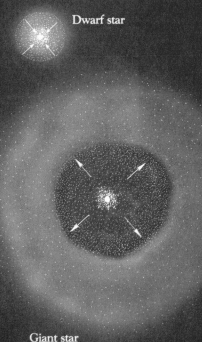

Dwarf star

Giant star

4. Stars die when appreciable amounts of their original cores have been used in nuclear reactions.
Small stars cool, contract, and become *dwarf stars*. Large stars expand and become *giant stars* because the nuclear processes in their cores change. After the giant stage, the stars become unstable. The smaller ones become dwarfs, but a larger one explodes as a *nova* or *supernova*.

Plate 16
THE STARS AND OUR FUTURE—
A NEW ASTROLOGY

From the dawn of recorded time to the present moment, many people believe that the stars control our destinies. This belief is called *astrology*—and the stars do hold our fates—but their control over us is in a form never dreamed of by the astrologers.

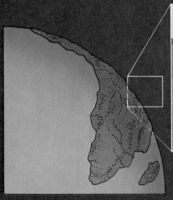

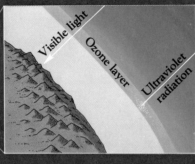

Ultraviolet energy is used to form ozone, which is unstable and reverts to ordinary oxygen, setting the stage for the process to continue.

Cosmic rays (high-energy particles) and ultraviolet are types of stellar radiation lethal to many forms of life, including humans. We are protected from them by the ozone layer in the atmosphere that absorbs much of the sun's ultraviolet rays and by our magnetic field that deflects most of the cosmic rays into the Van Allen belts.

Cosmic rays are deflected by the earth's magnetic field.

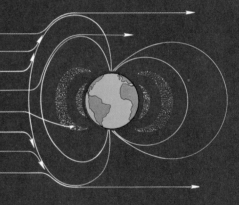

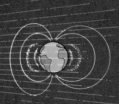

A supernova within about 30 light years of us would be a death star. It would put out so much of this lethal radiation that our feeble defenses would be overrun and most life on earth would be destroyed. Thus we live under the constant threat of a death star. *The same processes that created the elements and the dust cloud from which the earth formed can also destroy us.*

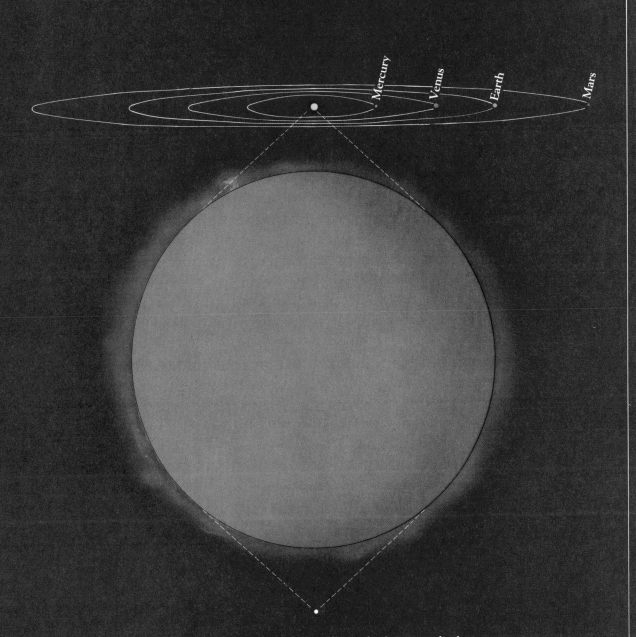

The sun itself, our source of life, will one day become a death star.
In about 5000 million years the sun will have used up its nuclear
fuel and will become a red giant star. It will expand, searing and
destroying the inner planets, its size possibly as great as earth's orbit.
It will expand rapidly, losing much of its mass to form a nebula. The
core will then contract, and the sun will become a white dwarf—
stable, very dense, and only about the size earth used to be.

PART I

GEOLOGY

Geology is the study of the solid rock earth. Our study begins with minerals in Chapter 2. The most common minerals are discussed here, and their identification by physical properties is described. Minerals are the building blocks of rocks, and rocks are the focus of Chapter 3. Here, the processes that form igneous, sedimentary, and metamorphic rocks are described. In a large measure geology is based on the recognition and interpretation of rocks. Mineral resources are also described. These resources occur at places where rock-forming processes have concentrated valuable materials.

Chapters 4 and 5 examine the forces deep within the earth that are responsible for the formation of igneous and metamorphic rocks. Chapter 4 explains the theory of plate tectonics and shows how it can account for such events as continental drift, the creation of new ocean floor, and the rise of mountains. Chapter 5 describes earthquakes and earthquake prediction. It also examines the internal structure of the earth (crust, mantle, and core) and explains how geologists can describe the deep interior, which no one has ever seen.

Chapter 6 returns to the surface of the earth to describe the erosive forces that sculpture the landscape. It covers the process of downslope mass movements, the role of rivers in cutting canyons and valleys and in carrying away the debris, the effects of ground water, the modification of the deserts by wind, and the types and effects of glaciers.

Chapter 7 explains how geologists are able to determine the approximate age of rocks using fossils and radioactive dating. It also provides a brief history of the last billion or so years of our nearly 5-billion-year-old earth. This history is founded on the achievements of geologists in the 19th and 20th centuries who painstakingly gathered the evidence for the immense age of the earth. Deciphering the details of the earth's history and processes is an ongoing task.

GEOLOGY: A BRIEF OVERVIEW

Observation and Interpretation

MINERALS

Identification of Minerals

COLOR

STREAK

LUSTER

HARDNESS

CRYSTAL FORM

CLEAVAGE

FRACTURE

SPECIFIC GRAVITY

Rock-Forming Silicate Minerals

Other Common Minerals

CHAPTER 2

GEOLOGY AND MINERALS

Oblique aerial photo of deformed rocks in southern British Columbia. (Photo from Geological Survey of Canada, G.S.C. #180345.)

GEOLOGY: A BRIEF OVERVIEW

Geology means science of the earth, but it is generally limited to the study of the solid rock earth. The other aspects of the earth are covered in the other parts of this book. The objective of geologists is to understand the processes that go on in and on our earth and to decipher the earth's history. Geology has had a profound influence on our thinking and continues to influence our daily lives. We will begin by considering a few of the intellectual and economic ideas that have come from the study of the earth.

Perhaps the most wide-reaching concept that has come from geology is the recognition of the immensity of the age of the earth. As geology developed, the accepted age of the earth has become greater and greater. At first it was thought that the hills, valleys, plains, and mountains that are the surface features of the continents were the original surface created when the earth formed. Once it was recognized that erosion sculptured the earth, it was realized that great amounts of time were required for these processes. Then the layers of sedimentary rocks that are formed by deposition of the debris of erosion were studied, and it was realized that even greater lengths of time are necessary for their accumulation. The last step was the use of the discoveries of the physicists to date rocks radioactively. This method gave us the currently accepted age of the earth of about 4600 million years.

The basis of modern geology is that the processes now active on the earth are the same processes that have always operated. This means that if we understand the processes—such as erosion, deposition, and volcanism—currently occurring, we can interpret how ancient rocks formed. This principle is called **uniformitarianism;** it is sometimes stated as "The present is the key to the past." Uniformitarianism and the great age of the earth go together in the sense that if one did not realize that great lengths of time are available for geologic processes, one might not appreciate the importance of slow processes like erosion.

Geology is also important politically and economically, because civilization is based on materials from the earth. Soil and water feed the living creatures of the earth, and our industrial society is based largely on metals and energy. Soils and water may be locally depleted by overuse, but in time they will be renewed. Metal mines and oilfields, on the other hand, cannot be renewed except in times measured in millions to hundreds of millions of years. The search for resources is an important function of geologists.

Intelligent use of the earth requires that geologic factors be taken into account in the design of dams, highways, and other structures—large and small. As population and urbanization increase, more and more such structures are needed, and many of the most desirable sites

Figure 2-1
A landslide in a
suburban part of
Oakland, California.

have been used already. Dams fail, and landslides destroy property
because the geology of the site was not considered (Figure 2–1). Even F 2-1
the more serious threats of volcanic eruption and earthquakes are not
always considered.

Observation and Interpretation

Geology and the other earth sciences differ from physics and chemistry in that observation, and not experimentation, is the main
method used. Geologists are not able to perform experiments on the
earth and so must depend upon observation. The observations are
generally interpreted using uniformitarianism as the cornerstone, but
geologists must also take into account the principles of all of the
other sciences. Some aspects of geology, such as the crystallization of
a volcanic rock, can be studied in the laboratory; but, of course, geologists cannot experiment with the volcano. The length of geologic
time cannot be simulated easily in experiments, and neither can the
high temperatures and pressures deep in the earth. Geologic observa-

Figure 2-2
The sedimentary beds in this part of Wyoming were deformed into an elongate dome. (Photo from U.S. Geological Survey.)

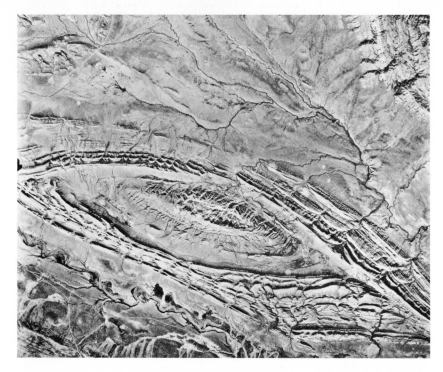

tion began with the naked eye and progressed to the microscope and then the X ray, as these new instruments were developed.

Most geologic work involves the study of an area, and so geologists generally plot their observations on maps. At first, geologists had to survey and make their own maps. Aerial photographs are commonly used as base maps today, making accurate geologic mapping much easier. Many geologic structures are easily seen on aerial photographs (Figure 2-2), and now satellites provide images covering wide areas. F 2-2 Radar images are used in some cases to enable us to detect surface features through thick tree cover (Figure 2-3). Infrared images detect F 2-3 heat and are useful in studying features as diverse as ground water and volcanoes.

The rocks below the earth's surface cannot be studied directly, but knowledge of them is very important in much geologic work, such as exploring for mineral resources. As a result, indirect methods of study, such as measuring the travel times of earthquake waves or waves from explosions, have been developed that reveal information about the deep rocks and their arrangement. We begin our study of geology with minerals, the components of the rocks of the earth.

MINERALS

Elements are the building blocks of chemistry and *are the most fundamental subdivisions of matter that can be made by ordinary chemical methods.* **Atoms** *are the smallest particles of an element that retain the chemi-*

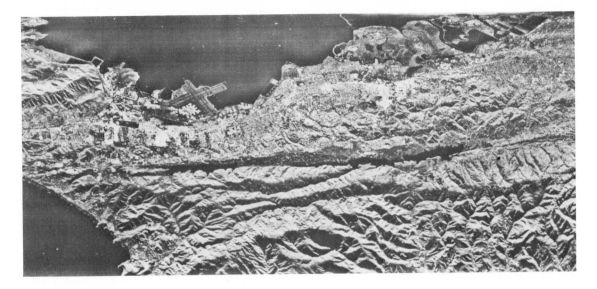

Figure 2-3
A radar image of part of the San Francisco peninsula in California. The San Andreas fault and other faults are clearly outlined. (Photo from U.S. Geological Survey in cooperation with NASA and Westinghouse Electric Co.)

cal properties of the element. In the earth, elements combine to form **minerals,** which are

- *naturally occurring,*
- *crystalline,*
- *inorganic substances, with*
- *definite chemical composition.*

Minerals, in turn, are the main constituents of **rocks,** which can be defined as *naturally occurring, firm material that forms part of the earth's crust.*

Identification of Minerals

Minerals are identified by their *physical properties,* such as color, hardness, and luster. This is fortunate because such properties are easily noted, and so we can learn to identify minerals without learning chemical analysis. In some ways physical properties are better for this purpose than chemical analysis. For example, diamond and graphite are both minerals composed entirely of the element carbon, but one is the hardest gem mineral and the other is soft, greasy feeling, and black. The difference is in the bonding of the carbon atoms (Figure 2-4). Mineralogists today use X rays to determine the internal structure of minerals. The commonly used physical properties are described below.   F 2-4

COLOR The color of a mineral is an obvious property. Generally, the color of an unaltered surface, best seen on a fresh break, is what is used, but in some cases the color of a tarnished surface may be dis-

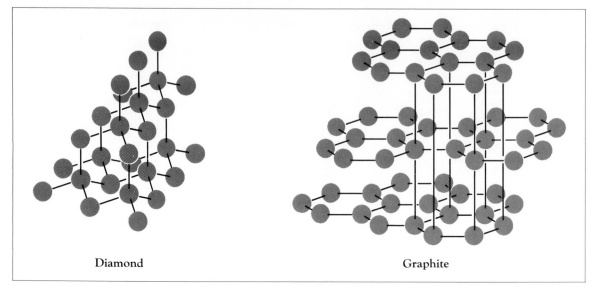

Diamond   Graphite

Figure 2-4
In the mineral diamond, each carbon atom is tightly bonded to four other carbon atoms. In graphite, the carbon atoms in each layer are tightly bonded, but the bonds between the layers are weak.

tinctive. For a few minerals color is diagnostic, but for most it is not; and a few minerals, such as quartz, may be nearly any color, generally because of minor impurities.

STREAK *The color of a powdered mineral* is called its **streak,** because it is best seen by rubbing the mineral on a piece of unglazed porcelain called a streak plate. For many minerals, especially ore minerals, the color of the streak may be quite different from the color of the mineral itself. Not all minerals have a streak. Those that are harder than the streak plate and those whose powder is white are said to have no streak. A few minerals, such as hematite, may have a wide variety of colors, but their streaks are all nearly the same.

LUSTER The **luster** of a mineral is *the way that it reflects light.* For most purposes only two lusters, metallic and nonmetallic, need to be distinguished, although mineral descriptions use many terms, mostly self-explanatory, such as glassy, waxy, greasy, silky, pearly, dull, and earthy. The differences between metallic and nonmetallic are learned most easily by looking at examples, because these terms are not easy to describe in words. A few mineral specimens may have characteristics of both, and they are called submetallic.

HARDNESS The **hardness** of a mineral is *a measure of its resistance to scratching or abrasion.* The hardness scale was devised to go from one to ten in fairly equal steps, but diamond, the hardest mineral, is very much harder than corundum, next to it. Hardness is measured by trying to scratch a fresh, unaltered surface on the unknown mineral with either minerals or other materials of known hardness. If a scratch is made on the unknown, it is softer than the scratcher. The test can be reversed, and the unknown can be used to scratch the materials of known hardness. The objective is to bracket the hardness of the unknown as closely as possible. After some experience, the hardness can be estimated by the ease with which a scratch can be made with a knife.

There are a number of problems with hardness tests. The shape of the point used to make a scratch and the amount of pressure used are variable. A softer material such as a knife (hardness 5.5) will leave a mark on quartz (hardness 7), just as softer chalk will leave a mark on a blackboard. Such a mark is commonly mistaken for a scratch.

A *standard scale of mineral hardness* is **Mohs hardness scale:**

1. talc   (softest mineral)
2. gypsum
3. calcite
4. fluorite
5. apatite
6. orthoclase
7. quartz
8. topaz
9. corundum
10. diamond   (hardest mineral)

A useful hardness scale referenced to Mohs is:

2.
   —fingernail
3.
   —copper cent (Use a bright, shiny one, or you will only test the hardness of the tarnish.)
4.
5.
   —knife blade or glass plate
6. —file
7. —quartz
8.

CRYSTAL FORM All **crystalline** materials form **crystals** in one of the six **crystal systems** (Figure 2-5). Not all minerals form actual crystals F 2-5 with smooth crystal faces, but they all have the organized internal structure of a crystal. If the crystal faces of a mineral are developed, the shape can aid in identification. To form crystals, a mineral generally has to form in an unrestricted space; this is why they are relatively rare (Figure 2-6). F 2-6

CLEAVAGE The crystalline internal structure of many minerals is such that the bonding between atoms is weaker in some directions

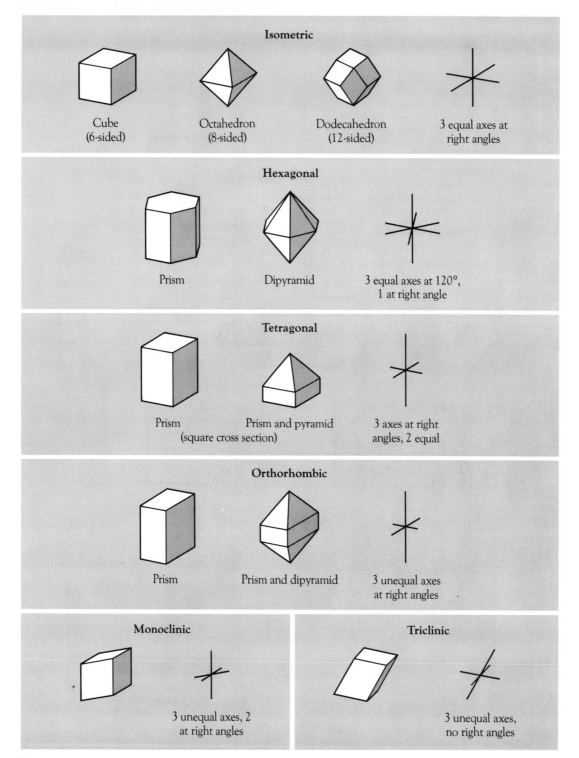

Figure 2-5
The six crystal systems.

A.

B.

C.

Figure 2-6
Crystals. **A.** Fluorite from Cumberland, England. **B.** Corundum from Montana. **C.** Three crystal forms of the mineral pyrite. **D.** Feldspar. (Photos **A** and **C,** from Field Museum of Natural History, Chicago; **B** and **D** from Ward's Natural Science Establishment, Inc., Rochester, N.Y.)

D.

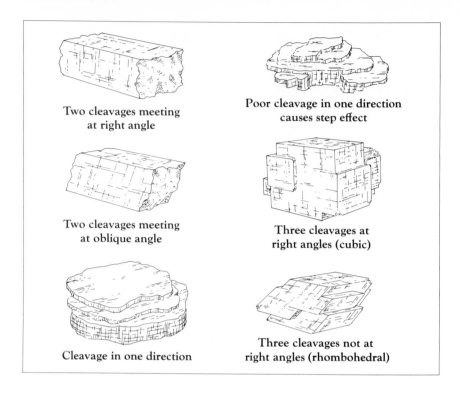

Two cleavages meeting
at right angle

Poor cleavage in one direction
causes step effect

Two cleavages meeting
at oblique angle

Three cleavages at
right angles (cubic)

Cleavage in one direction

Three cleavages not at
right angles (rhombohedral)

than in others. Thus, the mineral is weaker in certain directions than
in others. Such a mineral tends to break in these weak directions, and
these directions are called **cleavages.** Cleavages are recognized because
they are smooth plane surfaces that reflect light. Because cleavage is a
direction of weakness, in general one cleavage will result in two faces
on the mineral, one on each side. A mineral can have several cleavages (Figures 2-7 and 2-8). Depending on how much weaker the cleavage is than the rest of the mineral, the cleavage may be well
developed or very hard to find. Poor cleavages tend to form steps, as
in Figure 2-7. If there is more than one cleavage direction, the angle
between the cleavages is generally important in identification.

F 2-7
F 2-8

FRACTURE *The way a mineral breaks in a noncleavage direction is called its*
fracture. We use a number of terms to describe fractures, such as
hackly, fibrous, and splintery. **Conchoidal** *is the term given the rounded*
concave fracture that forms when glass is broken (see Figure 3-5, p. 63).

F 3-5

SPECIFIC GRAVITY **Specific gravity** *is the ratio of the weight of a mineral*
to the weight of the same volume of water. It is determined by weighing
the mineral in air and in water:

$$\text{Specific gravity} = \frac{\text{weight in air}}{\text{weight in air} - \text{weight in water}}$$

A.

B.

Figure 2-8
A. Mica crystal with one direction of cleavage. **B.** Halite crystal with three directions of cleavage at right angles. (Photos from Ward's Natural Science Establishment.)

Other physical properties of minerals include fluorescence, taste, magnetism, and radioactivity.

The mineral identification key (Appendix B) uses these physical properties to identify common minerals. Color (light or dark) and luster (metallic or nonmetallic) are used first; then hardness to narrow the choices to a small part of the key. In this way we limit the possibilities, and reading the descriptions and measuring the physical properties more closely enable us to identify the unknown.

Rock-Forming Silicate Minerals

The chemical composition of the continents is shown in Table 2-1. T 2-1 More than 99 per cent of the continental rocks are composed of only eight elements. Aluminum and iron are the only common metals among the eight. Copper, zinc, nickel, tin, chrome, and lead—metals that are in common use—do not appear on the list, and so they are among more than 80 other elements that make up less than 1 per cent of the continents. The ways in which the less abundant elements are concentrated are described in Chapter 3.

The eight abundant elements, as would be expected, form the minerals that compose most of the rocks that form the earth's crust. Fortunately for the student, there are only a handful of rock-forming minerals. Table 2-1 suggests why this is so. Oxygen and silicon are the most numerous atoms, and oxygen's large volume shows that it has a large atom. These two elements form a stable structure in which the small silicon atom is surrounded by four oxygen atoms (see Figure 2-9). Because these silicon-oxygen units can join in only a few stable F 2-9 forms (Figure 2-10), the number of rock-forming minerals is limited. F 2-10 The other abundant atoms fit into the spaces between the silicon-oxygen units. The rock-forming minerals and their relative abundances are shown in Figure 2-11. F 2-11

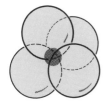

Figure 2-9
The stable structure of the silicate minerals is one silicon atom surrounded by four oxygen atoms.

Figure 2-10
Silicate mineral
structure.

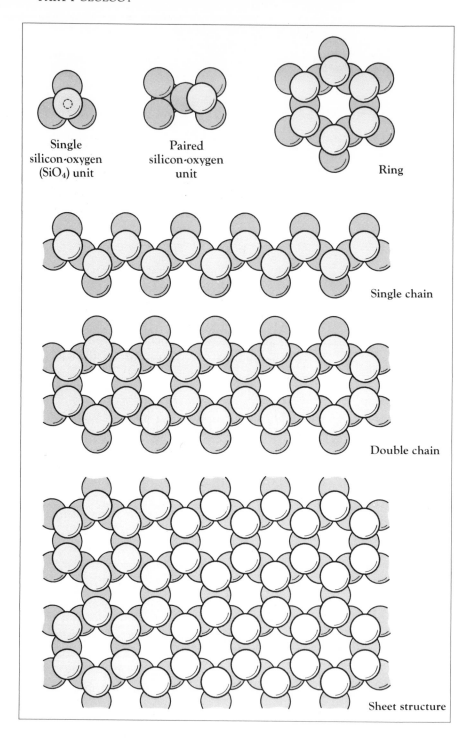

The minerals that form igneous rocks are described here, and we will see later how these primary minerals are modified by weathering to form some of the sedimentary rocks. Then we will consider how metamorphism changes the minerals in both igneous and sedimen-

Table 2-1   COMPOSITION OF THE CONTINENTS

Element	Symbol	Weight Per cent	Atom Per cent	Volume Per cent
Oxygen	O	46.40	62.17	94.05
Silicon	Si	28.15	21.51	.88
Aluminum	Al	8.23	6.54	.48
Iron	Fe	5.63	2.16	.48
Calcium	Ca	4.15	2.22	1.19
Sodium	Na	2.36	2.20	1.11
Magnesium	Mg	2.33	2.05	.32
Potassium	K	2.09	1.15	1.49
Totals		99.34	100.00	100.00

tary rocks. The rock-forming minerals described here are all **silicate minerals;** that is, they contain silicon and oxygen. All but two (olivine and quartz) contain aluminum as well. This leaves only five other elements to follow, and two of these, iron and magnesium, generally occur together.

Feldspars are the most important and most abundant family of minerals. There are two important feldspar minerals. **Plagioclase,** $(Ca,Na)(Al,Si)_4O_8$, is composed of calcium and sodium, as well as silicon, aluminum, and oxygen. It can contain all calcium, all sodium, or any mixture of the two. **Orthoclase,** $KAlSi_3O_8$, contains potassium as well as the common silicon, aluminum, and oxygen. Feldspars are all recognized by their light color, hardness (6), and two cleavages at right angles.

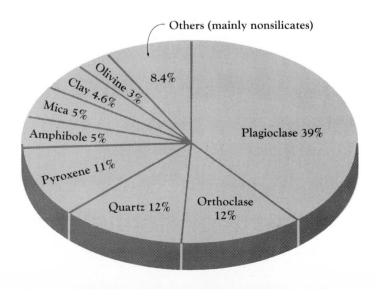

Figure 2-11
Abundance of minerals in the earth's crust. (Data from Ronov and Yaroshevsky, 1969.)

Figure 2-12
Cluster of quartz crystals. (Photo from Field Museum of Natural History, Chicago.)

Quartz, SiO_2, is the next most abundant mineral; it is composed of only silicon and oxygen. It is recognized by its hardness (7) and lack of cleavage. In most rocks quartz is generally clear, white, or light gray, but minor impurities can give it any color (Figure 2-12).

Olivine, $(Mg,Fe)_2SiO_4$, is the simplest of the iron-magnesium, or **ferromagnesian,** rock-forming minerals. It is generally recognized by its olive-green color. The gemstone peridot is a form of olivine.

The **micas** all have one perfect cleavage. There are three common types, which are easily distinguished by their colors. **Biotite,** $K(Mg,Fe)_3(AlSi_3O_{10})(OH)_2$, is black mica. **Muscovite,** $KAl_2(AlSi_3O_{10})(OH)_2$, is white mica. **Chlorite,** $(Mg,Fe,Al)_6(Al,Si)_4O_{10}(OH)_8$, is "green mica."

Augite, a **pyroxene,** $(Ca,Na)(Mg,Fe,Al)(Si,Al)_2O_6$, and **hornblende,** an **amphibole,** $Ca_2Na(Mg,Fe)_4(AlFe)(Al,Si)_8O_{22}(O,OH)_2$, are the common black or dark green minerals. Both have a hardness in the 5–6 range and two cleavages. The cleavage planes in augite are at right angles; in hornblende they are not at right angles, but at approximately 60 and 120 degrees.

Other Common Minerals

We should consider a few other relatively common minerals in addition to the primary igneous rock-forming minerals just discussed. They are mostly composed of the eight common elements. Some of these minerals result from modification of the primary silicates by weathering processes discussed in Chapter 3.

F 2–12

Calcite, $CaCO_3$, is the mineral of which limestone and marble are composed. It has a hardness of 3, is clear or white in color, and has three cleavages not at right angles. It is recognized by its effervescence with dilute hydrochloric acid.

Dolomite, $CaMg(CO_3)_2$, forms by the replacement of some of the calcium in calcite by magnesium. It has a hardness of 3.5 to 4; has three cleavages not at right angles; and is clear, white, pink, or brownish. It will not react with dilute hydrochloric acid unless it has been powdered, as by scratching with a knife; this distinguishes it from calcite.

Kaolinite, $Al_4Si_4O_{10}(OH)_8$, is a member of the complex family of **clay** minerals that result from weathering. It is very soft—hardness 2—is white, and has an earthy luster.

Chalcedony, SiO_2, is a very fine-grained quartz. It has a waxy luster and a hardness of 7; it may be any color.

Halite, NaCl, is rock salt and is distinguished by its salty taste. It has a hardness of 2.5, a glassy luster, and three cleavages at right angles; it is clear or white.

Gypsum, $CaSO_4 \cdot 2H_2O$, has a hardness of 2, glassy to pearly luster, and one good and two poor cleavages; it is clear to white or gray. Halite and gypsum both form from the evaporation of sea water or restricted saline lakes (Figure 2-13). F 2-13

Magnetite, Fe_3O_4, and **hematite,** Fe_2O_3, are iron oxides. Magnetite has a hardness of 6 and a metallic luster. It is black in color and has a black streak, and it is distinguished by being attracted by a magnet. Hematite has a hardness of 5–6.5, although the hardness may seem to be much lower. It usually has an earthy luster and is red to red-brown in color. It is distinguished by its red to red-brown streak (Figure 2-14). F 2-14

Figure 2-13
Some of the many forms of gypsum. (Photos from Field Museum of Natural History, Chicago.)

Figure 2-14
Hematite. **A.** Black, platy crystals of hematite with quartz crystals, Cumbria, England. **B.** Common, earthy hematite from iron mines in Ontario. (Photos from Field Museum of Natural History, Chicago.)

A.

B.

Pyrite, FeS_2, is commonly known as "fool's gold." It is commonly in cubic crystals, and it has a hardness of 6–6.5, a metallic luster, and a distinctive brass-yellow color and black streak (Figure 2-15).

The identification table in Appendix B covers these and some other minerals. About two thousand minerals have been discovered. Chapter 3 discusses the rocks that contain these minerals and some of the processes that form them, such as volcanism and weathering. Only a few minerals have economic value, and Chapter 3 also discusses how the rock-forming processes concentrate the relatively rare elements into valuable resources.

F 2-15

Figure 2-15
Intergrown pyrite crystals from Rio Tinto, Spain. (Photo from Field Museum of Natural History, Chicago.)

SUMMARY

- Geology is the study of the earth, its materials, processes, and history. The earth is about 4600 million years old.

- *Uniformitarianism* is the basis of geology; it is easily remembered as "The present is the key to the past."

- Geology depends much more on interpretation of observation than on experimentation.

- *Elements* are the most fundamental subdivision of matter that can be made by ordinary chemical methods.

- *Atoms* are the smallest particles of an element that retain the chemical properties of the element.

- *Minerals* are naturally occurring, crystalline, inorganic substances with definite chemical composition.

- *Rocks* are naturally occurring, firm material that forms part of the earth's crust.

- Minerals are easily identified by their *physical properties*.

- More than 99 per cent of the earth's crust is made of the eight elements listed in Table 2-1.

- Oxygen, the most abundant element, is also the largest of these eight common elements; it combines with silicon, the next most abundant element, to form the building block of the main rock-forming minerals—one silicon atom surrounded by four oxygen atoms.

- Because these one silicon and four oxygen units can only join together in a few ways, only a few silicate minerals form most of the rocks.

- The main rock-forming minerals are feldspars, quartz, olivine, micas, augite, and hornblende.

KEY TERMS

geology
uniformitarianism
element
atom
mineral
rock
streak
luster
hardness
Mohs hardness scale
crystalline

crystal
crystal systems
cleavage
fracture
conchoidal fracture
specific gravity
silicate mineral
feldspar
plagioclase
orthoclase
quartz

olivine
ferromagnesian
  mineral
mica
biotite
muscovite
chlorite
augite
pyroxene
hornblende
amphibole

calcite
dolomite
kaolinite
clay
chalcedony
halite
gypsum
magnetite
hematite
pyrite

REVIEW QUESTIONS

1. State the principle of uniformitarianism. Explain why this principle provides the key to the past.

2. Explain in your own words why geology depends more on observation than on experimentation.

3. To be considered a mineral, a substance must meet four requirements. List the four requirements and briefly explain what each means.

4. Below is a list of the physical properties that are most often used to identify minerals. Define each property and answer the questions.
   (a) Color.
   (b) Streak. Why is streak commonly a more reliable method of identification than color?
   (c) Luster.
   (d) Hardness.
   (e) Crystal. All minerals have the internal organization of a crystal (the atoms in a mineral are organized in a regular pattern). Why, then, are crystals relatively rare?
   (f) Cleavage. What causes cleavage?
   (g) Fracture.
   (h) Specific gravity. How would you go about determining a mineral's specific gravity?

5. (a) Describe a single silicon–oxygen unit. How many oxygen atoms does it have? How many silicon atoms?
   (b) Define a silicate mineral.

6. (a) Write the names and physical properties of each of the silicates discussed in this chapter.
   (b) Which of the silicates that you have described contains only silicon and oxygen? What is responsible for the various colors that this mineral can be?
   (c) What is a ferromagnesian mineral? Which of the silicates that you have described are ferromagnesian?

7. The following minerals will appear in the discussions of rocks presented in the next chapter. Learn the names of these minerals and answer the questions about them.
   (a) *Calcite.* Write the formula for this mineral. What simple test can you make to identify this mineral? What two rocks are composed of calcite?
   (b) *Dolomite.* What is the major chemical difference between calcite and dolomite? What is an easy way of determining whether you are holding a sample of calcite or of dolomite?
   (c) *Kaolinite* belongs to what family of minerals? What is its luster?
   (d) *Halite.* Write the formula. What is the common name for this mineral? How does this mineral form?
   (e) *Gypsum.* How does gypsum form?
   (f) *Magnetite* and *hematite* are both oxides of what? Which is attracted by a magnet? What other physical properties distinguish these two minerals from each other?
   (g) *Pyrite* is a sulfide of what?

SUGGESTED READINGS

Albritton, C. C. (ed.). *The Fabric of Geology.* Reading, Mass.: Addison-Wesley, 1963, 372 pp.

American Geological Institute. *Geology: Science and Profession.* Washington, D.C.: American Geological Institute, 1965, 28 pp.

Berry, L. G., and Brian Mason. *Mineralogy.* San Francisco: W. H. Freeman, 1959, 612 pp.

Dietrich, R. V., and B. J. Skinner. *Rocks and Rock Minerals.* Somerset, N.J.: John Wiley, 1979, 319 pp.

Ernst, W. G. *Earth Materials.* Englewood Cliffs, N.J.: Prentice-Hall, 1969, 150 pp. (paperback).

Geike, Archibald. *The Founders of Geology.* New York: Dover, 1962, 486 pp. (paperback). Originally published in 1905 by Macmillan.

Gillispie, C. C. *Genesis and Geology.* New York: Harper & Row, 1951, 306 pp. (paperback).

Hurlbut, C. S., Jr., and C. Klein. *Dana's Manual of Mineralogy,* 19th ed. New York: John Wiley, 1977, 532 pp.

Loeffler, B. M., and R. G. Burns. "Shedding Light on the Color of Gems and Minerals," *American Scientist* (November–December 1976), Vol. 64. No. 6, pp. 636-647.

Murray, R. C., and J. C. F. Tedrow. *Forensic Geology: Earth Sciences and Criminal Investigations.* New Brunswick, N.J.: Rutgers University Press, 1975, 217 pp.

Pough, F. H. *A Field Guide to Rocks and Minerals.* Boston: Houghton Mifflin, 1955, 349 pp.

Rapp, George, Jr. *Color of Minerals.* ESCP Pamphlet Series, PS-6. Boston: Houghton Mifflin, 1971, 30 pp.

Roy, C. J. *The Sphere of the Geological Scientist.* Washington, D.C.: American Geological Institute, 1962, 28 pp.

Short, N. M., and others. *Mission to Earth: Landsat Views the World.* NASA SP-360. Washington, D.C.: U.S. Government Printing Office, 1977, 459 pp.

U. S. Geological Survey. *Earth Science in the Public Service.* Professional Paper 921. Washington, D.C.: U.S. Government Printing Office, 1974, 73 pp.

IGNEOUS ROCKS

Texture and Occurrence

Volcanoes and Volcanic Rocks

TYPES OF VOLCANOES

Igneous Activity and Plate Tectonics

WEATHERING

Soils

SEDIMENTARY ROCKS

Clastic Sedimentary Rocks

Nonclastic Sedimentary Rocks

Features of Sedimentary Rocks

METAMORPHIC ROCKS

Foliated Metamorphic Rocks

Nonfoliated Metamorphic Rocks

Understanding Metamorphic Rocks

MINERAL RESOURCES

Exhaustibility of Resources

Types of Mineral Resources

IGNEOUS PROCESSES

METAMORPHIC PROCESSES

SEDIMENTARY PROCESSES

WEATHERING PROCESSES

ENERGY SOURCES

CHAPTER 3

ROCKS, WEATHERING, AND MINERAL RESOURCES

Mt. St. Helens in eruption, May 18, 1980. (Photo By R. M. Krimmel, U.S. Geological Survey.)

IGNEOUS ROCKS

We study rocks to learn how the earth works, and igneous rocks can tell us much about conditions below the earth's surface. **Igneous rocks** *are rocks that were once melted.* Volcanoes and volcanic rocks are obvious examples. From the study of active or recently active volcanoes, we have learned how to recognize older volcanic rocks at places where erosion has removed all traces of the volcano itself. Igneous rocks crystallize from magma; **magma** *is melted mineral material that may include suspended crystals and generally includes dissolved gases, mainly steam.* There are several major types of igneous rocks, and we should understand how each is emplaced. We will start by considering how igneous rocks are classified and then go on to see where each tends to form.

Igneous rocks are classified on the bases of composition and texture. Composition in this case means *mineral* composition—that is, the rock-forming silicate minerals that are in the rock. **Igneous texture** means the size of the mineral grains. In general, the size of the mineral grains is controlled by the rate of cooling. Large igneous bodies cool slowly and so have large crystals; smaller bodies cool more rapidly and so have small crystals.

The few igneous rock types shown in Figure 3-1 include most of the igneous rocks. Composition is shown horizontally and texture vertically. Textures are easy to recognize, and fortunately the overall color of an igneous rock generally is determined by its mineral composition. Thus, color and grain size, easily recognized features, are

F 3-1

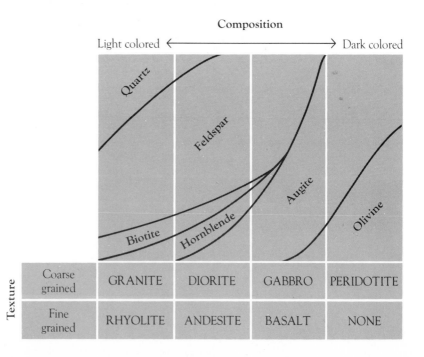

Figure 3-1
The classification of the igneous rocks.

A.

Figure 3-2
A. Granite is a light-colored, coarse-grained igneous rock. White feldspar, gray quartz, and black biotite crystals are easily seen. **B.** Basalt is a dark-colored, fine-grained igneous rock. Gas-bubble holes are filled with light-colored calcite and other minerals. (Photo **A** by A. Heitanen and **B** by J. R. Stacy; U.S. Geological Survey.)

B.

generally sufficient to identify an igneous rock. Two igneous rocks are much more abundant than the others. The most common coarse-grained igneous rock is **granite,** and the continents are generally underlain by granite. The ocean floors are composed of the volcanic rock **basalt,** and basalt volcanoes are also common on the continents (Figure 3-2).

F 3-2

Texture and Occurrence

Intrusive igneous rocks *are those that crystallize below the earth's surface* (Figure 3-3). Some intrusive rock bodies are **discordant;** that is, they F 3-3 *cut across the bedding of the earlier rocks into which they intrude.* Examples are **batholiths** and **dikes** (Figure 3-4). Batholiths are large bodies, so F 3-4 they crystallize slowly and thus have coarse textures. Most batholiths are composed of granite, and the cores of many mountain ranges are batholiths. **Concordant** *intrusive rock bodies are more or less parallel to the bedding planes;* **sills** and **laccoliths** *are examples.*

Figure 3-3
Intrusive igneous rock bodies. Sills and laccoliths are parallel, or concordant, with the older sedimentary rocks. Dikes and batholiths cut across the sedimentary beds and so are discordant.

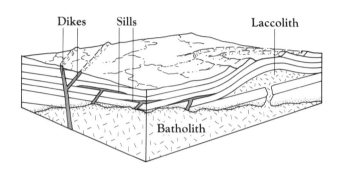

Figure 3-4
The light-colored rock is a dike cutting across the older rocks. (Photo by J. R. Stacy; U.S. Geological Survey.)

Figure 3-5
Obsidian or volcanic glass shows conchoidal fracture. Specimen is 15 centimeters (6 inches) long. (Photo by C. Milton; U.S. Geological Survey.)

Extrusive igneous rocks *crystallize at the earth's surface.* Volcanoes and lava flows are examples. Extrusive rocks, and the smaller intrusive bodies, cool rapidly and so are fine grained. Some volcanic rocks cool so rapidly that no crystals form, and such rocks are glass. **Obsidian** (Figure 3-5) is a common example of volcanic glass.

F 3-5

Volcanoes and Volcanic Rocks

Volcanic eruptions range from explosive to quiet, depending mainly on the viscosity of the magma and the amount of dissolved gases. The viscosity is determined for the most part by the composition of the magma. Basalt magma is quite fluid, and basalt volcanoes are generally quiet. Felsite (rhyolite) volcanoes are the most viscous and the most explosive; andesite volcanoes are intermediate. The amount of silica (SiO_2) in the magma controls the viscosity, because the silicon-oxygen (SiO_4) units mentioned in Chapter 2 tend to link together and so reduce the mobility of the melt. Basalt contains about 50 per cent SiO_2, and felsite over 70 per cent.

A melted rock is less dense than the same rock solidified because the rock expands when heated. This makes the melt buoyant, and so it tends to rise. If the magma reaches the surface, a volcano may result. All magmas contain volatiles, mainly water. At depth, where the magma is under the pressure of the weight of the overlying rocks, the water remains dissolved in the magma. When the magma nears the surface and the confining pressure is much less, steam separates from the molten rock. The situation is similar to what happens with beer, champagne, or pop. While the bottle is sealed, the pressure keeps the carbon dioxide dissolved in the fluid; but when the bottle is opened, the gas begins to bubble out of the liquid. The viscosity of the magma determines how the volatiles escape. If the magma is fluid, the gases escape quietly; but if it is viscous, the gas bubbles may explode.

The explosive products from a volcano are called **pyroclastic** *(fire-broken)* **rocks.** If the fragment is still viscous when ejected, it is called a volcanic **bomb.** Pyroclastic material up to sand size is called **ash,** and the rock formed from ash is called **tuff.** Larger fragments are called **volcanic blocks.** Hot ash that welds together when it lands is called **welded tuff.** Bubble holes formed by escaping gas are common in volcanic rocks. In some cases, the escaping gas forms a froth, much as champagne does when mishandled. The rock formed in this way is called **pumice,** and much pumice is light enough to float on water.

TYPES OF VOLCANOES   The quietest type of eruption is the **fissure eruption,** in which vast amounts of very fluid basaltic lava flow from fissures or dikes and cover large areas. The Columbia River Basalt of eastern Washington and Oregon is an example (Figure 3-6). There, F 3-6 about 1500 meters (5000 feet) of basalt covers an area of about 78,000 square kilometers (30,000 square miles).

If the basalt is somewhat less fluid and the magma flows from a single vent, a **shield volcano** may result. In this case the basalt forms a very gentle slope (Figure 3-7). The Hawaiian Islands are examples of F 3-7 shield volcanoes, and their most violent eruptions are fountains of hot, glowing lava.

At some places pyroclastic material is ejected from a central vent, and it builds up to form a steep-sided cone. These volcanoes are called **cinder cones,** and many of them are of basaltic composition. They are generally relatively small, under 300 meters (1000 feet) in height (Figure 3-8). They are commonly late features of basaltic vol- F 3-8 canism and may form small parasitic cones on the larger volcanoes. The basalt that forms cinder cones may be more viscous than most other basalt because it is cooler or because it contains less water.

Figure 3-6
Basalt flows on the Columbia Plateau in eastern Washington. Individual flows in this very fluid basalt can be followed for many tens of miles. (Photo from U.S. Department of the Interior, Bureau of Reclamation.)

Figure 3-7
An oblique aerial view of the summit of Mauna Loa volcano in Hawaii. In the distance the profile of 4200-meter (13,784-foot) Mauna Kea, a similar shield volcano, can be seen. (Photo by D. W. Peterson; U.S. Geological Survey.)

Figure 3-8
Vertical aerial photograph of Menan Buttes, two cinder cones, in southern Idaho. These cones have been somewhat modified by erosion. (Photo from U.S. Geological Survey.)

The classic, tall, cone-shaped large volcanoes, such as Fuji in Japan, are **composite volcanoes,** or **stratovolcanoes.** They are composed of both lava and pyroclastics. Their eruptions are explosive during a pyroclastic phase and relatively quiet when lava is erupted. The lava flows tend to strengthen the cone, so that these mountains may be very high. These composite cones are typically composed of andesite. The Cascade volcanoes of western North America are of this type (Figure 3-9).

As mentioned, the most explosive volcanoes are those whose composition is felsitic. Some of these are composite cones in which the youngest eruptions are of felsite. In such cases, the viscous felsite may

F 3-9

Figure 3-9
Mt. Adams, in the Cascade Mountains of Washington, is 3754 meters (12,307 feet) high and towers above the older Cascade Mountains upon which it formed. (Photo from Washington State Department of Commerce and Economic Development.)

plug the volcano, causing the pressure to build within it. When the pressure becomes great enough, an explosive eruption occurs (Figure 3-10). In other cases the volcano is a dome of viscous felsite that may prevent eruption until pressures are at the explosive level.

F 3-10

Igneous Activity and Plate Tectonics

The igneous activity just outlined does not occur randomly; much of it is associated with plate tectonics. At the mid-ocean ridges, basalt volcanism creates oceanic crust. The eruptions here are quiet. Submarine lava flows are common, and much of the igneous activity is probably intrusive. The probable source of the basalt is partial melting of the mantle, as the layer below the crust is called (Figure 3-11). The **mantle,** as we will see, is believed to be made of the rock **peridotite,** which is composed of olivine and pyroxene. (Augite is the common pyroxene that we met earlier.) When a rock is heated, the materials with the lowest melting points melt first, and if this early melt is separated from the unmelted rock, a magma of a composition different from the original rock is the result. We will address the problem of the source of the heat in the mantle in later chapters; but at depths between about 100 and 200 kilometers (60 to 120 miles), the temperature and pressure are such that the rocks are close to the melting point.

F 3-11

A.   B.

Figure 3-10 Mt. St. Helens. **A.** Before the eruption, the mountain was symmetrical and snow covered. Photo taken April 10, 1980 shows minor steam escaping. **B.** Summit during the eruption of May 18, 1980. **C.** Debris flow obliterates North Fork of Toutle River. Photo taken June 4, 1980. **D.** Summit after the eruption. (Photos from U.S. Geological Survey; **A** and **C** by Austin Post, **B** by R. M. Krimmel.)

C.

D.

Figure 3-11
Plate tectonics and
the origin of some
igneous rocks.

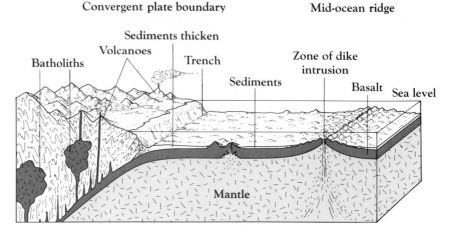

At convergent plate boundaries the volcanoes are andesite composite volcanoes (Figure 3-11). Apparently the basaltic ocean crust and overlying sediments are partially melted where they descend to depths where temperature and pressure are high. The sediments, especially those derived from the continents, may yield magma of felsite or granite composition when melted. This may account for the occasional explosive eruptions of these volcanoes. Because the felsite-granite magma is viscous, much of it probably crystallizes at depth, being unable to flow to the surface. This process may be the origin of batholiths, for they form near plate boundaries and are generally composed of granitic rocks; this may be the way that new continental crust is made.

On the plates themselves, the type of igneous rock depends on whether the crust is oceanic or continental. On the oceans the volcanic activity is simpler. Basalt volcanoes form submarine mountains and islands. It appears that basalt magma is generated at certain places, and the islands or submarine mountains form in lines as the plates move across these "hot spots" (Figure 3-12). On the continents F 3-12 basalt volcanism is also common, but many other types of volcanic and intrusive activity also occur. The differences in rock types may be the result of partial melting of the continental crust.

WEATHERING

A rock exposed at the earth's surface is, in general, subjected to conditions different from those under which it formed. The changes that take place in the rock as the result of these new conditions are called **weathering.** Weathering is the first step in both erosion (discussed in Chapter 6) and in the making of a sedimentary rock. Two types of weathering can be distinguished, but in nature they proceed hand in hand. **Mechanical weathering** *is the breaking up of an existing rock into*

Figure 3-12
As the Pacific plate moved over certain "hot spots," lines of volcanic islands were formed.

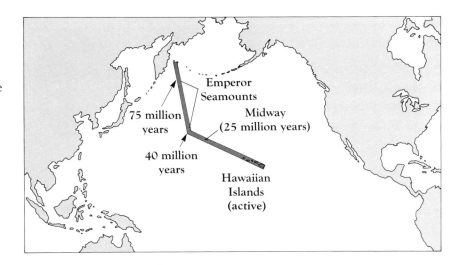

smaller fragments. In **chemical weathering** *the elements in the rock are rearranged to form new minerals.*

Mechanical weathering is the making of little rocks out of big rocks. Frost is the main agent of mechanical weathering. Water seeps into the tiny cracks in the rocks; and when it freezes, it expands, breaking the rock (Figure 3-13). Mechanical weathering tends to be most effective in cool, moist areas where the temperature goes through the freezing point many times a year. Some other agents of mechanical weathering are plant roots (Figure 3-14), burrowing animals, lightning, and human activities.

Chemical weathering is the reaction between the rock and its environment. Most rocks formed under conditions very different from those they encounter at the earth's surface. Water, carbon dioxide, and oxygen in the atmosphere react with the minerals in the rock to form new materials. Warm, moist climates favor chemical weathering,

F 3-13

F 3-14

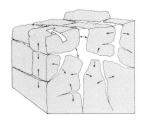

Figure 3-13
Mechanical weathering by frost action. The expansion of water when it freezes causes the breakup of the rock.

Figure 3-14
Sidewalk damaged by tree roots.

Figure 3-15
Mechanical weathering breaks the rock into smaller and smaller fragments, increasing the surface area. The increased surface area makes chemical weathering more effective.

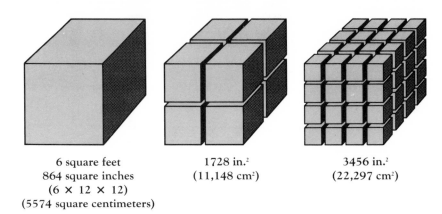

6 square feet
864 square inches
(6 × 12 × 12)
(5574 square centimeters)

1728 in.²
(11,148 cm²)

3456 in.²
(22,297 cm²)

as does the large surface area of the rock debris formed by mechanical weathering (Figure 3-15).

To understand chemical weathering, we need only consider what happens to the eight elements that make up most of the continents. Their behavior is summarized in Table 3-1. Oxygen, silicon, and aluminum in rocks combine with water to form clay and silica. The clay becomes the sedimentary rock *shale* and the silica *chert*. Iron in rocks combines with oxygen or oxygen and water at the earth's surface to form iron oxides. Magnesium, calcium, sodium, and potassium unite with carbon dioxide to form soluble products. The calcium forms calcite, and calcite is the mineral of which *limestone* is composed. Magnesium replaces some of the calcium in calcite to form *dolomite*. The

F 3–15

T 3–1

Table 3-1 A SIMPLIFIED CHART SHOWING THE REDISTRIBUTION OF ELEMENTS BY CHEMICAL WEATHERING

Element	Weathering Process			Product or Mineral	Sedimentary Rock
	H_2O	O_2	CO_2		
Oxygen (O) Silicon (Si) Aluminum (Al)	X			Clay $[(Al_2Si_2)O_5(OH_4)]$ Silica (SiO_2)	SHALE CHERT
Iron (Fe)	X	X X		*Iron Oxides* Hematite (Fe_2O_3) "Limonite" $[FeO(OH)]$	Sedimentary iron deposits
Magnesium (Mg)			X	Dolomite $[CaMg(CO_3)_2]$	DOLOMITE (replacement of limestone)
Calcium (Ca)			X	Calcite $(CaCO_3)$	LIMESTONE
Sodium (Na)			X	Goes to the oceans	
Potassium (K)			X	Most remains in soil	

Table 3-2   WEATHERING PRODUCTS OF COMMON MINERALS

	Weathering Products (see Table 3–1)					
Mineral	Clay	Silica	Iron Oxides	Dolomite	Calcite	Other
Olivine		X	X	X		
Augite	X	X	X	X	X	
Hornblende	X	X	X	X	X	
Biotite	X	X	X	X		Potassium to soil
Muscovite	X	X				Potassium to soil
Orthoclase	X	X				Potassium to soil
Plagioclase	X	X			X	Sodium to ocean
Quartz						Resists weathering, becomes sand grains

sodium goes into the oceans; much of the potassium stays in the soil attached to clay particles, and the rest goes into the oceans. The weathering products of the common minerals are summarized in Table 3-2.   T 3–2

Soils

Soil *is the material at the earth's surface that supports plant life* and so ultimately all life. The formation of soils involves both weathering of rocks and biologic activity. The rate of soil formation is quite variable and depends on rock type, climate, slope, and vegetation. At one extreme is the development of several inches of soil within a few years on newly erupted volcanic flows in some tropical areas; at the other is the lack of any soil after 10,000 years of exposure in some glaciated areas.

Most soils have three distinct *horizons*—the topsoil, the subsoil, and the partly decomposed bedrock (Figure 3-16). In many areas, es- F 3–16 pecially moist ones, humus is well developed and gives the upper soil a dark color. **Humus** *is plant material that has been partly decomposed by bacteria and molds in the soil.* Water passing through humus becomes acidic, and this promotes chemical weathering.

There are many types of soils, but only three major types are described here. Climate is the most important factor in soil formation, so the soils that develop in three different climates are discussed. In middle North America two soil types are well developed. The boundary between them is near the 64-centimeter (25-inch) rainfall line. In the east where the annual rainfall exceeds this amount, humus is well developed, and the acid soil water carries away the soluble materials.

Figure 3-16
A typical soil profile.

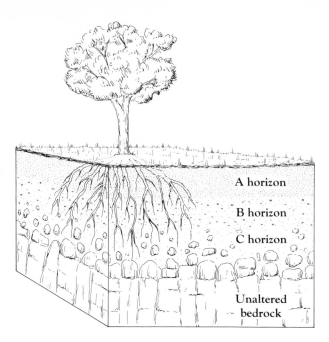

Aluminum and iron are deposited in the subsoil, and calcite and dolomite are completely removed. These soils are called **pedalfers** (*pedon* is soil, and *al* and *fe* are the symbols for aluminum and iron). These soils are acidic, and clays are well developed.

In the more arid west, less humus develops and there is less soil water, so the development of soil is slower. The soluble materials such as calcite are deposited in the subsoil. Such soils are termed **pedocals** (*pedon* is soil, *cal* is for calcite).

In tropical climates the soils are much different, and their formation is not completely understood. The heavy seasonal rainfalls leach everything except iron and aluminum from the soil. Iron oxides and aluminum oxides develop, and the iron oxides generally color the soil red. Such soils do not make good farmlands, but in some cases they are valuable ores, especially of aluminum.

SEDIMENTARY ROCKS

Sedimentary rocks form at the earth's surface, generally in layers or beds. The first step in the formation of sedimentary rocks is weathering, the mechanical and chemical breakdown of rock exposed to surface conditions. The weathered material is then eroded, transported, and finally deposited. Although sedimentary rocks cover three-quarters of the continental surface, they make up only a small fraction of the earth's crust (Figure 3-17). Sedimentary rocks are classified in two main groups—clastic and nonclastic.

F 3-1?

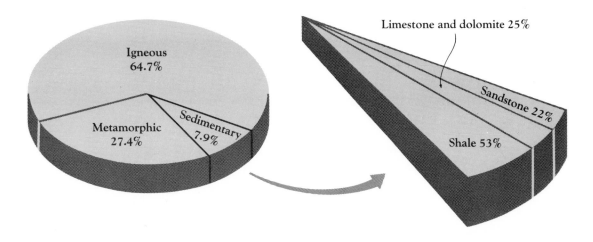

Figure 3-17
Sedimentary rocks
form only a small
amount of the earth's
crust, but they cover
much of the surface.
(Data from Ronov
and Yaroshevsky,
1969.)

Clastic Sedimentary Rocks

Clastic means broken, and **clastic rocks** *are composed of fragments broken from preexisting rocks.* They are classified by the size of the fragments without regard to the composition of the fragments. Clastic rocks are easily recognized by the their **clastic,** or fragmental, **texture** (Figure 3-18). The coarsest clastic sedimentary rock is called conglomerate. **Conglomerate** *is composed of rounded fragments, or clasts, larger than 2 millimeters (less than ⅛ inch).* If the fragments are angular, the rock is called **breccia. Sandstone** *is made of fragments between 2 millimeters and 1/16 millimeter.* Grains of the lower size limit are about the smallest that can be seen with a hand lens in bright sunlight. The fine-grained clastic sedimentary rocks are called siltstone (or mudstone) and shale. **Siltstone** *is similar to sandstone but much finer grained,* and **shale** *is composed of clay particles* that are extremely small. The clastic sedimentary rocks are summarized in Table 3-3.

   The shape and composition of the fragments may reveal the history of the rock. Angular fragments may have been abraded less by

F 3-18

T 3-3

Table 3-3 COMMON CLASTIC SEDIMENTARY ROCKS		
Rock	Fragment Size	Comments
Conglomerate	Over 2 mm	*Breccia* if fragments are angular
Sandstone	1/16 to 2 mm	
Siltstone	1/256 to 1/16 mm	Also called *mudstone.* Silt-size fragments give it a gritty feel
Shale	Less than 1/256 mm	Composed of clay. Distinguished from siltstone by lack of gritty feel if pure

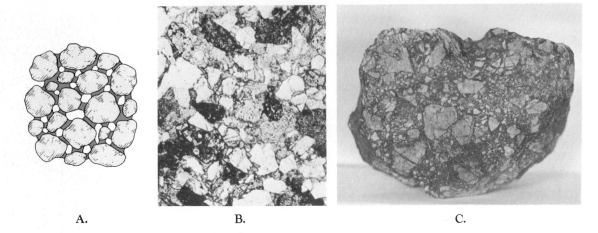

A.   B.   C.

Figure 3-18
Clastic texture. **A.**
The fragments are
cemented together by
material deposited
between the grains. **B.**
Photomicrograph of
sandstone showing
clastic texture. **C.**
Breccia. The larger
fragments are about
2.5 cm (1 in.) in
diameter. (Photo **B** by
L. Hoover and **C** by F.
S. Simons; U.S.
Geological Survey.)

stream transport than rounded fragments of the same composition.
Quartz is the mineral most resistant to weathering, so a quartz sandstone may be the product of deep weathering or long transportation.
Clastic rocks are commonly held together by cement deposited between the grains by ground water moving through these porous
rocks.

Nonclastic Sedimentary Rocks

Chemical weathering produces the materials that form the nonclastic
sedimentary rocks. The nonclastic rocks can be further subdivided
into biologic and chemical rocks, depending on how they are precipitated. The **biologic sedimentary rocks** *are those that are formed or precipitated by organisms.* Coal and coral reefs are two very different
examples. Most biologic sedimentary rocks are formed in the oceans
by organisms that take materials from the sea water and either precipitate the rock material as part of their life cycle or use it to build
shells or other structures that become rocks. The **chemical sedimentary rocks** *are those precipitated directly from sea water.* The evaporation
of sea water will cause the precipitation of halite, or rock salt. A rock
formed in this way is termed an **evaporite.**

The nonclastic sedimentary rocks are named on the basis of composition (see Table 3-4). The most common of them is **limestone** com-   T 3-4
posed of the mineral calcite. Most limestone is biologic in origin.
Chalk is a type of limestone composed of shells of very tiny organisms. **Coquina** is limestone made of much larger fossil shells. **Dolomite** is much less abundant and is formed mainly by replacement of
some of the calcium in calcite (limestone) by magnesium to form the
mineral and rock dolomite. **Chert** is composed of very fine-grained
silica (SiO_2) and so is chemically similar to quartz. Some chert is biologic in origin and some is a chemical precipitate. The latter type is

commonly associated with submarine volcanism. **Rock salt** and **gypsum** are common evaporites. **Diatomite** is made of the fossils of tiny organisms and looks very much like chalk, but because the fossils are composed of silica, diatomite, unlike chalk, does not react with acid.

Features of Sedimentary Rocks

Sedimentary rocks form in layers called **beds** or strata. In the field the beds can generally be seen, so the rock is recognized to be sedimentary (Figure 3-19). Hand specimens are, in many cases, too small to F 3-19 show bedding. It should be obvious that, if sedimentary rocks are undisturbed, the oldest rocks are at the bottom, and the rocks become younger toward the top.

    Fossils are commonly found in sedimentary rocks, especially the finer-grained rocks. They are evidence of life at the time the enclos-

Table 3-4   COMMON NONCLASTIC SEDIMENTARY ROCKS

Rock	Composition	Origin Biologic	Origin Chemical	How to Recognize
Limestone	Calcite ($CaCO_3$)	Common	Rare	Effervesces with dilute hydrochloric acid
Chalk	Calcite ($CaCO_3$)	Composed of microscopic fossils		A type of limestone
Coquina	Calcite ($CaCO_3$)	Composed of fossils		A type of limestone
Dolomite	Dolomite [$CaMg(CO_3)_2$]		Replacement of limestone	Effervesces with hydrochloric acid only if powdered
Chert	Silica (SiO_2)	X	X	Hardness 7; generally has waxy luster
Gypsum	Gypsum ($CaSO_4{\cdot}2H_2O$)		Evaporite	Hardness 2; shows cleavage if coarse grained
Rock salt	Halite ($NaCl$)		Evaporite	Hardness 2.5; salty taste; cubic crystals
Diatomite	Silica (SiO_2)	Composed of microscopic fossils		Looks like chalk but does not react with dilute hydrochloric acid
Coal	Carbon (C)	Plant material		Black; generally soft; burns

Figure 3-19
Sedimentary rocks
form in beds, or
layers. (Photo from
U.S. Geological
Survey.)

ing beds were deposited, and their study has enabled us to learn how
life on the earth has changed during geologic time. Fossils are also
used to date rocks, as we will see in Chapter 7. In addition, the type
of fossils can tell us a great deal about the environment in which the
beds were deposited.

Other features seen on bedding surfaces can disclose much about
the conditions under which a bed was deposited. **Mud cracks** (Figure   F 3-20
3-20) reveal periodic wetting and drying. **Ripple marks** (Figure 3-21)   F 3-21
show current action. Periods of erosion are shown by channels filled
with younger sediment (Figure 3-22). Careful study of these and   F 3-22
many other features of sedimentary rocks can reveal much about how
the rocks were formed.

Figure 3-20
Mud cracks form
when wet sediment
dries and shrinks.

Figure 3-21
Ripple marks indicate
water currents.

Figure 3-22
A channel indicates
localized erosion.

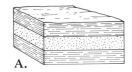
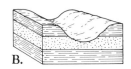
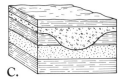

A.

B.

C.

METAMORPHIC ROCKS

Metamorphic rocks *are rocks that have been changed in mineralogy or texture or both by processes that occur without melting.* Heat, pressure, stress, shear, and active solutions are the main agents of metamorphism (Figure 3-23). The intensity of these agents determines **metamorphic grade.** Three types of metamorphism are recognized: dynamic, thermal, and regional.

Dynamic metamorphism occurs when rocks are broken, sheared, and ground by movement. This generally is caused by movement along a fault, as during an earthquake. The metamorphic changes are mainly mechanical, but some recrystallization may occur.

Thermal metamorphism is commonly called **contact metamorphism,** because it occurs at the boundary or contact of intrusive igneous bodies, such as batholiths, with the rocks that they intrude.

Regional metamorphism is metamorphism that occurs over a large area, or region. The rocks are recrystallized within the crust, and regional metamorphism generally accompanies widespread deformation. Such deformation occurs at times of mountain building and usually involves the collision of tectonic plates.

Metamorphic rocks are of two types. **Foliated rocks** *have a layered structure because they crystallized under stress or directed pressure.* Most regional metamorphic rocks are foliated rocks. **Nonfoliated rocks** *have no obvious directional character,* either because they did not crystallize under stress or because their minerals do not grow in layers.

F 3-23

Figure 3-23
Pressure, stress, and shear are important agents of metamorphism.

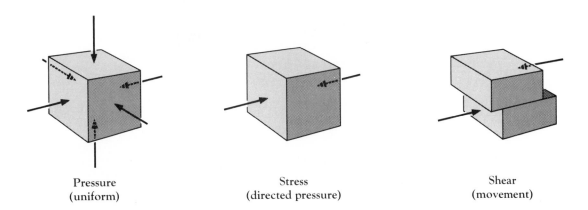

Pressure
(uniform)

Stress
(directed pressure)

Shear
(movement)

Foliated Metamorphic Rocks

A common sedimentary rock is shale, which is composed largely of clay. If shale is subjected to metamorphism, the first change that occurs is that the clay is recrystallized into mica. Because mica is a platy mineral, it grows in layers at right angles to the stress (Figure 3-24). In low-grade metamorphism—that is, at low temperature and pressure— shale becomes **slate**. In slate, the mica grains are too small to see, but they give the rock a sheen and a foliation along which it breaks easily. If the conditions are medium grade, the mica grains grow bigger, and they are easily seen in the metamorphic rock **schist**. If the metamorphism occurs at greater depth where high-grade metamorphic conditions are encountered, much of the mica is transformed into feldspar. The resulting rock generally has bands of feld-

F 3-24

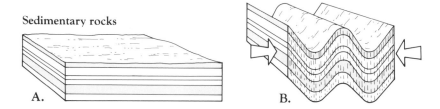

Sedimentary rocks

A.

B.

Figure 3-24 Foliation in metamorphic rocks. **A.** Before deformation and metamorphism. **B.** Foliation develops when platy minerals such as mica grow perpendicular to stress. **C.** Outcrop of folded slate showing foliation. (Photo by A. Keith; U.S. Geological Survey.)

C.

Figure 3-25
An exposure of gneiss near Gothenberg, Sweden.

spar alternating with darker minerals and is called **gneiss,** pronounced "nice," (Figure 3-25). These changes are summarized in F 3-25 Table 3-5. T 3-5

It should be obvious that in a sense metamorphism is the opposite of weathering. In weathering, the most common rock-forming mineral, feldspar, is changed into clay; and, in metamorphism, the clay changes into mica and finally to feldspar. Notice, too, that the mineral composition of gneiss is the same as that of granite; this is why it is suggested in the rock cycle (Chapter 1, Plate 7) that metamorphic rocks could be melted to form igneous rocks. Probably very few igneous rocks form in this way.

Nonfoliated Metamorphic Rocks

Marble is metamorphosed limestone. Limestone, as we have seen, is composed of the mineral calcite, and when it is subjected to metamorphic conditions, no new minerals form. The only change is that the calcite crystals get larger. In the recrystallization process, impurities are excluded from the calcite crystals in marble, so marbles tend to be white, and limestones may be gray or white.

Quartzite is metamorphosed quartz sandstone. Again, the quartz crystals in the sandstone become larger, and an interlocking texture develops, but no new minerals form. Quartzite is recognized by its sugary appearance.

Most rocks contain some impurities, so other minerals not mentioned here may be found in metamorphic rocks. In determining which minerals form in metamorphic rocks, only the overall chemical composition of the parent rock matters, so rocks of widely differing origins, such as shale and felsite tuff, can produce the same metamorphic rock.

Hornfels (singular) is the name given to thermal, or contact, metamorphic rocks. Any kind of rock can become a hornfels, and the degree of metamorphism may vary from intense to minor, depending on the distance from the intrusive body and its size and temperature. These factors make recognition of hornfels very difficult unless it is

Table 3-5   DEVELOPMENT OF FOLIATED METAMORPHIC ROCKS

Parent Rock	Low-Grade Metamorphism	Medium-Grade Metamorphism	High-Grade Metamorphism
SHALE ⟶	SLATE ⟶	SCHIST ⟶	GNEISS
Clay			
	Mica		
			Feldspar

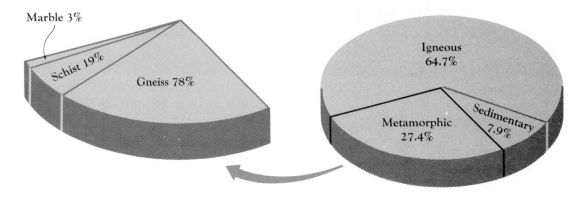

Figure 3-26
Abundance of
metamorphic rocks in
the earth's crust.
(Data from Ronov
and Yaroshevsky,
1969.)

seen in the field near the intrusive rock body. Pottery and bricks are made by heating clay and so are similar to hornfels.

Figure 3-26 shows the relative abundance of metamorphic rocks.   F 3-26

Understanding Metamorphic Rocks

Metamorphic rocks, especially the regional metamorphic rocks, are important because they can reveal the conditions, or environment, deep in the crust. Most metamorphic rocks are intensely deformed, making their interpretation difficult. As the smaller structures of metamorphic rocks were studied, their relationship to larger structures was better understood, and it was revealed that the belts of regional metamorphic rocks are really the eroded roots of mountain ranges. Study of the mineral groups in the metamorphic rocks also showed that the intensity, or grade, of metamorphism generally increases in the core of the ancient mountain ranges.

Mountain ranges are formed when the movement of tectonic plates causes collisions of the plates. In this process the rocks are deformed, and they are also moved into new environments where metamorphism occurs. Figure 3-27 shows where metamorphic rocks form   F 3-27
at convergent plate boundaries (volcanic island arcs). Metamorphism also occurs near mid-ocean ridges (Figure 3-27).

MINERAL RESOURCES

Mines are unusual places in the earth's crust. A mining area like Butte, Montana, is as unique as nearby Yellowstone National Park. About 30 per cent of the United States' copper production has come from a few square miles around Butte. The number and distribution of mines have many consequences—scientific, social, political, and economic. Modern industrial society is based on easily available metals and other mineral products. We will first examine the continual availability of mineral resources before turning to the various types of mineral resources.

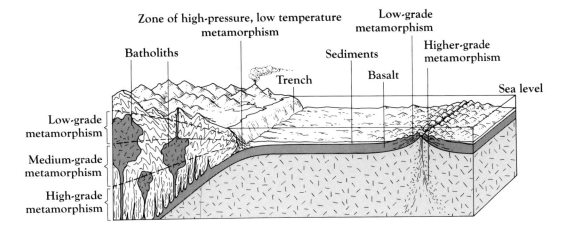

Figure 3-27
Metamorphism commonly occurs at mid-ocean ridges and at convergent plate boundaries.

Exhaustibility of Resources

A mine or a mining district has a finite amount of ore, and when the ore has been extracted, only a hole in the ground and a ghost town remain. This is the thesis of those who say that we will someday run out of certain resources, but is this necessarily so? A mine is a mine because some desired element is concentrated there to form ore. **Ore** *is material that can be mined at a profit.* Thus, the same concentration at another location may not be ore because of distance to market, difficult mining conditions, politics, or costly labor. It also follows that material that is valueless today may become ore later because of economic changes or new mining techniques. The mining of ore of lower and lower grade can go on until the limit of the average concentration of an element in the crust is reached. At that stage almost any rock would be ore, so no mines in the usual sense would exist. Thus, it is the biggest, easiest-to-mine, most economic deposits that will be mined out first; and, unless economics and technology intervene or new discoveries are made, we could run out of some materials. In a real sense, then, mines are finite, but many circumstances may prevent us from running out of a particular element.

Geologic resources can be placed into two categories—renewable and nonrenewable. The two geologic resources that are renewable are water and soil, which feed the world. Other resources such as sand, gravel, building stone, and similar bulk resources are so voluminous that they are inexhaustible, although locally they may be in short supply.

In this chapter nonrenewable resources are discussed. Metal mines are nonrenewable in the sense that the geologic processes that concentrated the elements operated for times measured in millions of years many millions of years ago. So, even though the same processes have gone on during the millions of years of geologic time, new mines are not being created in terms of human occupancy of the earth.

As soon as a resource is in short supply, economic factors come into play. As the price rises, use is reduced and substitutes are found. Recycling becomes economically feasible, and lower-grade deposits can be exploited economically. The amount of the reserve may also be increased by exploration, which is stimulated by higher prices. New theories and techniques may be developed. Plate tectonics has stimulated and focused the search for certain resources, and high-altitude satellite pictures have also broadened exploration to include the whole earth. New mining and especially new refining techniques can improve both the recovery of metals from ore and the grade of ore that can be profitably mined.

It should be clear that these factors may not come into play in all cases, and some resources may become truly scarce, with attendant social and economic problems. Although most metals may not run out, petroleum may. The future, however, is cloudy. It seems that in some cases the prophets of doom will be right; in general they may be crying wolf. Nevertheless, a blind belief that technology will solve all our problems seems unduly optimistic.

Types of Mineral Resources

Table 3-6 gives a classification of mineral deposits based on the means through which the desired elements are concentrated to form ore. Many mines have by-products, such as gold or silver, that enable them to operate with low-grade ore. The classification shows that the igneous, metamorphic, and sedimentary processes we have already examined may result in concentration of needed elements into economic deposits.

T 3-6

IGNEOUS PROCESSES
Deposits within Igneous Rock Bodies
Disseminated Magmatic Deposits In most deposits that originated in magma, some process concentrated the ore in a small part of the igneous body. If the mineral is valuable enough, it can be recovered economically even though it is scattered throughout the body. The South African diamond mines are such disseminated magmatic deposits; the diamonds are scattered throughout cylindrical or cone-shaped bodies similar in composition to peridotite. Because diamonds crystallize only under great pressure, these unusual rocks appear to have come from great depth.

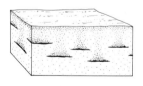

Figure 3-28
Chromite deposits form where heavy, early-formed chromite crystals sink and form layers in the still-fluid magma.

Crystal Settling Dense, early-formed crystals that sink to the bottom of a still-liquid magma body form this type of deposit. Typical examples are deposits of chromite and magnetite (Figure 3-28).

3-28

Table 3-6   CLASSIFICATION OF MINERAL DEPOSITS BY ORIGIN

Type of Ore Deposit	Example
Igneous processes	
Deposits within igneous rock bodies	
Disseminated magmatic	Diamonds: South Africa
Crystal settling	Chromite: Stillwater, Montana
Deposits from fluids associated with igneous rocks	
Late magmatic	Magnetite: Adirondack Mountains, New York
Pegmatite	Beryl, lithium: Black Hills, South Dakota
Hydrothermal	Copper: Butte, Montana
Metamorphic process	
Contact-metamorphic	Lead, silver: Leadville, Colorado
Sedimentary processes	
Evaporite	Potassium: Carlsbad, New Mexico
Placer	Gold: Sierra Nevada foothills, California
Weathering processes	Bauxite: Arkansas

Deposits from Fluids Associated with Igneous Rocks Almost all magmas contain at least a small percentage of water. Only a few of the common rock-forming minerals contain water (mica and hornblende), so, as crystallization proceeds, the water is concentrated in the remaining melt. Also concentrated are any elements other than the eight common elements because these atoms have the wrong size to fit into the crystal structures of the common rock-forming minerals. Thus, water solutions containing metals may be formed at the late stages of the crystallization of magma. Although the percentage of the potential ore elements may be quite small, the large volume of a batholith may provide substantial amounts of concentrated metals.

Late Magmatic Deposits If the valuable minerals are among the last to crystallize in a magma, they may be segregated. In some cases deformation after partial crystallization may separate the late fluids from the crystal mush. In other cases the late fluid may sink to the bottom and/or fill the spaces among the minerals that crystallized earlier.

Pegmatites Pegmatites are extremely coarse-grained rock bodies. Although most pegmatites have crystals a few centimeters long, in some instances the crystals reach several meters. Pegmatites form in both igneous and metamorphic rocks, but the processes are not completely understood. Fluids of late-igneous or metamorphic origin are apparently involved in the growth of these large crystals. Many pegmatites are simply very coarse-grained phases of their parent rocks, but some have unusual minerals. They are the prime sources for industrial mica, beryllium, lithium, columbium, tantalum, and rare earth elements, as well as gems and specimen crystals of common minerals.

Hydrothermal Deposits Hot-water solutions deposit sulfides and other ore minerals, usually in quartz veins. These deposits are called hydrothermal, and they are probably the most common source of metal ores. The mineralized hot waters may have come from the magmas of the batholiths with which many of these deposits are associated (Figure 3-29), or they may be circulating ground water that deposits metals F 3-29 from deep magmas or metals leached from surrounding rocks. Ground water may circulate through a great volume of rocks, so it can concentrate large amounts of metals even though the rocks through which it moves contain very little metal.

METAMORPHIC PROCESSES
Contact-Metamorphic Deposits In the metamorphosed zone around an intrusive igneous body, the magma and its fluids react chemically with the intruded rock, particularly if that rock is limestone, resulting in contact-metamorphic deposits of minerals replacing the preexisting rock (Figure 3-30).   F 3-30

SEDIMENTARY PROCESSES
Nodules containing manganese, nickel, copper, and other metals form on the deep-ocean floors at some places. These deposits, which may someday become economic to mine, are described in Chapter 8.

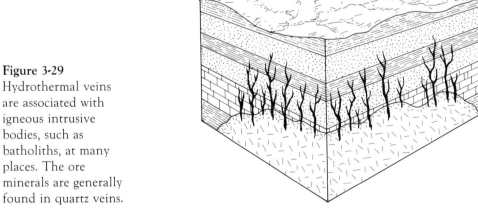

Figure 3-29
Hydrothermal veins are associated with igneous intrusive bodies, such as batholiths, at many places. The ore minerals are generally found in quartz veins.

Figure 3-30
Contact-metamorphic deposits may form where limestone, or other reactive rock, is intruded by an igneous rock body.

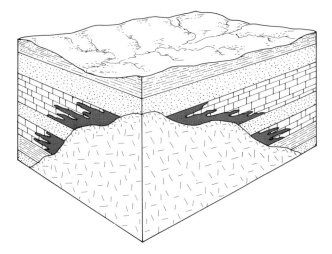

Evaporites The evaporation of salt lakes or of sea water in basins causes chemical precipitation of the dissolved salts in deposits known as evaporites. The usual sequence of precipitation from sea water is calcium and magnesium carbonate, then gypsum, then halite (rock salt), then potassium and bromine salts. Because of the low concentrations of dissolved material in salt water, very large quantities must evaporate to provide economic deposits, indicating that special conditions must contribute to the thickness of some evaporite beds.

Placer Deposits A river carries eroded fragments of many sizes and densities, as do ocean waves on beaches. For any given size, the lighter fragments are carried most easily. Thus, on the insides of curves in the river and other places where the current slows, the heavier fragments tend to be deposited (Figure 3-31). In this way, placer deposits are formed of dense minerals that are resistant to weathering. Native gold is a common placer mineral because gold is a very dense metal; diamonds and platinum also form placers.

WEATHERING PROCESSES
Weathering forms primary deposits of iron and aluminum, as well as other metals, and it may also further concentrate deposits formed in other ways. Extreme weathering may occur under tropical conditions

Figure 3-31
Placer deposits tend to form on the insides of curves where heavy minerals, such as native gold, are deposited.

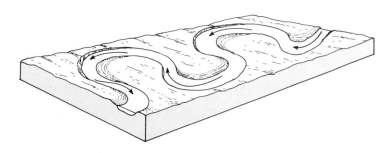

Figure 3-32
Ores may be
concentrated by
weathering, but such
deposits are generally
shallow.

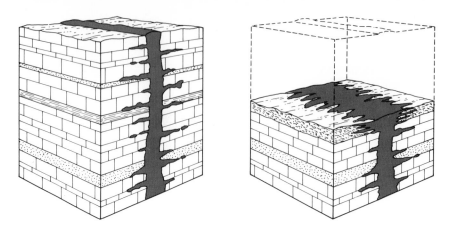

in which almost all other elements except iron and/or aluminum are removed from the soil by leaching. The aluminum ore called **bauxite** is formed in this way, and it is extensively mined.

Enrichment of mineral deposits by weathering is also common and has created many bonanzas, but it has also misled many prospectors and investors. A simple case of residual concentration is shown in Figure 3-32. Here, solution of a soluble rock such as limestone has F 3-32 concentrated the ore minerals. In the more common case, weathering of sulfide minerals produces sulfuric acid and soluble sulfates in a series of complex chemical reactions. The soluble materials move slowly downward to the water table where they are precipitated (Figure 3-33). Such enriched deposits may be valuable, especially if gold F 3-33 is involved, but one should not be fooled into thinking that the deposit is large because the primary unenriched ore may not be worth mining. Weathering does color the outcrop of a vein red or red-brown and so may help the prospector to find a prospect.

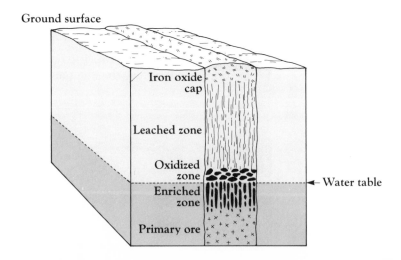

Figure 3-33
Sulfide mineral veins
may be enriched by
weathering processes.

ENERGY SOURCES

Many of our energy resources are of sedimentary origin. The fossil fuels, coal and petroleum, are formed by altered organic material buried with sediments. Uranium minerals are concentrated by surface processes, although they are also found in veins of hydrothermal origin.

The uranium deposits of the western United States are mainly found in sandstone, and the uranium was deposited in the sandstone by ground water. The uranium was probably removed from igneous rocks by weathering processes and was transported by ground water. In many cases the uranium was deposited in the sandstone at places where organic material, mainly plants, had accumulated.

Energy is also obtained from other sources. **Geothermal energy** is heat from within the earth. Hot springs are an example, but for them to be valuable for energy, temperatures must be higher than in most hot springs. The source of the geothermal heat is generally from deeply circulating ground water or from nearby intrusive rock bodies (Figure 3-34). Waterfalls are used to generate **hydroelectric energy,** F 3-34 and dams are built for the same purpose. **Solar energy** is obtained directly or indirectly from the sun.

Figure 3-34
Geothermal power plant at The Geysers in California. (Photo from Pacific Gas and Electric Company.)

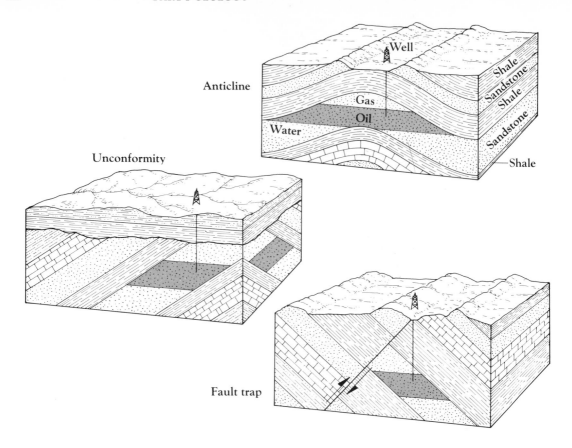

Figure 3-35
Types of petroleum traps.

Fossil Fuels. Coal originates from plant material that is buried with sedimentary rocks. Under most conditions, plant material decomposes before it is covered or buried, and coal does not form. Coal beds can develop if the organic material is quickly buried before it decomposes, as might happen in a swamp. The stagnant or nearly stagnant water in the swamp slows the decaying processes and allows the plant material to accumulate at the bottom of the swamp as **peat,** the first step in the formation of coal. With further compaction, the peat becomes **lignite** and then **bituminous** (*soft*) **coal.** Most coal is bituminous, although in relatively rare cases further compaction and heating produce *hard coal,* or **anthracite.** The original plant material is reduced to 5 per cent or less of its original thickness in the process of turning into coal. The swamps in which coal forms are unusual places, for some coal beds are tens of meters thick.

Petroleum forms mainly in organic-rich marine rocks. Time and moderate heating from burial are apparently required to form petroleum. Dark organic-rich shale is generally the source rock in which petroleum forms. To form an economic oil field, the petroleum must migrate from the shale into a more permeable reservoir rock from which the oil can be recovered. A reservoir rock must be both porous and permeable; that is, it must have space between the grains **(porosity)**, and these spaces must be interconnected **(permeability)** so that the fluid can be recovered (see Figure 6-16, p. 159). The third requirement for an oil field is a trap that prevents the escape of the petroleum from the reservoir rock (Figure 3-35). The oil shales of western United States are fine-grained rocks that accumulated in lakes. Their organic material has not completely changed into petroleum.

F 3-35

SUMMARY

- *Igneous rocks* are rocks that were once melted.

- *Magma* is melted mineral material that may include suspended crystals and dissolved gases, especially steam.

- Igneous rocks are classified on the bases of texture and composition.

- The continents are underlain by granite, and the oceans by basalt.

- *Intrusive rocks* are those that crystallize below the surface.

- *Dikes* and *batholiths* are *discordant* intrusive bodies.

- *Sills* and *laccoliths* are *concordant* intrusive bodies.

- *Pyroclastic rocks* include *bombs* and *tuff.*

- Volcanoes range from quiet shield volcanoes to explosive volcanoes.

- Basalt volcanism at the mid-ocean ridges forms the crust under the oceans. The basalt probably comes from partial melting of the mantle, the layer below the crust.

- At convergent plate boundaries, andesite volcanics and batholiths probably originate from melting of the downward-moving slab of ocean floor.

- *Mechanical weathering* is the breaking of an existing rock into smaller fragments.

- *Chemical weathering* is the reorganization of elements into materials more in equilibrium with their new environment.

- *Soil* is the material at the earth's surface that supports plant life.

- In humid areas *pedalfers*, soils rich in aluminum and iron, form.

- In more arid areas *pedocals*, soils rich in calcium, form.

- *Clastic sedimentary rocks* are made of fragments of other rocks. They are classified as shale, sandstone, and conglomerate, based on the size of the fragments.

- *Nonclastic sedimentary rocks* are precipitated either by chemical or biological means.

- *Chemical sedimentary rocks* are generally formed by evaporation of sea water. *Halite* or *rock salt* is an example.

- The most common *biologic rock* is *limestone*.

- The shape and size of fragments, the fossils, and other features of sedimentary rocks can reveal much about their history and origin.

- *Metamorphic rocks* are rocks that have been changed in either texture or mineralogy, or both, without melting.

- *Dynamic metamorphism* occurs when rocks are broken and sheared, as in a fault zone.

- *Thermal metamorphism* is *contact metamorphism* and occurs at the boundary or contact of igneous rock bodies. Such rocks are called *hornfels.*

- *Regional metamorphism* occurs in large areas and typically produces foliated rocks.

- *Foliated rocks* are those that crystallize under

stress and so have a directional property. Typical foliated rocks are *slate, schist,* and *gneiss.*

- *Nonfoliated metamorphic rocks* are *marble* (from limestone or dolomite) and *quartzite* (from quartz sandstone).

- *Ore* is material that can be mined at a profit.

- Mines are finite.

- Mineral deposits may be associated with igneous rock bodies as disseminated minerals, or they may settle to the bottom (or top), forming a layer. Fluids from intrusive rocks may also form deposits.

- *Contact-metamorphic deposits* form when fluids from an intrusive body react with nearby rocks.

- Hot water containing quartz and other minerals may form *hydrothermal veins,* a very common type of mineral deposit.

- Evaporation of sea water may concentrate some elements and form economic deposits.

- Heavy resistant minerals, such as gold, may form *placer deposits.*

- Weathering may concentrate materials, but in some cases the concentrate is quite small.

- *Geothermal energy* is heat from within the earth.

- *Coal* originates from plant material that was buried.

- *Petroleum* originates from organic material buried, generally, in marine rocks.

KEY TERMS

igneous rock	composite volcano, or	chalk	marble
magma	stratovolcano	coquina	quartzite
igneous texture	mantle	dolomite	hornfels
granite	peridotite	chert	ore
basalt	weathering	rock salt	disseminated magmatic deposit
intrusive igneous rock	mechanical weathering	gypsum	
discordant intrusive		diatomite	crystal settling
body	chemical weathering	bed	late magmatic deposit
batholith	soil	fossil	pegmatite
dike	humus	mud crack	hydrothermal deposit
concordant intrusive	pedalfer	ripple mark	contact-metamorphic
body	pedocal	metamorphic rock	deposit
sill	sedimentary rock	metamorphic grade	placer
laccolith	clastic rock	dynamic metamorphism	bauxite
extrusive igneous rock	clastic texture	phism	geothermal energy
obsidian	conglomerate	thermal, or contact,	hydroelectric energy
pyroclastic rock	breccia	metamorphism	solar energy
bomb	sandstone	regional metamorphism	coal
ash	siltstone	phism	peat
tuff	shale	foliated metamorphic	lignite
volcanic block	biologic sedimentary	rock	bituminous coal
welded tuff	rock	nonfoliated metamorphic rock	anthracite
pumice	chemical sedimentary	phic rock	porosity
fissure eruption	rock	slate	permeability
shield volcano	evaporite	schist	
cinder cone	limestone	gneiss	

REVIEW QUESTIONS

1. Define the following: igneous rocks, sedimentary rocks, metamorphic rocks.
2. Of what is magma composed?
3. Igneous rocks are classified on the basis of what two characteristics?
4. Examine Figure 3-1 and answer the following questions:
   (a) What is the difference between the igneous rocks granite and rhyolite? Which rock cooled faster?
   (b) Which of the igneous rocks shown in Figure 3-1 forms the continents?
   (c) Which of the igneous rocks forms the ocean floors?
5. (a) What is the difference between an intrusive igneous rock and an extrusive igneous rock?
   (b) Under what conditions would you expect a rock to be coarse grained? Under what conditions would you expect it to be fine grained? What is volcanic glass?
   (c) Explain how dissolved gases affect the explosiveness of a volcanic eruption.
6. For each of the following describe the structure, explain how it forms, state the type of magma usually involved, and describe the explosiveness of the eruption:
   (a) Shield volcano.
   (b) Cinder cone.
   (c) Composite volcano (stratovolcano).
7. For each of the following sites state what type of igneous rock is formed and what is believed to be the source of its magma:
   (a) Mid-ocean ridges (diverging plate boundaries).
   (b) Composite volcanoes at convergent plate boundaries.
   (c) Batholiths at convergent plate boundaries.
   (d) Middle of oceanic plate.
8. (a) Which of the sites listed in Question 7 may be the source of new continental crust?
   (b) Which of the sites listed in Question 7 is the source of new oceanic crust?
   (c) Which of the sites in Question 7 creates strings of islands lying in a straight line?

9. (a) What is weathering?
   (b) What is the difference between mechanical weathering and chemical weathering?
   (c) List some agents of mechanical weathering. Of these agents, which is by far the most important?
   (d) Why is mechanical weathering likely to be more extensive in a cool, moist climate?
   (e) What materials in the atmosphere react with the rocks exposed at the earth's surface?
   (f) What two conditions favor chemical weathering?
10. (a) Turn to Figure 3-1 and write down the minerals that compose granite.
    (b) Turn to Table 3-2 and make a list of the weathering products of granite.
    (c) As granite moves through the rock cycle, what sedimentary rocks might form?
11. Granite is a hard igneous rock; shale is a soft sedimentary one.
    (a) Which rock do you think is more susceptible to chemical weathering? Why?
    (b) Which rock do you think might be more susceptible to abrasion by pebbles in a stream or by wind-driven sand?
12. (a) What is humus?
    (b) Which has the most effect on the type of soil formed, climate or rock type?
    (c) The term *pedalfer* comes from what three words? In what areas do pedalfers form? Describe the formation of a pedalfer. Why are pedalfers acidic?
    (d) From what words is the term *pedocal* derived? In what areas do pedocals form? Describe the development of a pedocal.
13. (a) What does the word *clastic* mean? What are clastic rocks?
    (b) How are the fragments in a clastic rock held together?
    (c) Clastic rocks are classified by what property?
    (d) Arrange the following names so that those with larger rock fragments come before those with smaller fragments: silt-

stone (mudstone), shale, conglomerate, and sandstone.

14. (a) What is the difference between a biologic and a chemical rock?

(b) What is an evaporite? Is an evaporite a biologic or a chemical rock?

(c) Of what mineral is limestone composed? Is most limestone deposited by marine organisms or by nonliving processes? Name and describe two types of limestone.

(d) How does dolomite differ from limestone? How does dolomite form? Is the formation a biologic or a chemical process?

(e) Name two nonclastic sedimentary rocks that are composed of silica, and describe the ways in which they form. One of these rocks looks very much like chalk; what test can you perform to distinguish it from chalk?

(f) If large amounts of plant material are buried and compressed before they have a chance to decay, the sedimentary rock _____ forms. What element composes this rock?

15. (a) Explain how pressure, shear, and stress differ from one another.

(b) What is dynamic metamorphism? Where does it occur? Which is the principal agent in dynamic metamorphism?

(c) What is thermal, or contact, metamorphism? Which two agents of change seem to be predominant during contact metamorphism?

(d) What is regional metamorphism? Where does it occur?

16. (a) What are foliated rocks?

(b) True or false: Most regionally metamorphosed rocks are foliated.

(c) When shale is subjected to metamorphism, its _____ minerals are recrystallized into a group of platy minerals called _____.

(d) Describe the differences among shale, slate, schist, and gneiss, and state the conditions under which slate, schist, and gneiss are formed.

17. (a) What do granite, shale, and gneiss have in common?

(b) Are many igneous rocks believed to be formed from melted metamorphic rock? How is most granite believed to be actually formed?

18. (a) What conditions are absent during the formation of a nonfoliated metamorphic rock?

(b) What changes occur when limestone is metamorphosed to marble?

(c) What changes occur when quartz sandstone is metamorphosed to quartzite?

(d) Hornfels is a metamorphic rock and should not be confused with the mineral hornblende. What is hornfels?

19. If a sedimentary rock and an igneous rock contain the same minerals, can they be changed into the same metamorphic rock?

20. What is an ore? Why are metal mines classified as nonrenewable resources?

21. (a) Explain the difference between crystal settling and a late magmatic deposit in which the fluid settled to the bottom.

(b) Hydrothermal deposits are the most common sources of ore. Describe how they form.

(c) What is a pegmatite?

22. How does contact metamorphism create ore at some places?

23. (a) What is an evaporite? What minerals are found in evaporite beds?

(b) Explain the following statement: the process that creates a placer deposit does not form new ore but mechanically concentrates previously existing ore.

24. (a) Intensive weathering has produced many of the world's deposits of aluminum ore. What is this ore called? Explain the role of weathering in its formation.

(b) Explain why weathering has caused more than one investor to go bankrupt.

25. If you wished to prospect for uranium in the western United States, what kind of rock would interest you? From where did the uranium originate, and how was it deposited?

26. What is geothermal energy? From what sources is it usually tapped? California and Iceland both lie on plate boundaries. Do you think geothermal energy is likely to solve the energy problems of the American Midwest?

27. (a) Describe the formation of peat and coal.
(b) Where does petroleum usually form? In what kind of rock does it usually form?
(c) State the conditions required for the formation of an economic oil field.

SUGGESTED READINGS

Barnea, Joseph. "Geothermal Power," *Scientific American* (January 1972), Vol. 226, No. 1, pp. 70–77.

Bonatti, Enrico. "The Origin of Metal Deposits in the Oceanic Lithosphere," *Scientific American* (February 1978), Vol. 238, No. 2, pp. 54–61.

Brobst, D. A., W. P. Pratt, and V. E. McKelvey. *Summary of United States Mineral Resources.* U.S. Geological Survey Circular 682. Washington, D.C.: U.S. Government Printing Office, 1973, 19 pp.

Cheney, E. S. "The Hunt for Giant Uranium Deposits," *American Scientist* (January–February 1981), Vol. 69, No. 1, pp. 37–48.

Decker, Robert, and Barbara Decker. "The Eruptions of Mount St. Helens," *Scientific American* (March 1981), Vol. 244, No. 3, pp. 68–80.

Decker, Robert, and Barbara Decker, *Volcanoes.* San Francisco: W. H. Freeman, 1981, 244 pp.

Dietrich, R. V., and B. J. Skinner. *Rocks and Rock Minerals.* Somerset, N.J.: John Wiley, 1979, 319 pp.

Eaton, G. P., and others. "Magma beneath Yellowstone National Park," *Science* (May 23, 1975), Vol. 188, No. 4190, pp. 787–796.

Ernst, W. G. *Earth Materials.* Englewood Cliffs, N.J.: Prentice-Hall, 1969, 150 pp. (paperback).

Heiken, Grant. "Pyroclastic Flow Deposits," *American Scientist* (September–October 1979), Vol. 67, No. 5, pp. 564–571.

Hsü, K. J. "When the Mediterranean Dried Up," *Scientific American* (December 1972), Vol. 227, No. 6, pp. 26–36.

Keyfitz, Nathan. "World Resources and the Middle Class," *Scientific American* (July 1976), Vol. 235, No. 1, pp. 28–35.

Kittleman, L. R. "Tephra," *Scientific American* (December 1979), Vol. 241, No. 6, pp. 160–177.

Kuenen, P. H. "Sand," *Scientific American* (April 1960), Vol. 202, No. 4, pp. 94–106. Reprint 803, W. H. Freeman, San Francisco.

Laporte, L. F. *Ancient Environments.* Englewood Cliffs, N.J.: Prentice-Hall, 1968, 116 pp. (paperback).

Menard, H. W. "Toward a Rational Strategy for Oil Exploration," *Scientific American* (January 1981), Vol. 244, No. 1, pp. 55–65.

Moore, J. G. "Mechanism of Formation of Pillow Lava," *American Scientist* (May–June 1975), Vol. 63, No. 3, pp. 269–277.

Park, C. F., Jr. *Earthbound: Minerals, Energy, and Man's Future.* San Francisco: Freeman, Cooper, 1975, 279 pp.

Peck, D. L., T. L. Wright, and R. W. Decker. "The Lava Lakes of Kilauea," *Scientific American* (October 1979), Vol. 241, No. 4, pp. 114–128.

Pettijohn, F. J., and P. E. Potter. *Atlas and Glossary of Primary Sedimentary Structures.* New York: Springer-Verlag, 1964, 370 pp.

Sheets, P. D., and D. K. Grayson (eds.). *Volcanic Activity and Human Ecology.* New York: Academic Press, 1979, 644 pp.

Skinner, B. J. *Earth Resources,* 2nd ed. Englewood Cliffs, N.J.: Prentice-Hall, 1976, 152 pp. (paperback).

ÉVIDENCE FOR CONTINENTAL DRIFT

THE EARTH'S MAGNETISM

SEA-FLOOR SPREADING

PLATE TECTONICS

CONTINENTS

Deformed Rocks

FOLDS

FAULTS

Larger Features of Continents

CHAPTER 4

The San Andreas fault cuts through California. (Photo by R. E. Wallace, U.S. Geological Survey.)

PLATE TECTONICS AND THE EARTH'S SURFACE STRUCTURE

EVIDENCE FOR CONTINENTAL DRIFT

The idea that the continents may have moved, or drifted, is a very old one. The shapes of Africa and South America suggest that they may have been joined at one time and have since moved apart. Late in the 19th century, much evidence for this latter point was gathered by geologists working in widely scattered places in the southern hemisphere. Those studies were largely ignored by the majority of geologists who lived and worked in Europe and North America, where the evidence for continental movement is not obvious. Another reason that continental drift was not taken seriously by many geologists until after World War II is that no reasonable mechanism was proposed that could move continents across the face of the earth.

The evidence that the southern continents were once parts of a much larger continent called **Gondwanaland** goes far beyond the jigsaw fit obvious on maps. Similarities include rock types, ages of the rocks, fossils in the rocks, and rock structures, such as folds and faults. These matches are not perfect because erosion has removed some of the rocks, deposition has covered some, and later deformation has changed some. This is like putting together a jigsaw puzzle with some of the pieces missing and some deformed. The difficulties are really worse than that because the many layers of rocks mean that it is really a multilayer jigsaw puzzle with many pieces missing or deformed.

One of the most compelling matches is the glaciation found on all of the southern continents. In the 19th century, glacial deposits and glacially eroded bedrock were found in Africa, South America, India, and Australia (Figure 4-1). Later, glaciation of the same age was F 4-1 found on Antarctica. These glacial rocks are all about 300 million years old. On each of the continents, the evidence is clear that the

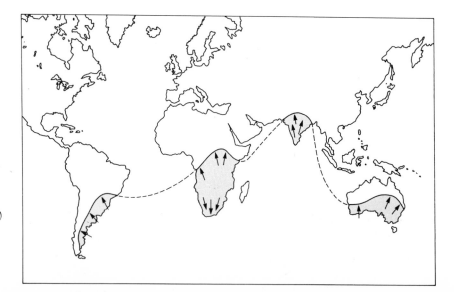

Figure 4-1
Areas of ancient glaciation (about 300 million years ago). Arrows show the direction of ice movement.

Figure 4-2
Gondwanaland reconstructed, showing the areas of ancient glaciation.

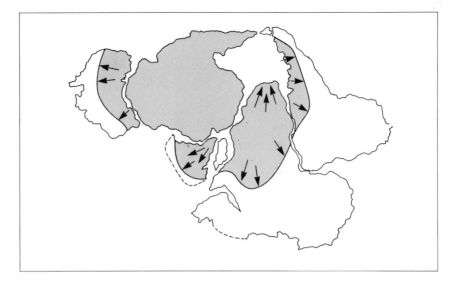

glaciation was done by continental glaciers, not mountain glaciers. The distribution of glacially deposited rocks and the scratched and striated bedrock indicate the direction of movement of the ice.

The problem presented by these glacial rocks is that their present distribution is from well north of the equator to near the south pole. It is difficult to imagine continental glaciers forming at the same time in such widely differing latitudes. Even if the rotational axis of the earth, the south pole, is moved into the middle of the glaciated areas, some of the glacial deposits would still be within about 10 degrees of the equator. The problem disappears if the continents are brought together in a reconstruction of Gondwanaland (Figure 4-2). The glaciated areas meet, and the directions of the ice movement are in agreement.

Fossils, too, suggest that the southern continents once formed the single continent Gondwanaland. On each of the continents the beds above the glacial deposits contain a distinctive fossil flora called the **Glossopteris flora** for its most abundant genus. It seems unlikely that the same plants would be found on continents widely separated by oceans as they are today. It is equally unlikely that the same plants would grow in both equatorial and polar climates. Fossils of the swimming reptile *Mesosaurus* are also found on most of the southern continents. Again, it seems very unlikely that an animal about 50 centimeters (18 inches) long that lives in fresh water would swim across an ocean.

THE EARTH'S MAGNETISM

The next type of evidence for continental drift came from studies of the very weak magnetism impressed on rocks by the earth's magnetic field. Although these studies had begun earlier, it was in the 1950s that they reopened the whole question of continental movement and then provided critical evidence. Before considering rock magnetism, we will consider the earth's magnetic field and its origin.

The simplest evidence for the earth's magnetic field is that a compass needle points toward the earth's magnetic pole. The earth's north and south magnetic poles are not at the same places as the geographic, or rotational, poles. The magnetic north pole is north of central Canada, about 11½ degrees away from the north pole. Because a compass points toward the magnetic north pole, it points to the west of true north in eastern North America and to the east of true north in western North America. A simple compass is a magnetic needle, free to turn in the horizontal plane. If instead of using a simple compass we use one that is also free to pivot in the vertical plane, we can learn more about the earth's magnetic field. The needle in such a compass will align itself with the earth's magnetic field. The needle will still point toward the magnetic north pole, but its vertical angle will change with latitude. At the equator, it will be horizontal; at a magnetic pole, it will be vertical; and at points in between, the angle will vary in a regular way (Figure 4-3). The earth's magnetic F 4-3 field can be mapped with such an instrument, and the form of the field is similar to that of a bar magnet (Figure 4-3). The resemblance is not exact, and the earth's south magnetic pole is not at the point opposite to the north magnetic pole.

The earth's magnetic field is believed to be caused by electrical currents moving in the earth's fluid, nickel-iron core. Although the magnetic field is similar to that of a bar magnet, the temperatures within the earth are too high for such a magnet to exist. (The evidence for the earth's internal structure is described in Chapter 5.) It is believed that the fluid, conducting core acts much as a dynamo does; all that is needed is an electrical current to flow in the core. An electrical current causes a magnetic field, and a conductor moving in a magnetic

Figure 4-3
The earth's magnetic field. Note that the magnetic poles do not coincide with the geographic poles. The vertical angle at which a compass free to pivot in the vertical direction would point is also shown.

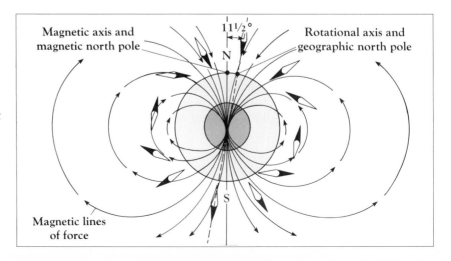

Figure 4-4
Paleomagnetic data show only the direction to the magnetic pole and the distance from the pole, or latitude.

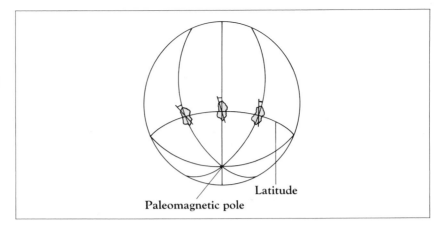

Latitude

Paleomagnetic pole

field has an electric current induced into it. These well-known effects can work together in the core to sustain a magnetic field. The situation is as if a dynamo produced an electric current by moving a conductor through a magnetic field, and the current is used to run an electric motor that turns the dynamo. Friction prevents such a perpetual-motion machine from working, so additional energy must be added. In the earth, the additional energy comes from motion of the fluid core caused by the earth's rotation.

Weak as the earth's magnetic field is, it can magnetize rocks. Basalt is a common volcanic rock, and its magnetism is easily measured. As an igneous rock such as basalt cools, the earth's magnetic field magnetizes the rock when it cools to a certain temperature called the **Curie point.** The Curie point temperature of basalt is about 500°C (900°F), a temperature at which basalt has already hardened because its melting temperature is much higher. When a rock cools through its Curie point, the earth's magnetic field is frozen into it. A much stronger magnetic field or heating to the Curie point is required to change this magnetism. The magnetism frozen in a rock is called **fossil magnetism** or **remanent magnetism** (not remnant).

Fossil magnetism records both the direction of the north (or south) magnetic pole and the latitude of the rock at the time the rock cooled. The horizontal component of the magnetism shows the direction of the pole, and the vertical angle indicates the latitude (Figure 4-4). Notice that this information does not reveal where at that lati- F 4-4 tude the rock formed.

One way to use fossil magnetism to show the movement of continents is to plot on a globe the locations of the north magnetic pole for each continent for rocks of a given age (geologic period). For each continent the pole will apparently be in a different place, showing that the continents must have moved relative to one another. The magnetic pole was only at one place at any time; so if we move the plotted continents and their magnetic poles so that their apparent

Figure 4-5
A. Locations of the magnetic poles in the geologic past for Africa and South America are plotted relative to present positions. **B.** If South America is moved into contact with Africa, the two plots coincide, showing they were once joined. Numbers are approximate ages in millions of years.

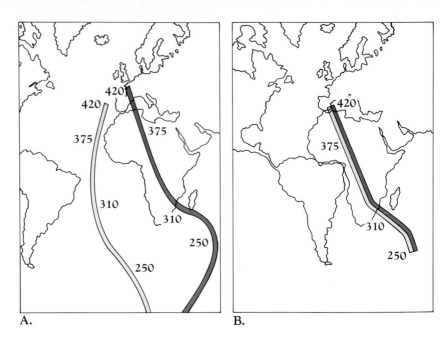

A.

B.

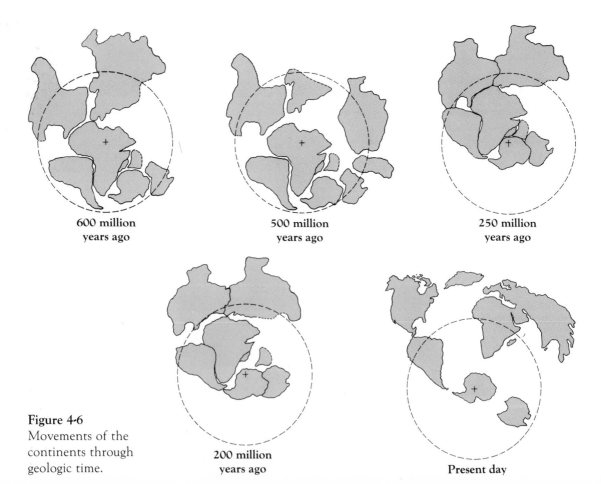

600 million years ago

500 million years ago

250 million years ago

200 million years ago

Present day

Figure 4-6
Movements of the continents through geologic time.

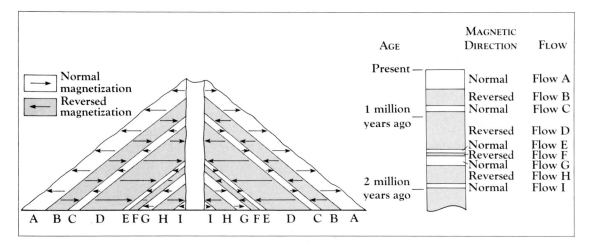

Figure 4-7
Dating the lava flows on the volcano enables construction of a time scale based on magnetism.

poles are at the same place, we have reconstructed the positions of the continents relative to each other at that time. In Figure 4-5, the pole locations for two continents have been plotted for several geologic periods, and the positions coincide if the continents are moved back to their earlier locations. The movements of the continents for the last 250 million years are shown in Figure 4-6. F 4-5 F 4-6

Studies reveal that the earth's magnetic field has reversed its polarity many times in the geologic past. These magnetic reversals are revealed when the fossil magnetism in a thick series of basalt flows is studied. In some flows the magnetism is **normal**—that is, the same as the earth's present field. In others the poles are **reversed**—that is, the fossil north direction is toward the present south pole (Figure 4-7). Dating of the basalt flows shows that these reversals were worldwide, and their times are well known. These magnetic reversals are very important in understanding the ocean floors, and they are the key to knowing that the continents have moved over the face of the earth. F 4-7

SEA-FLOOR SPREADING

The floors of the deep parts of the oceans are nearly flat and are geologically simple, with surprisingly uniform structures. Seismic studies reveal that beneath about 2 kilometers (1.2 miles) of interlayered sedimentary rocks and basalt, the crust underlying the oceans consists of 4.8 kilometers (3 miles) of basalt (Figure 4-8). The only interruptions to this uniform structure are submarine mountains that are basalt volcanoes. At places these volcanoes reach the surface of the oceans and form islands, such as the Hawaiian Islands. F 4-8

If we move closer to the center of the oceans, the situation changes drastically. The ocean floor rises, reducing the water depth, the bottom topography becomes very irregular, and volcanic activity is abundant (Figure 4-9). These areas are called the **mid-ocean ridges,** or **rises,** and they form a continuous ridge through all of the oceans (see Figure 4-15, p. 107). At a few places the mid-ocean ridge is high enough F 4-9

Figure 4·8
The structure of the ocean floors.

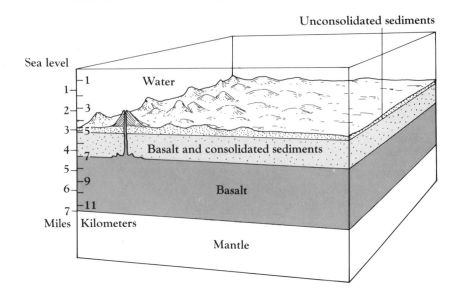

that parts of it form islands. Iceland is the largest of these islands, and there the ridge's features can be seen. Iceland is made of basalt, and its volcanoes, hot springs, and geysers show that it is geologically young. In late 1963 a new volcanic island, Surtsey, was formed by eruption about 19 kilometers (12 miles) to the southwest, and in 1973 an eruption began on Heimaey, just southwest of Iceland.

Submarine studies on the mid-ocean ridge reveal similar volcanic activity. At most places a valley, apparently caused by faulting, is at the center of the ridge, and a similar feature is also found in Iceland. The ridge itself is made of many ridges parallel to the main ridge. Some of the peaks are volcanoes, and some are formed by faulting.

The significance of these features is revealed when the age of the ocean floors is studied. The volcanic activity on the mid-ocean ridges suggests that these features are young, and dating of their rocks confirms that they are. Rocks can be dated using either the radioactive minerals or the fossils that they contain; the methods of geologic dating are described in Chapter 7. As one moves away from the ridges in either direction, the age of the ocean crust is found to be older, and this age increases regularly with distance from the ridge crest. Of course, the uppermost sediments on the ocean floor are all of the same current age. The rocks that are older with distance from the crest are those below the sediments, and those rocks form the basaltic crust of the oceans. It is necessary to drill through the sediments to obtain samples of the oceanic crustal rocks. These observations suggest that volcanic activity at the mid-ocean ridges forms new crust, which pushes aside the older crust. This process is called **sea-floor spreading.**

Magnetic studies also show sea-floor spreading. For many years it was known that the magnetic intensity measured over the oceans

Figure 4-9
Mid-ocean ridge.

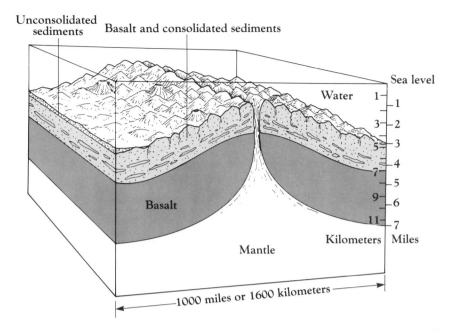

forms stripes of alternating high and low magnetic intensity. (These measurements are made by a magnetometer, airborne or towed by a ship.) These magnetic stripes were a puzzle because they are unlike anything found on the continents. The magnetic reversals noted in the last section proved to be their cause. The magnetic stripes are parallel to the mid-ocean ridges. As each portion of the mid-ocean ridge crest is formed by the cooling and solidifying basalt, the earth's magnetism is impressed in that basalt. At times when the earth's magnetic field is in the orientation that it is now, the rocks have normal magnetism. The parts of the mid-ocean ridge crest that formed when the earth's magnetic field was in the opposite sense are said to have reversed magnetism. Reverse-magnetized rocks have their north pole at the present south pole.

The magnetic stripes result from oceanic crust with stripes of normal and reverse magnetism. The magnetic intensity measured over the oceans is the combination of the magnetism of the rocks and the earth's magnetic field. Where the rocks are normally magnetized, their field adds to the earth's field to form a high magnetic intensity. Where the crust is reverse magnetized, its field is opposite to the earth's and so subtracts from the earth's field, forming a stripe of reduced magnetic intensity (Figure 4-10).

F 4–10

This concept can be tested because the timing or age of the magnetic reversals is known from studies of volcanic rocks on the continents. Where samples have been recovered by drilling, the age of the rocks and their magnetic direction are in agreement. Another test is the symmetry of the magnetic stripes. Figure 4-11 shows the magnetic intensity as measured across a mid-ocean rise and shows a mirror im-

F 4–11

Figure 4-10
The magnetic stripes in the oceans are formed by reversals of the earth's magnetic field. The earth's magnetic field is implanted in the newly formed ocean crust at mid-ocean rises.

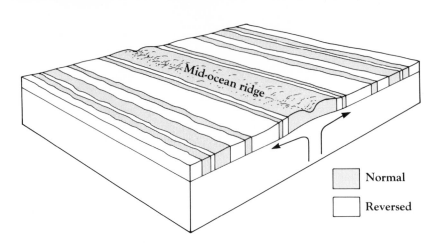

Normal

Reversed

Figure 4-11
The magnetic stripes are symmetrical about a mid-ocean rise, as shown by the observed and reversed magnetic intensity plots. A similar plot is obtained by calculation, assuming a rate of sea-floor spreading. (After F. J. Vine.)

age of that profile. The similarity is obvious, showing that the magnetic intensity is symmetrical across the mid-ocean ridges. The test can be carried further. If we assume a reasonable rate of sea-floor spreading, we can calculate the magnetic profile because we know the times of magnetic reversal. The calculated profile is very close to the one actually measured. Thus, it can be said that the basaltic crust of the ocean acts like a tape recorder: the basaltic crust is the tape, sea-floor spreading moves the tape, and the earth's magnetism is recorded on the tape. From the width of the magnetic stripes, the rate of sea-floor spreading is generally shown to be a few centimeters a year, up to about 10 centimeters per year (4 inches per year).

One problem remains. If new crust is being created at the mid-ocean ridges, what happens to the old crust? If new crust is continually created and nothing happens to the old crust, the earth

Profile reversed

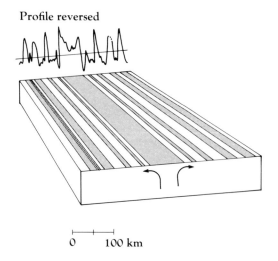

0   100 km

Observed profile

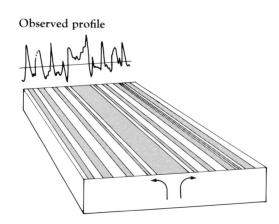

would be expanding like a balloon. We must look for where the oceanic crust is consumed or returned into the earth. The margins of the oceans are likely places, and two types of ocean-continent boundary occur.

At many coasts a **continental shelf** lies between the continent and the ocean. Continental shelves have an average water depth of only about 60 meters (200 feet) and are about 70 kilometers (42 miles) wide on the average. The transition between the shallow shelf and the deep ocean is the **continental slope.** The east coast of North America is an example of this kind of coastline. The shelf-slope coastlines are geologically quiet, with no volcanic activity or earthquakes. (These quiet coasts are described in more detail in the chapters on oceanography.) Other coasts do have such activity, and they are probably the places where oceanic crust is absorbed.

The active margins of the oceans are called *convergent boundaries.* Volcanoes and deep-ocean trenches are associated with these margins. At many places convergent boundaries are marked by **volcanic island arcs.** They are called volcanic island arcs because they are strings of islands made by active volcanoes, and they form curves or arcs on a map (Figure 4-12). The arcs are also areas of many earthquakes, and F 4-12 they are associated with the deep-ocean **trenches** that are the deepest places in the oceans. Convergent boundaries are where the ocean crust moves downward into the earth (Figure 4-13). The downward- F 4-13

Figure 4-12
Volcanic island arcs of the Pacific Ocean. The trenches associated with the island arcs are shown.

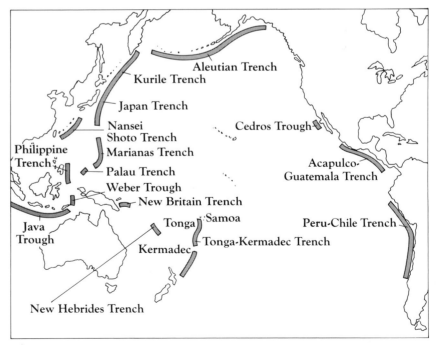

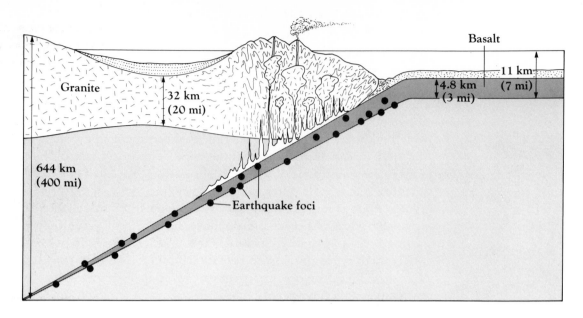

Figure 4-13
Cross-section of a convergent boundary, showing volcanic islands, trench, and deep earthquake foci.

moving slab of crust apparently causes the earthquakes as it moves past the other rocks because the earthquakes occur on a plane sloping below the volcanoes. The deepest earthquakes on the planet occur in this zone, and they are as deep as 700 kilometers (420 miles). It is probable that at that depth the temperature and pressure are such that the slab of oceanic crust is no longer rigid. The volcanic rocks are apparently formed by partial melting of the descending slab of oceanic crust and sediments. The volcanoes are andesite and so are compositionally different from the basalt volcanics of the oceans. The deep trench on the oceanic side of a convergent boundary is apparently also caused by the downward-moving slab of oceanic crust. Most island arcs, especially those on the west side of the Pacific Ocean, have shallower seas between them and the continent (Figure 4-13). Along the west coast of Central America and South America, the trench is at the edge of the continent, and the andesite volcanoes, such as the Andes Mountains, are on the continent itself.

PLATE TECTONICS

Sea-floor spreading shows how the floors of the oceans move. We saw previously that there is much evidence to indicate that the continents have also moved. **Plate tectonics** brings these ideas together, showing that large parts of the earth's surface—the plates—move as units and that these movements cause many of the structures of the earth's crust. The moving units are called plates because they are quite thin in relation to their surface area. The term *tectonics* means the defor-

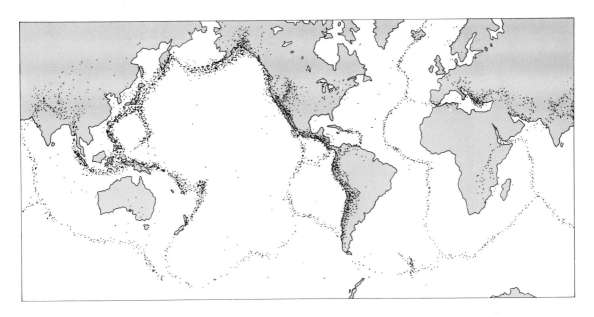

Figure 4-14
Epicenters of all earthquakes for a 7-year period. The epicenters outline the plates shown in Figure 4-15.

mation of rocks and the structures that result. The outlines of the plates can be seen from study of earthquakes because earthquakes result from movement of the plates past each other. Figure 4-14 shows a plot of all of the earthquakes that occurred over the whole earth in a 7-year period. The plate outlines are clear, and there are fewer than a dozen plates.

The boundaries of the plates can be determined in Figure 4-15.

Figure 4-15
The crustal plates.

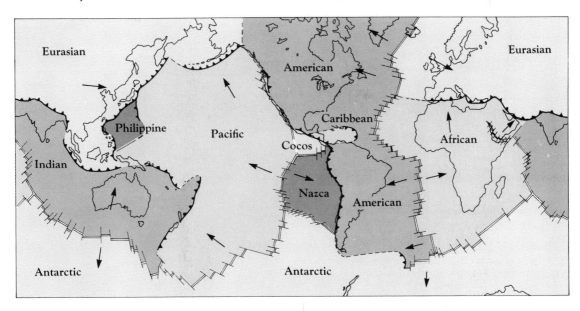

F 4-14

F 4-15

Figure 4-16
Plate tectonics diagram. Plates form at mid-ocean ridges, are consumed at convergent plate boundaries, and pass each other on transform faults.

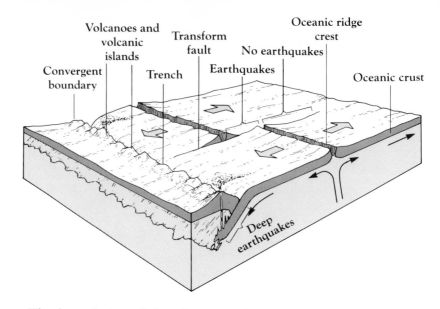

The boundaries of the plates can be determined in Figure 4-15. Each plate begins at a mid-ocean ridge and moves away from the ridge toward a trench or island arc, where it is consumed. The third and last type of boundary is the *transform fault,* where one plate moves past another. The three types of plate boundaries and the earthquakes associated with each are summarized in Figure 4-16. We will now consider each of these types of boundary and the structures associated with each.

F 4-16

Figure 4-17
Photograph from space looking northeast toward the Sinai peninsula. A spreading center in the Red Sea to the right split the Sinai peninsula from Africa. (Photo from NASA.)

Figure 4-18
Ocean-continent
plate collision.

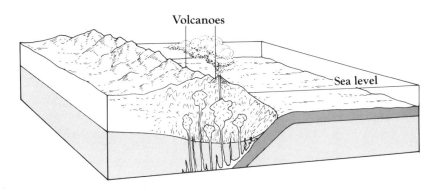

A mid-ocean ridge is a **divergent boundary** because the two plates in contact are moving away from each other. Divergent boundaries can also form under continents—for example, at the Red Sea (Figure 4-17). The Red Sea apparently formed because a spreading center F 4–17 similar to a mid-ocean ridge formed on the border between Africa and the Arabian Peninsula. As the continent split, the Red Sea formed. An earlier stage of the same process seems to be the cause of the rift valleys of eastern Africa (Figure 4-15). These African rift valleys are downfaulted valleys that are volcanically active.

Convergent boundaries are places where plates come together or collide; there are three types. An *ocean-continent convergence* is shown in Figure 4-18. Continental rocks are less dense than the ocean crust, F 4–18 so they tend to remain at the surface, and the ocean plate moves downward. A trench develops at the coast, and andesite volcanoes form on the continent. The west coast of South America is an example of such an ocean-continent plate collision. The place where the ocean plate is moving downward is called a **subduction zone.** Generally the original continental margin is a continental shelf and slope composed of sedimentary rocks. These rocks are deformed by folding and faulting as the plates collide and the subduction zone develops. These rocks are also metamorphosed and intruded by the magmas that cause the andesite volcanoes. These processes transform the original continental margin into mountains.

The second type of convergent boundary is the *ocean-ocean plate convergence*, shown in Figure 4-19. When two oceanic plates meet, one F 4–19 of them moves downward or is subducted. In this case a volcanic island arc forms above the subduction zone, with an area of ocean between the island arc and the continent. Japan and much of the western Pacific area are examples of this type of plate convergence.

The third type of convergence occurs where two continental plates collide. In this case, both plates remain at the surface and no new subduction zone forms. Notice that an initial subduction zone is nec-

Figure 4-19
Ocean-ocean plate collision.

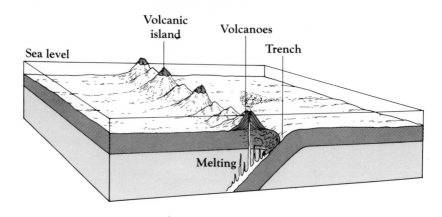

Figure 4-20
Continent-continent plate collision.

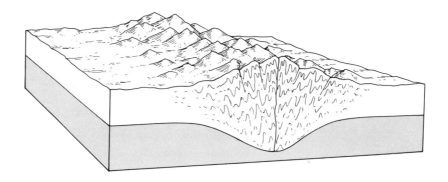

essary to allow the two continents to collide (Figure 4-20). A *continent-* F 4-20
continent collision deforms the rocks and makes continents thicker near
the zone of convergence. An example is India's collision with Asia;
the resulting Himalaya Mountains are the highest on earth.

Transform faults, the third type of plate boundary, are the places
where one plate moves past another (Figure 4-16). A map showing
the mosaic of plates (Figure 4-15) reveals that transform faults are at
places where the mid-ocean ridges are offset. Where plates move past
each other in opposite directions, shallow earthquakes occur; but
where the plates on each side of the transform fault are moving in the
same direction, no earthquakes occur. Notice in Figure 4-16 that the
direction of movement of the plates is the opposite of movement on a
simple fault that would cause the offset of the mid-ocean ridge. (Look
ahead to Figure 4-27, p. 116, which shows how an ordinary fault
causes the offset of a feature.) The San Andreas fault in California is
a transform fault; that fault has been the cause of many destructive
earthquakes. The San Andreas fault is about 1000 kilometers (620
miles) long, and the total movement has been estimated at about 560
kilometers (350 miles).

CONTINENTS
We know more about the rocks on the continents than we do about those under the oceans. On the continents we can study the minor structures in the rocks as well as the larger structures, but, because we must use indirect methods in studying rocks under the oceans, we see only larger structures. We will begin with the smaller features that can be seen in a single outcrop or in a small area such as a mountain.

Deformed Rocks

The continents are underlain by granitic rocks, but their surfaces are largely covered by sedimentary rocks. These sedimentary rocks, such as sandstone, limestone, and shale, were originally deposited in flat layers, or beds. Where these beds are tilted, bent, or broken, their deformation is obvious. At other places, the sedimentary rocks have been intruded by igneous rocks or have been metamorphosed, and these processes also deform the rocks. We begin by considering folding and faulting. When deforming forces are applied to rocks, they either bend or fold, or they break or fault. If the rocks are brittle, they tend to fault; if they are weak, they will probably be folded. The temperature and pressure affect the strength of rocks, so faulting is favored near the surface, and deeper rocks are more likely to respond by folding.

FOLDS Folded rocks generally record compressive forces, and, as we will see, compressive effects are common in continental rocks. Figure 4-21 shows how compression can cause **folding** and that folded rocks F 4-21 occupy less area than they did before they were folded. Folds may be open and upright as in Figure 4-22A, or they may be tight as in Figure F 4-22 4-22B. In some cases the folds are overturned as in Figure 4-22C. A fold is said to be overturned if the beds on one limb are inverted— that is, if the older rocks are above the younger rocks.
   Basin-like folds are called **synclines** (Figure 4-23B), *and hill-like folds* F 4-23 *are* **anticlines** (Figure 4-23A). Notice that eroded synclines have young rocks in their centers and that eroded anticlines have older rocks in their cores. In tight folds it is better to use the relative ages of the

Figure 4-21
Folds are generally caused by compression, and folded rocks take up less area than they did before they were folded.

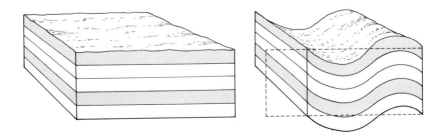

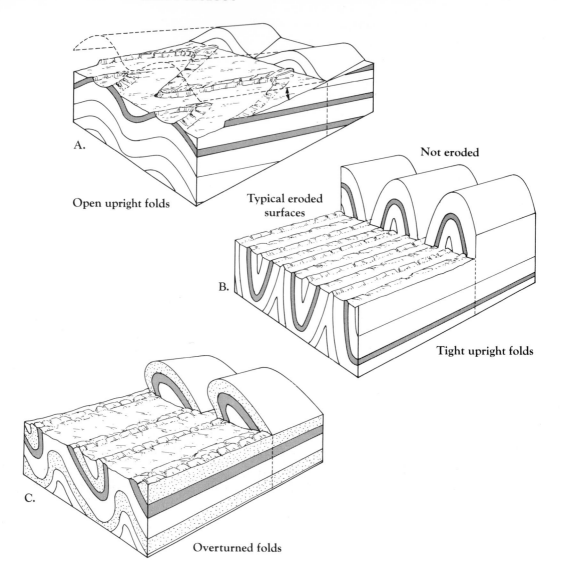

A.

Open upright folds

Not eroded

Typical eroded surfaces

B.

Tight upright folds

C.

Overturned folds

Figure 4-22
Types of folds. Erosion exposes older rocks in the cores of anticlines and younger rocks in the centers of synclines.

beds to define anticline and syncline. In an undisturbed sequence of sedimentary rocks, the oldest beds are at the bottom.

FAULTS **Faults** *are breaks in rocks along which movement has occurred.* Earthquakes are caused by fault movements. Faults can be classified by the type of movement that has occurred on them. Figure 4-24 **F 4-24** shows how faults can be caused by either compression or tension. **Reverse faults** are caused by compression, and the area occupied by the rocks is reduced. **Normal faults** are tensional, and the area of the rocks is enlarged. Reverse faults are much more common than normal faults. This prevalence and the compression recorded by folds

Figure 4-23
Folds. **A.** Anticline.
B. Syncline. **C.**
Overturned fold. **D.**
Anticline in
sandstone, Maryland.
E. Syncline in thin-
bedded limestone,
Texas. (Photo **D** by G.
W. Stose and **E** by C.
C. Albritton, Jr.; U.S.
Geological Survey.)

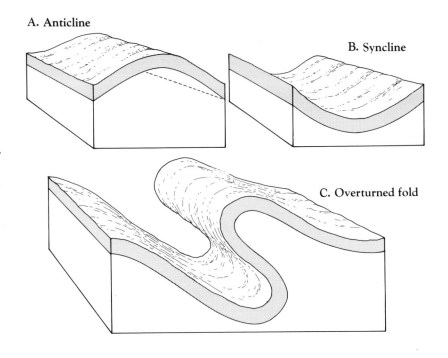

A. Anticline

B. Syncline

C. Overturned fold

D.

E.

suggest that compression is the more common type of force in the
earth's crust. Most normal faults are caused by the stretching associated with uplift rather than by tension.

Thrust faults are a special type of reverse fault. They differ from
most reverse faults in that the fault plane is nearly horizontal (Figure
4-25). The amount of movement on many thrust faults is very large. **F 4-25**
In most cases the displacement on reverse faults is measured in
meters, but on some thrust faults it can be measured in kilometers.

Faults that break the surface initiate erosion that in turn tends to
obliterate the surface break, as shown in Figure 4-26. The figure also **F 4-26**
shows how the beds are generally bent near a fault plane. In the ab-

Figure 4-24
Faults. **A.** Reverse faults are caused by compression and shorten the width of the block. Normal faults are tensional and lengthen the block. **B.** Normal (vertical) fault in sandy shale, Tennessee. (Photo by A. Keith; U.S. Geological Survey.)

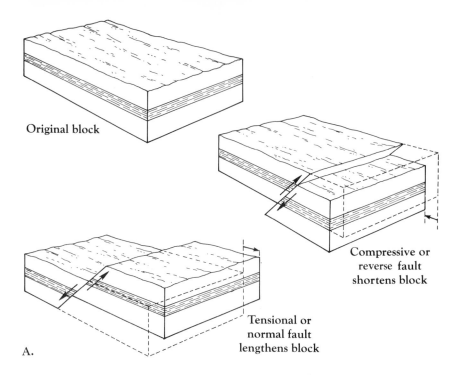

Original block

Compressive or reverse fault shortens block

Tensional or normal fault lengthens block

A.

B.

sence of a feature that is cut by the fault and can be seen on both sides of the fault, the actual amount and direction of movement on a fault cannot be determined. As might be expected, the movement on most faults is diagonal.

Some fault planes are nearly vertical, and the displacement is horizontal along the fault plane. Another type of fault with horizontal

Figure 4-25
A. Thrust faults are reverse faults with shallow fault planes. Movements on some are measured in kilometers. **B.** Thrust fault in limestone that is bent at the fault plane, Chile. (Photo by K. Segerstrom; U.S. Geological Survey.)

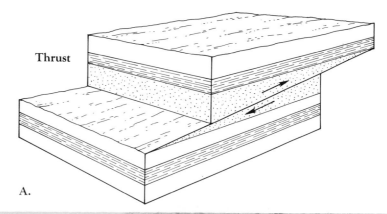

B.

movement is the *transform fault*. This type of fault is associated with sea-floor spreading, and was described earlier. Figure 4-27 shows the F 4-27 difference in movements between an ordinary fault and a transform fault.

Larger Features of Continents

The folds, faults, and other evidence of rock deformation that we have been considering are generally concentrated in mountain

Figure 4-26
Erosion modifies a
surface fault.

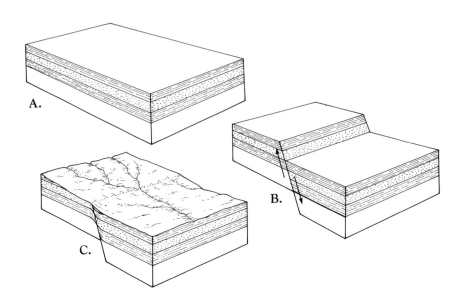

A.

B.

C.

Figure 4-27
Movement on a
transform fault is in
the opposite sense to
the movement on an
ordinary fault.

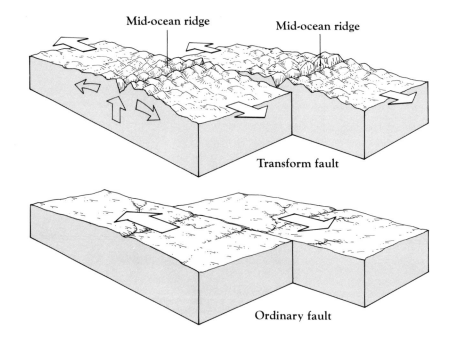

Mid-ocean ridge   Mid-ocean ridge

Transform fault

Ordinary fault

ranges. Mountain ranges with folded and faulted rocks are unique, as far as we know, because they are not found on any other planet. The earth's mountains are of several types (Figure 4-28), but the moun- F 4–28 tains of interest here are those with folded and faulted sedimentary rocks and cores of igneous and metamorphic rocks. Mountain ranges of this type are found on both the east and west coasts of North America and elsewhere.

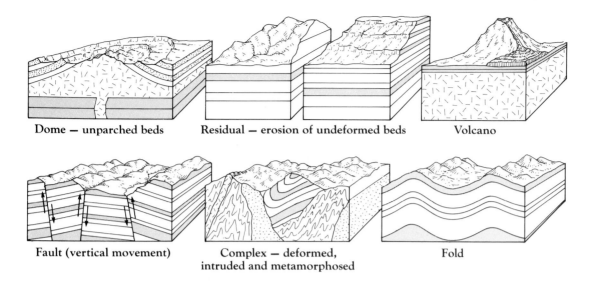

Dome — unparched beds   Residual — erosion of undeformed beds   Volcano

Fault (vertical movement)   Complex — deformed, intruded and metamorphosed   Fold

Figure 4-28
Types of mountains.

The first recognition that mountains are different from the rest of the continents came in 1859, when James Hall noted that the sedimentary rocks in the Appalachian Mountains were much thicker than those farther west. Hall was studying fossils, and he observed that rocks of a given age and containing the same fossils thickened and contained more sandstone and shale in the mountains, and they were thinner with more limestone in the interior of the continent. In 1873 James Dwight Dana noted that this thickening occurred throughout the Appalachian Mountains, and he realized that the thicker sedimentary rocks meant that a subsiding basin the size of the Appalachian Mountains had once existed. He gave the name **geo-syncline** to this huge elongate basin. The geologic history of the continents has been largely the development of geosynclines. We now recognize that geosynclines are closely related to sea-floor spreading and plate tectonics. We will begin with the history and development of a typical geosyncline.

A geosyncline begins as an area of subsidence in which sediments are deposited. The features in geosynclinal sedimentary rocks, such as ripple marks, mud cracks, and fossils, all indicate that the water was shallow. This implies slow subsidence over a long time. Geosynclines form at the margins of continents, and so continental slopes and volcanic island arcs are probably geosynclines at the sedimentation phase. During this phase, geosynclines typically acquire between 10,000 and 15,000 meters (30,000 and 45,000 feet) of sedimentary rocks. Rocks of the same age farther inland are typically only about one-tenth this thick. Sedimentation continues for hundreds of millions of years. The sedimentary rocks on the ocean side of a geo-

Figure 4-29
Deformation of
geosynclines. **A.**
Geosyncline develops
on a continental shelf
and slope. **B.** Volcanic
island arc develops
and deforms the
geosyncline. **C.**
Continental collision
causes a second
deformation of the
geosynclinal rocks.

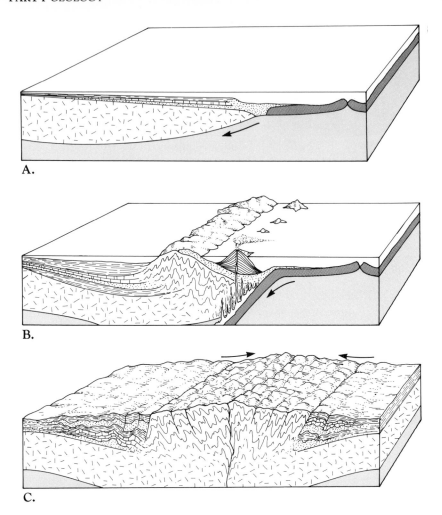

syncline typically contain more sandstone and shale than those
farther inland, and volcanic rocks are also common on the oceanic
side.

The sedimentation phase ends when the rocks are deformed, probably caused by collision with another plate. More than one time of
deformation is recorded in the rocks of many mountain ranges. The
first deformation may occur when a subduction zone develops (Figure
4-29B). The resulting volcanic island arc can account for the clastic   F 4-29
and volcanic rocks on the oceanic side of the geosyncline. The final
phase of deformation may occur when the two continents collide
(Figure 4-29C). Intrusion of batholiths and metamorphism are associated with these collisions, and these deep-seated rocks are seen in the
cores of mountain ranges where the overlying folded and faulted sedimentary rocks have been removed by erosion. The granitic and metamorphic rocks formed during this deformation are similar to the

Figure 4-30
Age of the
continental crust
rocks of North
America (in billions
of years).

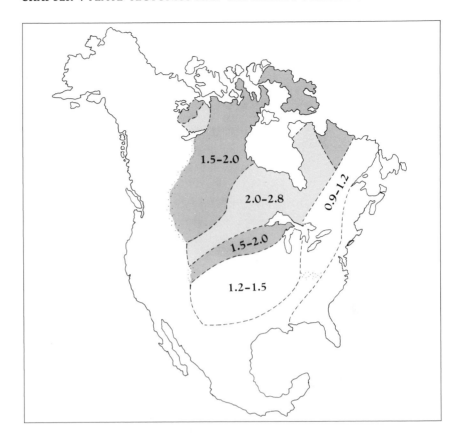

rocks that make up the continents, and so this may be the process by which the continents are formed and are enlarged.

The interiors of most continents are relatively low, flat areas, covered at places with a thin layer of flat or slightly deformed sedimentary rocks. Under the sediments, and exposed at the surface at many places, are igneous and metamorphic rocks of granitic composition. These rocks are the continental crust. They appear to be very ancient rocks because they are deformed and metamorphosed, but young rocks can also be metamorphosed, so their appearance may be deceiving. These rocks have been studied because at places they contain rich metal mines and because very old rocks may hold clues to the earth's early history. Radioactive minerals can be used to date these rocks that, because of their metamorphism, contain no fossils. Such dates reveal that rocks of similar ages form large belts (Figure 4-30). F 4-30 Further study of these belts reveals that their rocks have structures very much like those seen in the cores of present-day mountain ranges. Thus, these belts of old rocks may be the deeply eroded stumps of ancient mountains. The sizes and shapes of these belts are similar to those of present mountain ranges, and, in general, the oldest rocks are at the centers of the continents, with progressively younger rocks toward the edges. All of this suggests that mountain-building may be the process by which the continents are formed.

The concept of the geosyncline came from study of mountain ranges. We now see that geosynclines are continent–ocean borders and that the theory of plate tectonics explains the accumulation of sediments, their later deformation, and the building of new continents. Throughout geologic time, continents have collided, welded together, and later split apart as new spreading centers formed. We will attempt to trace these movements in Chapter 7.

SUMMARY

- The evidence for continental drift includes:
  the shapes of the continents and how they fit together
  the match of the rock types
  the match of geologic structures
  similarity of climates—glaciers
  similarity of fossils
  paleomagnetic data

- The earth's magnetic field is believed to be caused by electrical currents in the earth's liquid core.

- The earth's magnetic field is frozen into a rock such as basalt when the rock cools. This occurs at a temperature well below the melting point.

- The *fossil* or *remanent magnetism* reveals the latitude and north direction at the time the rock cooled.

- The polarity of the earth's magnetic field has reversed many times in the geologic past, and these reversals are the cause of the magnetic stripes of the oceans.

- At the mid-ocean ridges or rises (also called *divergent plate boundaries*), new ocean floor is created by basalt moving upward from the earth's interior.

- The ocean floors are progressively older as one moves away from the ridges in either direction.

- *Sea-floor spreading* is the concept that the oceanic crust is created in this way and moves outward from the mid-ocean rises.

- As new crust is created at the mid-ocean ridges, a similar amount is consumed at *convergent plate boundaries*.

- At convergent boundaries, the descending slab causes the deepest earthquakes, and partial melting produces magma. *Subduction* is a term used to describe this process. Volcanoes of andesitic composition, in many cases forming volcanic island arcs, are the result.

- *Plate tectonics* is the concept that thin (about 100-kilometer—60-mile) slabs or plates created by sea-floor spreading move across the earth's surface. Some of the plates are all ocean, and some are both ocean and continent.

- Movement of the plates is the cause of continental drift.

- *Rock deformation* occurs where two plates collide; and if one of the plates is made of oceanic crust, it will descend or be subducted.

- A *transform fault* is where two plates move past each other.

- *Folding* is generally caused by compression.

- Basin-like folds with younger rocks in the center are called *synclines*.

- Hill-like folds with older rocks in the center are called *anticlines*.

- *Faults* are breaks in rocks along which movement has occurred.

- *Normal faults* involve extensions of the rocks.
- *Reverse faults* are caused by compression.
- *Thrust faults* are a type of reverse fault with a very shallow or flat fault plane.
- *Geosynclines* are large, elongate areas of subsidence and sediment accumulation that are later deformed and uplifted to form mountain ranges. In this way new parts of continents are created.
- Geosynclines tend to form at continent-ocean boundaries, and their origin and later deformation are caused by plate tectonics.

KEY TERMS

continental drift	reverse remanent	trench	anticline
Gondwanaland	magnetism	plate tectonics	fault
Glossopteris flora	mid-ocean ridge, or	divergent boundary	reverse fault
Curie point	rise	convergent boundary	normal fault
fossil, or remanent,	sea-floor spreading	subduction zone	thrust fault
magnetism	continental shelf	transform fault	geosyncline
normal remanent	continental slope	folding	
magnetism	volcanic island arc	syncline	

REVIEW QUESTIONS

1. State two reasons why the idea of continental drift was rejected by most geologists until relatively recently.
2. Geologists think that the continents of the southern hemisphere were once joined to form the supercontinent Gondwanaland. How does the following evidence support this theory?
   (a) Glacial deposits and striations.
   (b) Fossils of the *Glossopteris* flora and of *Mesosaurus*.
3. (a) What is the difference between earth's geographic poles and its magnetic poles? Where are they located?
   (b) What is fossil magnetism? At what point in a rock's history is the magnetism imprinted?
   (c) What does the horizontal component of a rock's fossil magnetism record? What does the vertical component record?
   (d) Explain how fossil magnetism demonstrates past continental movement.
4. Describe the process called sea-floor spreading.
5. How has dating of the basalt rocks of the oceanic crust provided evidence of sea-floor spreading?
6. Alternating stripes of high and low magnetic intensity exist over the ocean floors.
   (a) What causes the high magnetic intensity? What causes the low intensity?
   (b) The stripes are parallel to what feature of the ocean floor?
   (c) Explain how magnetic studies of the ocean floor support the theory of sea-floor spreading. (Use your answers to Questions 6a and 6b and what you have learned about fossil magnetism.)
7. Describe the two types of ocean-continent boundary. Which is geologically active?
8. Where is old oceanic crust returned to the interior of the earth? Describe or make a labeled drawing of this process.
9. Compare the composition of the volcanic is-

land arcs with the composition of the other volcanoes in the ocean. How is the rock of the volcanic island arcs believed to form?

10. (a) What is a divergent plate boundary?
    (b) In what continent has a divergent boundary started to form?
    (c) What occurs at a divergent boundary?

11. List the three kinds of convergent boundary and describe what happens at each.

12. What happens along a transform fault? What determines whether a transform fault is the site of much earthquake activity?

13. Which process (folding or faulting) is more likely to occur deep within the earth's crust? Which is more likely to occur on or near the surface? Give a brief reason for your answers.

14. Folding is most often a response to what kind of force? Why?

15. (a) When the sedimentary rocks in an area have been left undisturbed, where are the oldest sedimentary layers found?
    (b) What is a syncline? Where are the oldest rocks located in a syncline?
    (c) What is an anticline? Where would you find the oldest rocks of an anticline?

16. (a) Draw a reverse fault. What kind of force causes this fault? Does reverse faulting increase or decrease the area occupied by the rocks?
    (b) Draw a normal fault. What is the cause of most normal faults?
    (c) How does a thrust fault differ from a reverse fault? Compare the amount of movement of the rocks along a reverse fault with the amount of movement along a thrust fault.
    (d) Draw a transform fault. What causes transform faults?

17. Describe the history of a geosyncline by answering the following questions:
    (a) Where do geosynclines form?
    (b) Describe the deformation that may occur. What kinds of rock may result from this deformation?

18. Why do geologists suspect that mountain-building may be the process by which continents are formed? (Answer the question by describing some of the evidence.)

SUGGESTED READINGS

Cook, F. A., L. D. Brown, and J. E. Oliver. "The Southern Appalachians and the Growth of Continents," *Scientific American* (October 1980), Vol. 243, No. 4, pp. 156–188.

Dewey, J. F. "Plate Tectonics," *Scientific American* (May 1972), Vol. 226, No. 5, pp. 56–68.

Dietz, R. S. "Geosynclines, Mountains and Continent-Building," *Scientific American* (March 1972), Vol. 226, No. 3, pp. 30–38.

Dietz, R. S., and J. C. Holden. "The Breakup of Pangaea," *Scientific American* (October 1970), Vol. 223, No. 4, pp. 30–41.

Elders, W. A., and others. "Crustal Spreading in Southern California," *Science* (October 6, 1972), Vol. 178, No. 4056, pp. 15–24.

Heezen, B. C., and I. D. MacGregor. "The Evolution of the Pacific," *Scientific American* (November 1973), Vol. 229, No. 5, pp. 102–112.

Heirtzer, J. R., and W. B. Bryan. "The Floor of the Mid-Atlantic Rift," *Scientific American* (August 1975), Vol. 233, No. 2, pp. 78–91.

James, D. E. "The Evolution of the Andes," *Scientific American* (August 1973), Vol. 229, No. 2 pp. 60–69.

Macdonald, K. C., and B. P. Luyendyk. "The Crest of the East Pacific Rise," *Scientific American* (May 1981), Vol. 244, No. 5, pp. 100–116.

McKenzie, D. P., and Frank Richter. "Convection Currents in the Earth's Mantle," *Scientific American* (November 1976), Vol. 235, No. 5, pp. 72–89.

McKenzie, D. P., and J. G. Sclater. "The Evolution of the Indian Ocean," *Scientific American* (May 1973), Vol. 228, No. 5, pp. 62–72.

Molnar, Peter, and Paul Tapponnier. "The Collision between India and Eurasia," *Scientific American* (April 1977), Vol. 236, No. 4, pp. 30–41.

Moorbath, Stephen. "The Oldest Rocks and the Growth of Continents," *Scientific American* (March 1977), Vol. 236, No 3, pp. 92–104.

Oliver, Jack. "Exploring the Basement of the North American Continent," *American Scientist* (November–December 1980), Vol. 68, No. 6, pp. 676–683.

Sclater, J. G., and C. Tapscott. "The History of the Atlantic," *Scientific American* (June 1979), Vol. 240, No. 6, pp. 156–175.

Tokosöz, M. N. "The Subduction of the Lithosphere," *Scientific American* (November 1975), Vol. 233, No. 5, pp. 88–101.

Valentine, J. W., and E. M. Moores. "Plate Tectonics and the History of Life in the Oceans," *Scientific American* (April 1974), Vol. 230, No. 4, pp. 80–89.

Vine, F. J. "Sea-floor Spreading—New Evidence," *Journal of Geological Education* (February 1969), Vol. 17, No. 1, pp. 6–16.

Wyllie, P. J. *The Dynamic Earth: Textbook in Geosciences.* New York: John Wiley, 1971, 416 pp.

EARTHQUAKES

Earthquake Prediction and Control

INTERNAL STRUCTURE OF THE EARTH

Deep Interior

Crust

CAUSE OF PLATE TECTONICS AND SEA-FLOOR SPREADING

CHAPTER 5

EARTHQUAKES AND THE EARTH'S INTERIOR

Three-story reinforced concrete building astride a fault was totally destroyed in the 1972 Managua, Nicaragua, earthquake. (Photo by R. D. Brown; U.S. Geological Survey.)

In this chapter we examine the interior of the earth and search for the energy source that causes plate tectonics. All studies of the deep interior must be done by indirect methods; the deepest drill holes are about 10 kilometers (6 miles). The study of earthquake waves that pass through the earth have revealed much about the interior, so we will consider them first. It is perhaps ironic that earthquakes, which bring misery to many people, are also the best window we have to the interior of our planet.

EARTHQUAKES

An earthquake occurs when two parts of the earth's crust move past each other. The break in the rocks along which this movement occurs is called a fault. A fault or break in the crust is caused when the forces within the earth exceed the strength of the rocks. Once a fault exists, movement on it occurs when the internal forces overcome the friction on the fault plane (Figure 5-1). Most earthquakes are the re- F 5-1 sult of movement along an existing fault, but the initial movement on a newly formed fault would also cause an earthquake. *An* **earthquake** *is the vibrations caused by movements along a fault. The study of vibrations in the earth caused by earthquakes or explosions is* **seismology.**

An earthquake causes three main types of seismic wave:

1. **P wave.** The P stands for primary because P waves travel fastest and so are the first to arrive at a point. They are also called P waves

Figure 5-1
Slow deformation continues until the rocks break suddenly, causing an earthquake.

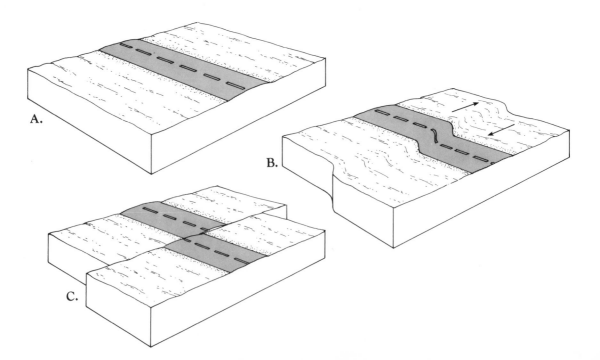

Figure 5-2
Ground motion
caused by P waves.

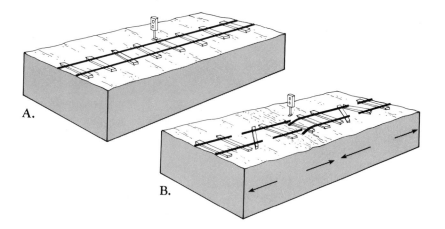

because they are push waves, like sound waves, and cause compression and rarefaction as they move. The ground motion that they cause is shown in Figure 5-2.

F 5–2

2. **S wave.** The S stands for secondary because S waves are the second waves to arrive at a point. They are also called S waves because they are shear waves, and their ground motion is shown in Figure 5-3.

F 5–3

3. **Surface wave.** These waves travel along the surface with a motion much like that of waves in water (Figure 5-4). They are the slowest of the waves.

F 5–4

P waves and S waves vibrate more rapidly than do surface waves, but their amplitude, or amount of movement, is much less. P waves and S waves travel through the earth and tell us much about the interior. Surface waves with their larger amplitude cause most earthquake damage.

When an earthquake is felt by people or is detected on a seismograph, one of the first questions asked is Where was the quake? *The*

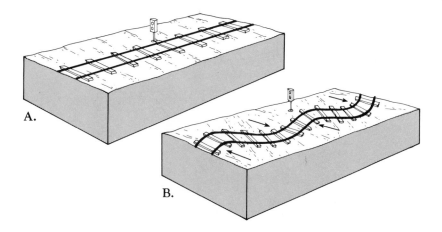

Figure 5-3
Ground motion
caused by S waves.

Figure 5-4
The ground motion caused by a surface wave is similar to the wave produced by a rock thrown into a pond.

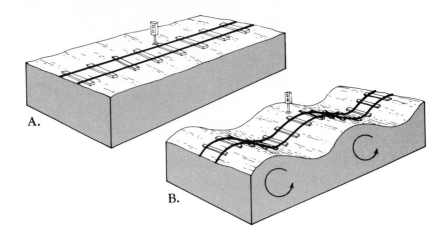

A.

B.

point in the earth where the earthquake occurs is called the **focus.** *The place on the surface directly above the focus is called the* **epicenter** (Figure 5-5). **F 5-5** The distance to the epicenter is proportional to the time difference between the arrival of the P wave and the arrival of the S wave. Because these waves travel at different velocities, the farther away the earthquake is, the longer the S minus P time. At a single seismic station, the S − P time locates a circle, the radius of which is the distance to the epicenter. The circles of S − P distance from two seismic stations will generally intersect at two points, one of which is the epicenter. If three stations are used, the circles will mutually intersect at only one point, which is the epicenter (Figure 5-6). **F 5-6**

The next thing people want to know is how big the earthquake was. The size of an earthquake can be measured in two ways. *The* **intensity** *of an earthquake is a measure of how much damage it caused.* Table 5-1 is a typical intensity scale. A more fundamental measure of **T 5-1** the size of an earthquake is *the amount of energy released,* and this *is called the* **magnitude.** The magnitude is measured by the amount of movement of a standard seismometer corrected for distance to the

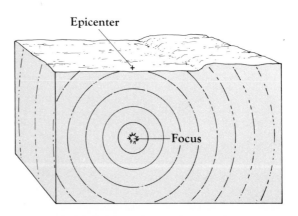

Epicenter

Focus

Figure 5-5
Focus and epicenter of an earthquake.

Figure 5-6
Locating an epicenter. At each seismic station, the S–P time tells the distance to the epicenter. Three stations are needed to locate the epicenter.

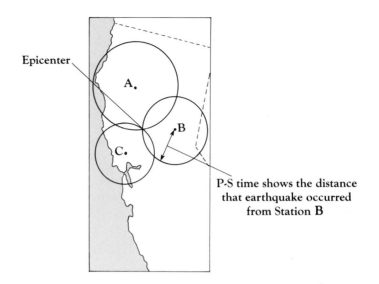

Epicenter

P-S time shows the distance that earthquake occurred from Station **B**

epicenter. The location and magnitude of an earthquake are what is usually reported soon after it occurs. The intensity, or damage, depends on both magnitude, or energy released, and the depth of the focus (Table 5-2). A shallow, low-magnitude earthquake may have a higher intensity—that is, cause more damage—than a deep, higher-magnitude earthquake.

T 5-2

Earthquake damage to structures is generally caused by the surface, or ground, waves because of their higher amplitude or movements. As these waves pass, the ground may move several inches so that the foundation of a building is deformed. This deformation can damage the building (Figure 5-7). A rigid building will collapse if its strength is exceeded by the deformation. Earthquake-resistant structures are designed to be limber, or elastic. Such buildings are deformed, but they should not break; they return to their original positions. The upper floors of tall buildings may sway many feet during an earthquake. The foundation of a building is very important in determining its behavior during an earthquake because the underlying rocks control the amount of ground movement. If the building's foundation is on bedrock, there will be little ground motion and, consequently, less damage. On the other hand, if the rocks are not well consolidated, the ground wave will cause more movement. For this reason, structures on natural or poorly engineered artificial fill tend to sustain more damage than buildings on bedrock. Structures astride the fault may be totally destroyed by the fault movement. Landslides triggered by earthquakes also cause much damage. In the past, fires started by earthquake rupture of gas mains or power lines have caused much destruction because water mains and transportation routes are also broken, making fire fighting very difficult.

F 5-7

Table 5-1	THE MERCALLI EARTHQUAKE INTENSITY SCALE (Adapted from C. F. Richter, 1956)
I.	Not felt. Some objects may sway.
II.	Felt by some people at rest and on upper floors.
III.	Felt by most people indoors but may not be recognized. Hanging objects swing.
IV.	Felt by everyone indoors but still may not be recognized. Windows, doors rattle; wood structures creak. Standing autos rock. Dishes rattle.
V.	Felt and recognized by everyone indoors, most outdoors. Sleepers wake. Doors swing, windows rattle. Pendulum clocks affected. Liquids may spill. Unstable objects upset.
VI.	Felt by everyone; many frightened, walk unsteadily. Weak plaster and masonry (like adobe) crack. Windows break. Crockery broken. Furniture moved. Trees shaken visibly, rustle.
VII.	Difficult to stand. Noticed by auto drivers. Weak chimneys break at roofline. Architectural ornaments fall. Unreinforced masonry damaged. Concrete ditches damaged. Waves on ponds, water muddied. Small slides on sand banks. Furniture broken. Hanging objects quiver.
VIII.	People thrown down, frightened. Masonry damaged or collapsed unless of resistant design. Elevated structures, tanks, twist or fall. Unbolted frame houses move on foundations. Springs and wells change. Wet ground cracked. Auto steering affected. Tree branches broken.
IX.	Panic. Reinforced masonry destroyed or seriously damaged. Foundations damaged. Unbolted frame houses shifted off foundations, frames damaged. Reservoirs seriously damaged. Underground pipes broken. Cracks in ground. Ejection of sand or water from soft ground.
X.	Panic. Most structures and foundations ruined. Dams seriously damaged. Railroads slightly bent. Large landslides. Water thrown on banks of rivers. Sand and mud shifted horizontally.
XI.	Panic. General destruction of buildings. Pipelines unusable. Railroads greatly bent.
XII.	Damage nearly total. Rock masses displaced. Lines of sight distorted. Objects thrown in air.

Table 5-2   THE APPROXIMATE RELATIONSHIP BETWEEN MAGNITUDE AND INTENSITY OF EARTHQUAKES. The distance from the epicenter and the number of earthquakes of each magnitude expected each year are also shown. The intensity (damage) depends on the magnitude (energy) as well as the depth of the focus, the distance, and the local geologic setting, and so it appears as a range in this table.

Magnitude (Richter scale)	Maximum Intensity (Modified Mercalli scale)	Distance Felt	Number Expected per Year
3.0–3.9	II–III	24 km (15 mi.)	49,000
4.0–4.9	IV–V	48 km (30 mi.)	6,200
5.0–5.9	VI–VII	112 km (70 mi.)	800
6.0–6.9	VII–VIII	200 km (125 mi.)	120
7.0–7.9	IX–X	400 km (250 mi.)	18
8.0–8.9	XI–XII	720 km (450 mi.)	1

SOURCE: Adapted from Essa, 1969.

Figure 5-7
A. Surface break or fault scarp caused near Hebgen Lake, Montana, in the 1959 earthquake. **B.** Building in Anchorage, Alaska, damaged in the 1964 earthquake. (Photo **A** by J. R. Stacy, U.S. Geological Survey; photo **B** from NOAA.)

A.

B.

Earthquake Prediction and Control

Earthquakes cause much damage and take many lives each year, and, if they could be predicted, much of this heartache could be prevented. The first step in prediction is to see where earthquakes occur. This is shown by plate tectonics (see Figure 4-14, p. 107), because the plates are outlined by earthquake epicenters. Seismic-risk maps (Figure 5-8) show where experience suggests that future earthquakes are likely. Figure 5-8 indicates that much of California has a high seismic risk, so one way of reducing earthquake hazards would be to avoid

F 5-8

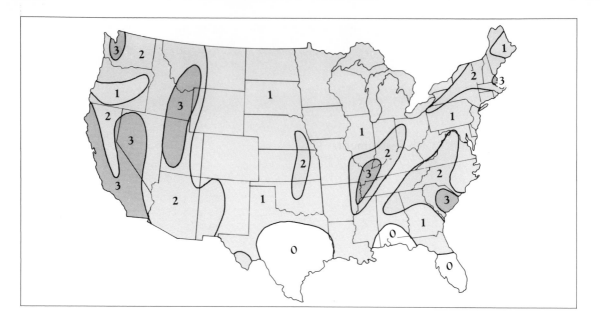

Figure 5-8
Seismic risk map.
(From ESSA.)

that state. Such a solution is not realistic, especially because serious earthquakes are, in general, restricted to a few faults. Most earthquake-prone areas use zoning laws to avoid placing structures on or near active faults. In this way active earthquake faults can be avoided, although the fact that no historic earthquakes have occurred on a fault does not necessarily mean that the fault is inactive.

Before an earthquake occurs, forces in the earth must build up; this deforms the rocks slightly. An earthquake results when the forces have built up enough to overcome the friction on the fault plane and movement occurs. The deformation that precedes many earthquakes can be detected in a number of ways. Sensitive tilt meters detect tilting and swelling of the surface. At some places swarms of minor earthquakes, detectable only on seismographs, may accompany tilting. Such events have been used to predict earthquakes in Japan. On some faults forces build up over many years and culminate in destructive earthquakes. Such faults are said to be **locked.** At other places very small movements, measured in millimeters, occur from time to time, causing **creep** on the fault (Figure 5-9). These movements, amounting to a few centimeters a year, seem to dissipate energy before it builds up enough to cause a serious earthquake. Parts of the San Andreas fault in California appear to b′ locked—and thus potentially dangerous—and other parts are active.

One of the more promising methods of earthquake prediction involves the slowing of the P-wave velocity in an area before an earthquake in that area. Apparently as forces build up and deform the rocks, microscopic cracks form in the rocks. Ground water moves into these cracks, and P waves from distant earthquakes moving through the rocks are slowed. The length of time the slowing occurs is proportional to the magnitude of the impending earthquake. Just before the earthquake, the P-wave velocity returns to normal.

The minor deformation that results from the buildup of forces before an earthquake causes other changes, too. The electrical and magnetic properties of rocks change, and the changes can be detected with instruments. Levels in water wells may also rise or fall prior to an earthquake. Other changes not readily detected by humans or by instruments apparently also occur, and these effects are sensed by certain animals. In spite of all these promising possibilities, much research will be necessary to develop a reliable method of earthquake prediction.

Once the forces have built to the point that an earthquake is possible, what can trigger the movement? Human intervention can also trigger earthquakes. Beginning in 1962, Denver, Colorado, experienced a number of earthquakes, and some of them caused damage. Before that time the area had had very few earthquakes. The increase was eventually traced to the pumping of waste water from the Rocky Mountain Arsenal into deep wells. Apparently the water reduced the friction on a fault plane and so allowed movement to occur. Pumping water into oil fields to increase the flow of petroleum from other wells has also caused earthquakes at a few oil fields.

In other cases, newly created lakes have caused earthquakes in the same way. Building a dam creates a lake, and water from the lake infiltrates into the earth. After Lake Mead was formed by Hoover Dam, a number of earthquakes occurred there. In 1967 at Koyna, India, an earthquake of magnitude 6.5 on the Richter scale caused about 200 deaths. The lake believed to be responsible was formed in 1962 and 1963 by the construction of a dam. Earthquakes believed to have been triggered by dam-impounded lakes have occurred at a number of other places as well. It must be understood, however, that most dam-impounded lakes do not cause any earthquakes.

Control of earthquakes seems to be a possibility because we know a few ways of triggering them. It might be possible to trigger a few small earthquakes instead of waiting for the forces to build up to a major earthquake. This could be done by pumping water down drill holes into the fault plane. The problems with such plans are that the behavior of faults under such circumstances is somewhat unpredictable and that the social and legal aspects are equally unpredictable. A larger earthquake than anticipated might occur. People would object to experimenting with faults near their homes. All sorts of damage would be blamed on the resulting earthquake. At present, prediction, realistic zoning and land use, and earthquake-resistant design and construction are the best defenses against earthquake damage.

Figure 5-9
Curb in Hayward, California, offset by fault creep. The slight change in direction of the curb was caused by very slow movements along the Hayward fault.

INTERNAL STRUCTURE OF THE EARTH

The earth consists of a number of layers, and their presence is deduced from the travel times of seismic waves. Seismic studies can also give us information about the nature of the layers. We will begin by considering the deep layers of the earth and then explore the crust.

If the earth were homogeneous, seismic waves would travel in straight lines; and if the layers were homogeneous, the seismic waves would travel in straight lines in each layer. The density of the layers increases with depth, and so the seismic velocity also increases, making the seismic-wave paths curve. Seismic waves behave much as light does. At each interface where the velocity changes, the wave is refracted. At these layer boundaries, some of the wave is also reflected; though these reflected waves are detected by seismographs, we will not consider them here.

Deep Interior

We will consider the behavior of the P wave first. In Figure 5-10 we **F 5-10** will assume that an earthquake occurs at the top of the cross section of the earth, and that we have seismic stations at various distances from the epicenter. The travel time of the P wave increases in a regular way with distance from the epicenter until a point is reached about 11,430 kilometers (7100 miles) from the epicenter. From that point to 15,940 kilometers (9900 miles) from the epicenter no P wave is received. This **P-wave shadow zone** is caused by refraction at the boundary between the **mantle** and the **core.** Between 15,940 kilome-

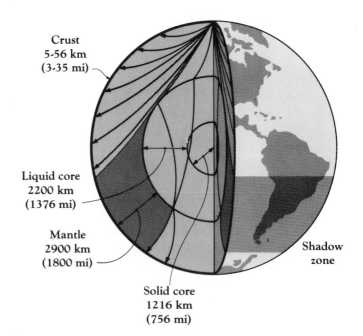

Crust
5-56 km
(3-35 mi)

Liquid core
2200 km
(1376 mi)

Mantle
2900 km
(1800 mi)

Solid core
1216 km
(756 mi)

Shadow zone

Figure 5-10
Paths of P waves through the earth.

Figure 5-11
Paths of S waves through the earth.

Shadow zone

ters and the bottom of the cross section, P waves are again received; these P waves traveled through the core. The P waves that pass through the core are slowed by the core. Notice in Figure 5-10 that the waves traveling through the inner part of the core are slowed less than those that move entirely through the outer core. Thus, the core has two distinct parts.

The behavior of the S wave is much simpler. At distances up to the same 11,430 kilometers, its travel time increases regularly; from that point on, it disappears. Thus the **S-wave shadow zone** is the entire area past 11,430 kilometers from the epicenter (Figure 5-11). The S wave is a shear wave, and shear waves cannot pass through a liquid, so the disappearance of the S wave shows that the outer core is liquid. The liquid outer core also slows the P wave, and the faster travel time of P waves in the inner core suggests that it is solid. The sizes of the deep layers are shown in Figures 5-10 and 5-11.

The composition of these deep layers must also be determined indirectly. We assume that the earth formed at the same time and from the same material as the nearby planets. Meteorites are believed to be fragments of small planets that also formed at that time. Occasionally meteorites fall to earth, and large ones cause impact craters such as Meteor Crater in Arizona (see Figure 17-40, p. 461). Meteorites, then, may be similar in composition to the earth. (The origin of meteorites is discussed in more detail in Chapter 17.)

Meteorites are of two distinct types, iron and stony, as well as mixtures of the two. Iron meteorites are composed mainly of iron with some nickel, and stony meteorites are made of olivine and pyroxene (peridotite is the name given to rocks of this composition). The earth's core is believed to be composed of iron and nickel, and the mantle of peridotite. At the temperature and pressure deep in the earth, rocks of such compositions would have the observed seismic velocities. An iron-nickel core would also be liquid.

The earth's magnetic field is believed to originate in the core. Electrons moving through the metallic conducting core generate a magnetic field; and a conductor, such as the core, moving in a magnetic field has a current induced in it. The current, in turn, causes a magnetic field, starting the process again. Although the process sounds like a dog chasing its own tail, calculations show that motions of the liquid core caused by the earth's rotation could sustain a magnetic field in this way.

Crust

A pronounced change in seismic velocity separates the crust from the mantle. This velocity change was discovered and recognized as important early in the 20th century, so it was used to define the crust. We now know that the movements associated with sea-floor spreading and plate tectonics occur in the upper mantle.

The change in seismic velocity that separates the crust from the mantle is called the **Mohorovičić discontinuity,** *or the* **Moho,** *or the* **M-discontinuity,** because it was discovered in 1909 by a Yugoslavian named Andrija Mohorovičić. A seismic station closer than about 800 kilometers (500 miles) to an earthquake receives two P waves and two S waves. This occurs because one set of P waves and S waves travels directly through the crust, and the other set travels partly in the crust and partly in the mantle (Figure 5-12). Modern, more sensitive seismographs actually detect three sets of P and S waves, indicating that there are two layers of different velocities in the crust. The layers in the crust detected seismically are below the surface covering of sedimentary rocks. The depth of the Mohorovičić discontinuity varies. Under the continents its depth is between 20 and 60 kilometers (12 and 37 miles), with the thickest parts under mountain ranges. Beneath the oceans its depth is more uniform at about 11 kilometers (7

F 5–1

Figure 5-12
Determination of the depth of the Mohorovičić (Moho) discontinuity. At station A, a single set of direct P and S waves are received. At station B, two sets of P and S waves are received: one set travels directly through the crust; the other set travels partly through the mantle. At distances greater than 100 to 150 kilometers (60 to 90 miles), the waves traveling through the mantle arrive first at station B.

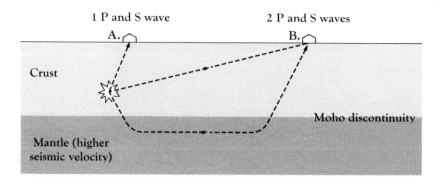

Figure 5-13
The crusts of the continents and the oceans are different in both rock type and thickness.

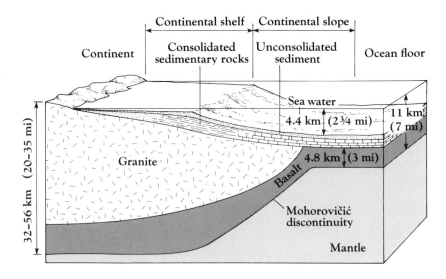

miles) below sea level. At the scale of Figures 5-10 and 5-11, the outer line labeled "crust" is close to the thickness of the crust.

The structure of the crust is shown in Figure 5-13. The continents F 5-13 are underlain by rocks of granitic composition, and the oceans by basalt. The basalt layer also extends below the continental granite. The rock composition of the continents is known from direct observation in deeply eroded areas, where erosion has stripped away the surface sediments. Drilling on the ocean floors reveals the basaltic composition there, and the seismic velocities are in agreement with these observations.

The velocity change at the Moho is probably caused by a change in composition between the crust and the mantle. The crust is composed of granite and basalt, and the mantle is believed to be made of the olivine rock peridotite. A more important zone, not so easily detected by seismic methods, exists in the upper mantle. At depths between 100 and 200 kilometers (60 to 120 miles) is a zone where the seismic waves are slowed. This is the **low-seismic-velocity zone** (Figure 5-14), and it is believed to exist because the rocks there are near their F 5-14 melting point. Above that zone the temperature is not high enough to cause melting; below there, the pressure causes the melting-point temperature to increase faster than the actual temperature increases. Apparently, then, the temperature and pressure are near the rocks' melting point only in the low-seismic-velocity zone.

The layer above the low-seismic-velocity zone is called the **lithosphere.** The rocks of the lithosphere are rigid, and the plates that move across the face of the earth are believed to extend through the lithosphere. *The rocks in the upper mantle below the lithosphere are not rigid, and this layer is called the* **asthenosphere** (Figure 5-14). The movement of plates

Figure 5-14
The lithosphere is the
crust and upper
mantle. Below the
lithosphere is the low-
seismic-velocity zone.

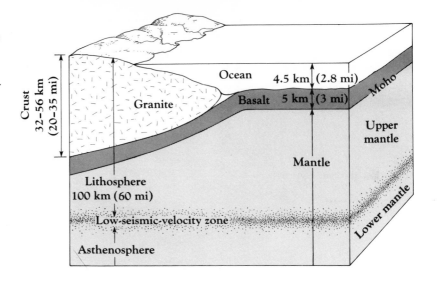

Figure 5-14
The lithosphere is the
crust and upper
mantle. Below the
lithosphere is the low-
seismic-velocity zone.

is believed to occur in the asthenosphere, and magma probably is also generated in this layer.

The entire lithosphere appears to float in the asthenosphere, and this may account for the difference in elevation between the continents and the oceans. The granitic rocks of the continents are thicker and less dense than the basalt of the oceans. Thus, the continents are higher than the oceans, just as a less dense material floats higher than a denser material (Figure 5-15). The continental granite is thicker under mountains than elsewhere, and this may explain why mountains are higher than the rest of the continents. *The concept that the lithosphere floats in the asthenosphere is called* **isostasy.** Isostasy may also explain how mountains are eventually leveled by erosion. As the mountains are eroded, they continue to rise until their crust is the same thickness as the rest of the continent (Figure 5-16). Isostasy is shown by the moat created by depression due to the weight of some volcanic islands.

F 5-15

F 5-16

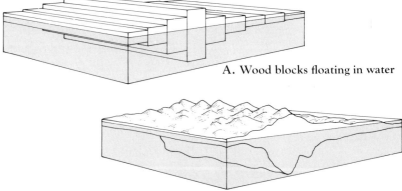

A. Wood blocks floating in water

Figure 5-15
Isostasy is the concept
that the lithosphere
floats on the
asthenosphere and
explains why
mountains have deep
roots.

B. Mountain range floating in upper mantle

Figure 5-16
Isostasy and erosion cause the roots of mountain ranges to be reduced as erosion reduces the height of the mountains.

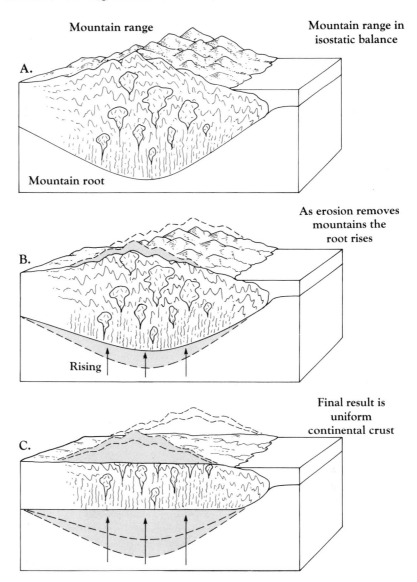

Mountain range

Mountain range in isostatic balance

A.

Mountain root

As erosion removes mountains the root rises

B.

Rising

Final result is uniform continental crust

C.

CAUSE OF PLATE TECTONICS AND SEA-FLOOR SPREADING

The surface structures that we see on the earth are caused largely by plate tectonics, and the movements of plates are caused by forces within the earth. We can deduce the nature of movements in the mantle that cause plates to move. Material moves upward at the mid-ocean ridges where new oceanic crust is being created. The volcanic activity, the age of the rocks, the high heat flow, the earthquakes, and the symmetrical magnetic stripes all show that new oceanic crust is created at the mid-ocean ridges. It seems likely that the mantle rocks move upward at the mid-ocean ridges and that partial melting of the

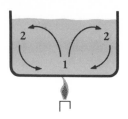

Figure 5-17
Convection. Fluid heated at the bottom (1) rises, and cool fluid from above (2) falls to take its place. Such a flow is called convection.

mantle peridotite is the origin of the basaltic ocean crust created there (see Figure 4-16, p. 108).

The ocean crust moves away from the mid-ocean ridges, making room for new crust to form at the ridge crests. Because new material is added to the oceanic crust, somewhere the crust must return to the mantle. The return occurs at the subduction zones—the volcanic island arc–deep sea trench areas. At these places, the ocean crust moves downward into the mantle. The downward motion causes deep earthquakes as well as shallow ones; and the ocean crust and sediments are melted, causing the volcanic activity (Figure 4-16).

Thus, oceanic crust is created at the mid-ocean ridges and consumed at subduction zones. Clearly, material leaves the mantle at the ridges and is returned at the subduction zones by this process. There must be movement within the mantle, or else it would become thinner at the ridges and thicker at the subduction zones. *Circulation of rising warm material and descending cool material is called* **convection,** and this is probably the process that occurs in the mantle. Convection can be understood from study of Figure 5-17. The liquid is F 5-17 heated at the bottom, and the bottom fluid becomes warm and expands. The expansion makes the warm fluid less dense than the cooler fluid, so the warm fluid rises. As the warm fluid rises, cooler, more dense fluid from the top sinks to take the place of the rising warm fluid. Possible convection currents in the mantle are shown in Figure 5-18.   F 5-18

The rate of sea-floor spreading is measured in centimeters per year, and this is the rate of movement of mantle convection. It is possible to move convectively at such rates and still transmit the much faster earthquake P waves like a solid. Tar and Silly Putty © are both substances that respond like solids to rapid movements but slowly flow under their own weight.

Figure 5-18
Convection currents in the asthenosphere could cause plate tectonics.

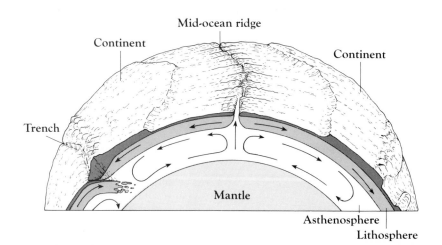

If convection occurs in the mantle, there must be a source of heat energy deep in the earth to cause it. Natural radioactive elements are present in the earth; and, as such elements decay, they release energy. Enough heat is generated in this way to cause geologic processes.

Another possible source of heat energy is the energy stored in the liquid core of the earth. As the core slowly crystallizes, energy is released. Heat energy would be released in this way at the bottom of the mantle where it is needed to cause convection. The central part of the core is believed to be solid, suggesting that the core may be slowly crystallizing. An interesting aspect of this idea is that, of the planets we have explored in the space program, only earth has mountain belts of folded and faulted rocks, and only earth has a strong magnetic field. The magnetic field is believed to originate in the liquid metallic core, so perhaps only earth has folded mountains because only earth has a liquid conducting core. This idea is discussed in more detail in Chapter 17.

SUMMARY

- An *earthquake* is the vibration caused by movement of a fault.

- *P waves* travel fastest and can pass through a fluid.

- *S waves* travel slower and cannot pass through a fluid.

- *Surface waves* cause most damage.

- The point where an earthquake occurs is called the *focus,* and the surface above is the *epicenter.*

- The distance to the epicenter is proportional to the S-P time, and three seismic stations are needed to locate an epicenter.

- The *intensity* of an earthquake is a measure of the damage, and the *magnitude* is a measure of the energy released.

- Several methods of earthquake prediction are currently under study.

- Earthquakes can be triggered by humans by creating large lakes and by pumping fluid into the earth.

- The earth has three main layers: *crust, mantle,* and *core.* Their dimensions, revealed by the behavior of earthquake waves, are shown in Figure 5-10.

- The core has two parts, and the outer core is a fluid because S waves cannot pass through it.

- The mantle is probably composed mainly of olivine and pyroxene (peridotite).

- The core is probably composed mainly of iron and nickel.

- The *Mohorovičić discontinuity,* or *Moho,* or *M discontinuity,* is a change in seismic velocity that separates the crust from the mantle.

- The more important break in the upper part of the earth is the *low-seismic-velocity zone* about 100 to 200 kilometers (60–120 miles) below the surface. Here the upper mantle rocks are near their melting point.

- The layer above the low-seismic-velocity zone is called the *lithosphere,* and this is the layer that moves in plate tectonics.

- Below the lithosphere is the *asthenosphere.*

- The concept that the lithosphere floats in the asthenosphere is called *isostasy.*

- *Convection currents* in the mantle may be the cause of plate tectonics.

- The source of the heat energy that causes plate tectonics and other geologic processes may be natural radioactivity, or it may come from crystallization of the core.

KEY TERMS

earthquake	intensity	S-wave shadow zone	lithosphere
seismology	magnitude	crust	asthenosphere
P wave	locked fault	Mohorovičić discon-	isostasy
S wave	creep	tinuity, or Moho,	convection
surface wave	P-wave shadow zone	or M-discontinuity	
focus	mantle	low-seismic-velocity	
epicenter	core	zone	

REVIEW QUESTIONS

1. What is an earthquake?
2. (a) Name and describe the three principal waves that earthquakes generate.
   (b) Which of the three waves causes most of the damage? Why?
   (c) Explain or draw a sketch to show how three seismic stations can locate the epicenter of an earthquake.
3. Explain the difference between an earthquake's intensity and its magnitude. The intensity depends on what two factors?
4. You are planning to build a house in an area through which a potentially active fault runs. What precautions do you plan to take?
5. Explain the difference between locked and creeping faults.
6. The rock deformation that precedes an earthquake can sometimes be detected. Explain how the following methods of prediction work:
   (a) detection of tilting
   (b) detection of slowing of P waves
7. Describe some ways in which human activities have triggered earthquakes.
8. The behavior of seismic waves has given geologists information about the interior of the earth.
   (a) What does the behavior of S waves tell us about the states of the mantle and the outer core?
   (b) What is the S-wave shadow? Explain why this shadow locates the boundary between the mantle and the core.
   (c) Describe the P-wave shadow. What causes it?

   (d) What effect does liquid have on the velocity of P waves?
   (e) Use your answers to Questions (c) and (d) above to explain what P waves tell us about the internal structure of the earth.
9. (a) Why are meteorites considered a clue to the composition of the earth's interior?
   (b) State the composition of two types of meteorite.
   (c) Of what material is the earth's mantle believed to be composed?
   (d) Of what material is the earth's core believed to be composed?
10. Fill in the blanks:
    (a) On the continents, a thin layer of sedimentary rocks lies over the continental crust made of the igneous rock ____. Underneath the continental crust lies another igneous rock, ____.
    (b) A thin sedimentary layer lies over the oceanic crust composed of the igneous rock ____.
11. What is the Moho? What is believed to cause this change in velocity?
12. Answer the following questions about the low-seismic-velocity zone:
    (a) Where is this zone located?
    (b) What is believed to be the cause of this zone?
    (c) Name the layer that lies between the top of the zone and the surface of the earth.
    (d) Are the rocks above the zone rigid or nonrigid?
    (e) Are the rocks of the zone rigid or nonrigid?

(f) Locate the asthenosphere in relation to the zone.

13. (a) Define isostasy.
    (b) How might isostasy account for the fact that the earth's surface has highlands (the continents) and basins (the oceans)?
    (c) Use isostasy to describe what happens as mountains are eroded.

14. One of the reasons why geologists did not accept the theory of continental drift until recently was that no reasonable mechanism had been proposed for moving rigid continents over what was believed to be an entirely rigid mantle.

    (a) The discovery of a low-velocity zone in the mantle eliminated part of the objection to continental-drift theory. Which part of the objection did it eliminate?
    (b) Convection requires heat. What two sources of heat exist in the interior of the earth?

SUGGESTED READINGS

Anderson, D. L. "The San Andreas Fault," *Scientific American* (November 1971), Vol. 225, No. 5, pp. 52–68.

Bolt, B. A. "The Fine Structure of the Earth's Interior," *Scientific American* (March 1973), Vol. 228, No. 3, pp. 24–33.

Boore, D. M. "The Motion of the Ground in Earthquakes," *Scientific American* (December 1977), Vol. 237, No. 6, pp. 68–78.

Christensen, M. N., and G. A. Bolt. "Earth Movements: Alaskan Earthquake, 1964," *Science* (September 11, 1964), Vol. 145, No. 3627, pp. 1207–1216.

Grantz, Arthur, and others. *Alaska's Good Friday Earthquake, March 27, 1964.* U.S. Geological Survey Circular 491. Washington, D.C.: U.S. Government Printing Office, 1964, 35 pp.

Hammond, A. L. "Dilatancy: Growing Acceptance as an Earthquake Mechanism," *Science* (May 3, 1974), Vol. 184, No. 4136, pp. 551–552.

Hansen, W. R., and others. *The Alaska Earthquake, March 27, 1964: Field Investigations and Reconstruction Effort.* U.S. Geological Survey Professional Paper 541. Washington, D.C.: U.S. Government Printing Office, 1966, 111 pp.

Iacopi, Robert. *Earthquake Country.* Menlo Park, Calif.: Lane, 1964, 192 pp. (paperback).

Kate, R. W., and others. "Human Impact of the Managua Earthquake," *Science* (December 7, 1973), Vol. 182, No. 4116, pp. 981–990.

Kerr, R. A. "Quake Prediction by Animals Gaining Respect," *Science* (May 16, 1980), Vol. 208, No. 4445, pp. 695–696.

National Science Foundation and U.S. Geological Survey. *Earthquake Prediction and Hazard Mitigation Options for USGS and NSF Programs.* Washington, D.C.: U.S. Government Printing Office, 1977, 76 pp.

Plafker, George. "Tectonic Deformation Associated with the 1964 Alaska Earthquake," *Science* (June 25, 1965), Vol. 148, No. 3678, pp. 1675–1687.

Press, Frank. "Earthquake Prediction," *Scientific American* (May 1975), Vol. 232, No. 5, pp. 14–23.

Tocher, Don. "Earthquakes and Rating Scales," *Geotimes* (May–June 1964), Vol. 8, No. 8, pp. 15–18.

Walker, Jearl. "How to Build a Simple Seismograph to Record Earthquake Waves at Home," *Scientific American* (July 1979), Vol. 241, No. 1, pp. 152–161.

Wallace, R. E. *Goals, Strategy, and Tasks of the Earthquake Hazard Reduction Program.* U.S. Geological Survey Circular 701. Washington, D.C.: U.S. Government Printing Office, 1974, 27 pp.

Wyllie, P. J. "The Earth's Mantle," *Scientific American* (March 1975), Vol. 232, No. 3, pp. 50–63.

TYPES OF DOWNSLOPE MASS MOVEMENT

GEOLOGIC ROLE OF RIVERS

Transportation

Erosion

Deposition

Graded Rivers

Development of Valleys

Development of Landscapes

GROUND WATER

Geologic Role of Ground Water

DESERTS AND WIND

Wind

GLACIERS

Mountain Glaciers

Continental Glaciers

CHAPTER 6

EROSION SCULPTURES THE SURFACE

Huge gully in Georgia. (Photo by USDA Soil Conservation Service.)

Figure 6-1
The hydrologic cycle.

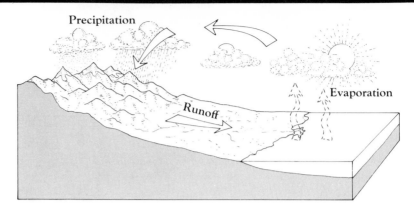

The earth's internal processes, such as plate tectonics, add material to the continents and uplift them. The wearing down of the continental surface is, of course, called **erosion,** and it is probably the most obvious surface process. Its effectiveness can be seen from the measurement of sediment carried by the principal rivers. These measurements suggest that at the present rate of erosion it would require only about 12 million years to complete the erosion of the continents. Twelve million years is a very long time by human standards, but the earth is at least 4600 million years old. It is clear, then, that the continuing internal processes causing the buildup of continents prevent erosion from completing its task of wearing the continents to smooth plains.

The sun's energy powers the hydrologic, or water, cycle (see Figure 6-1), which causes most erosion. This energy evaporates water and moves the resulting water vapor in the atmosphere. This water vapor is returned to the surface as rain or snow. The precipitation that falls on the continents is our interest. Some of the water infiltrates the ground and becomes ground water; some is used by plants, and much of this water is returned to the atmosphere as water vapor; some evaporates directly; the rest forms running water. These rivers are agents of erosion and the most important agents of transportation; they transport the eroded material ultimately to the oceans. They also transport weathered material brought to them by other erosive processes such as landslides.

In addition to the sun, the other force involved in surface processes is gravity. It causes not only the downhill movement of surface waters but also landslides and other such land movements.

F 6-1

TYPES OF DOWNSLOPE MASS MOVEMENT

Downslope movement, or **mass movement,** are terms used to designate certain erosion processes. They are descriptive in the sense that they refer to downhill movement commonly involving massive amounts of material. A more common term for these processes is *landslide,* but, as we will see, landslide is used by geologists in a more restricted sense.

Downslope movements range from almost imperceptibly slow to dramatically fast and deadly. The classification depends upon the type of movement, the speed of movement, and whether **bedrock** or weathered material is involved. **Overburden,** or **regolith,** is the name given to the unconsolidated surface layer, generally consisting of soil and weathered bedrock, that develops on all but the steepest slopes.

The three types of movement are falls, slides, and flows. As the names imply, **falls** are rapid movements through the air by free fall, bouncing, and rolling. **Slides** move by sliding, generally on a shear surface. **Flows** behave like viscous fluids; overburden is generally involved in flows, and the amount of water influences the speed of motion. Many mass movements involve more than one type of movement and are not easily classified.

The most common flow is **creep** (Figure 6-2). This very slow down-slope movement of overburden or overburden and bedrock can be recognized by tilted fenceposts and trees. In some cases the movement is rapid enough that not even shrubs can take root. Creep can occur on very gentle slopes. With increasing amounts of water the flow becomes more rapid, and **earthflow** occurs when a definite scarp appears at the head and a tongue moves out on the surface. With even more water, a rapidly moving **mudflow** results. Mudflows can move long distances, and thick mud can carry large boulders (Figure 6-3). F 6-3

Solifluction is the type of flow that occurs when frozen ground melts at the surface. The resulting mud flows down even gentle slopes, forming the smooth, treeless slopes commonly seen in high mountain ranges and in arctic areas.

Slides are rapid downslope movements that result when rock or overburden breaks off on a shear surface (see Figure 6-4). A slide is F 6-4

F 6-2

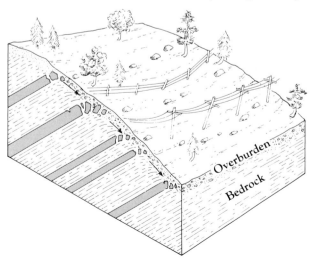

Figure 6-2
Creep is the slow movement of surface material.

Figure 6-3
Damage caused by a mudflow at Big Sur, California. This mudflow occurred at the first rain after a forest fire had destroyed the vegetation. (Photo from Caltrans.)

termed a **landslide** if both bedrock and overburden are involved; it is a **debris slide** if only overburden is involved; and it is a **rockslide** if only bedrock is involved. In all of these slides, the slide material breaks into many fragments. If the bedrock remains intact and is only rotated and moved downslope, the slide is called a **slump** (see Figure **F 6-5** 6-5).

Falls move through the air and so can only occur on very steep slopes or cliffs. Only bedrock is involved because overburden cannot

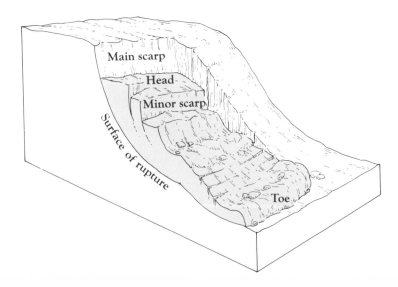

Figure 6-4
Landslide.

Figure 6-5
Slumps occur if the rock remains intact.

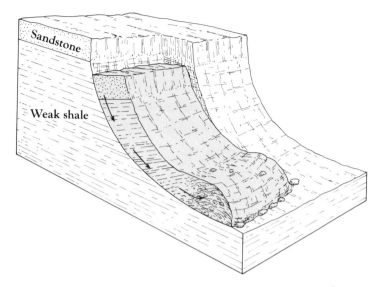

accumulate on very steep slopes. A typical **rockfall** occurs at a cliff, and the surface formed by the accumulation of fallen fragments is called a **talus** (see Figure 6-6).

F 6–6

Downslope movements are the most important processes in the erosion of the landscape. Most people think of rivers as the main agents of erosion, but, of course, a river can erode only its immediate bed. The erosion of the valley sides is done by downslope movement of material to the river, and the river carries this material away. Rivers are the most important agents of transportation. The downslope processes that we are discussing are the common, ordinary, ex-

Figure 6-6
Rock fragments at the foot of a cliff form a talus. (Photo by C. B. Hunt; U.S. Geological Survey.)

pectable hillside erosion processes, and they occur to a greater or lesser extent on all hillsides. These processes must be considered when any buildings or other engineered structures are designed on any hillsides.

GEOLOGIC ROLE OF RIVERS

Rivers obviously carry the surface water of the lands to the ocean. The running water also erodes, transports, and deposits sediment. Generally these processes go on slowly so one is not conscious of them, even though they shape the land on which we live (Figure 6-7). F 6-7 At times, however, these processes are rapid and become hazards to life and property.

Transportation

Carrying sediment is the most important geologic role of a river. Although rivers are agents of erosion themselves, most of the material carried by rivers is brought to them by other processes. A river can only erode its bed, but valley-side processes, such as creep and rain wash, deliver much weathered material to the river (Figure 6-8). An- F 6-8 other way to see the importance of transportation by rivers is to consider that they form only a very tiny percentage of the surface area, but they must carry away almost all of the weathered material of the whole landscape. If this were not true, landscapes could not change.

The methods of river transportation depend on the size of the particles and the velocity of the water. Three distinct loads can be considered. The **dissolved load** is material carried in solution and is, in fact, the soluble products of weathering. The **suspended load** is the fine particles that, if abundant, give the water a brown, muddy look. The **bed load** is the larger particles that are pushed or rolled along the bottom or rise as a result of impact and are carried a short distance by the flowing water. The size of the particles that compose either the bed load or the suspended load depends on the velocity of flow at a particular moment.

Erosion

Stream erosion is also caused by three processes. Actual erosion of bedrock can occur only where bedrock is exposed in the stream bed. This generally occurs only at places where the stream is rapidly downcutting, such as near its headwaters. **Solution** can occur only in soluble rocks, and limestone is the only common soluble rock. **Abrasion** is the wearing or grinding of the bedrock by the movement of the bed load. **Direct lifting** is the lifting and carrying of small particles a distance downstream by turbulent water. This process is probably more

Figure 6-7
These two photos of the Colorado River, taken about 100 years apart, show very few differences. **A.** 1871; **B.** 1968. (Photo **A** by J. K. Hillers and **B** by E. M. Shoemaker; U.S. Geological Survey.)

A.

B.

Figure 6-8
The material transported by a river is largely delivered to it from the valley sides.

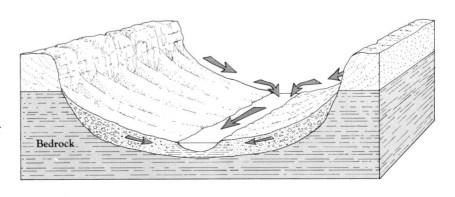

Bedrock

important in moving the bed load and the suspended load than in actual erosion of bedrock.

The elevation of the mouth of a river is called its **base level,** *and this is clearly the lowest elevation to which it can erode* (see Figure 6-11, p. 154). The ultimate base level is sea level for a river that flows into the ocean. For a tributary stream, base level is the elevation at which it joins the main stream; and because this elevation changes as the main stream develops, it is a local or temporary base level.

Deposition

Deposition of sediment occurs where the flow of a river slows. Typically this happens where there are wide flood plains and where the river meets the ocean. In the latter case, a **delta** results if the ocean waters are relatively calm or the river load is great; but if waves and storms prevail, the river's load is carried away by the ocean waters. The Mississippi Delta (Figure 6-9) and the Nile Delta are examples, F 6-9 and their characteristic shape is the result of changing positions of the main channel in the delta. Deltas can be excellent agricultural areas, but flooding and changing channel locations are problems.

Alluvial fans form where rivers flow out of mountain ranges into relatively flat plains. At such places the slope of the river changes abruptly from steep in the mountains to gentle on the plains. The result, of course, is that the river can no longer transport its load, so deposition occurs. In many cases the river's water infiltrates and becomes ground water below the plain. Alluvial fans develop conical shapes, apparently because the position of the main distribution

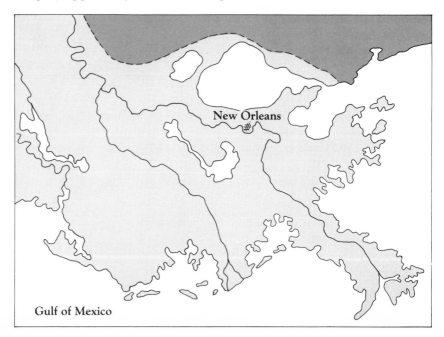

Figure 6-9
The delta of the Mississippi River.

New Orleans

Gulf of Mexico

Figure 6-10
Alluvial fans at the mouths of streams entering Death Valley, California, are shown in this vertical aerial photograph. (Photo from U.S. Geological Survey.)

channel changes (Figure 6-10). Alluvial fans, although somewhat sim-   F 6–10
ilar to deltas in origin and shape, differ in that the load of a mountain
stream is generally much coarser than that of a river like the Mississippi that crosses plains. Some alluvial fans have gentle slopes, and
the presence of such fans would not be suspected by most people who
had not studied a map. This can lead to problems because homes and
housing developments have been built on fans. The problems develop, especially in drier climates, when the position of the channel
changes or when mountain storms cause mudflows.

Graded Rivers

One of the most interesting and far-reaching discoveries made about
rivers is that most are delicately balanced and respond quickly to
changes. The balance is between load and ability to transport the

Figure 6-11
Profile of a graded river.

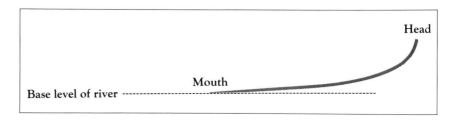

load. The ability of a stream to transport load depends on slope, amount of water, velocity, and shape of channel. One aspect of the stream's balance is its longitudinal profile (along the length of the stream). A stream is steeper near its headwaters where the flow is least, and more gently sloping near the mouth where the amount of water is greater and the load generally finer grained (Figure 6-11). At F 6-11 each tributary, the influx of load and water changes the longitudinal profile. The most obvious change a graded river makes is in its slope. If the amount of water flow increases or the load decreases, the river erodes its bed and the slope is reduced; if the load is increased or the flow decreases, deposition occurs on the river bed and the slope is steepened. Such changes occur seasonally on most rivers, but the term **graded river** implies a long-term balance over a number of years. Thus, the variations in climate of wet and dry years make precise definition of a graded river difficult.

A dam on a river illustrates the behavior of a graded river. The dam changes the original river into two rivers. The headwaters now flow into the lake above the dam, where the load is deposited. This shorter journey of the load does not require the steeper slope of the original river, and deposition occurs on the river bed, reducing the slope. Below the dam, the situation is reversed. The water that is released has no load. The original river bed is steeper than needed, so it is lowered by erosion. These changes above and below a dam occur not only on the main river but also on all of the minor tributaries. These predictable changes, in many cases, are the cause of law suits brought by landowners whose fields are either covered with silt or gullied.

Development of Valleys

River valleys vary from narrow canyons with nearly vertical walls to wide, almost flat valleys. Examples of all stages from narrow to broad valleys can be found in nature, and this apparently is the progression as erosion continues. We will consider the changes that occur in the life of a river, recognizing that in nature conditions may not remain static long enough for all of the changes to occur.

The first stage in the development of a river valley is called youth, and it begins when an area is subjected to erosion for the first time. Uplift of an area probably starts the cycle. Rain water runs down the low places, and rills or gullies soon develop. The gully that collects the most water will be eroded deepest, so that it gathers even more water. Thus the river and its valley are born. This process can be seen on a small scale on new road cuts and other excavations.

During youth the valley is characteristically narrow and has a **V** shape. The **V** shape develops because the river is actively downcutting, and the valley sides are shaped by downslope movements (see Figure 6-12A). The river carries away the downslope material. This F 6-12 phase ends when downcutting of the river slows. As might be expected, resistant rocks are eroded slower than softer rocks, so streams in youth may have waterfalls, rapids, or even lakes.

The mature stage begins when downcutting slows and the river begins to widen the valley. During youth, the river's course tends to be much straighter than during maturity. The river valley is widened in maturity by erosion on the outside of a curve (Figure 6-13). Once a F 6-13 curve is started, the flow of the river causes more curves as the water careens from one bank to the other. The resulting curve patterns are called **meanders.** At the mature stage, the meanders occupy the width of the valley (Figure 6-12B).

Figure 6-12
Stages in the development of a river valley.

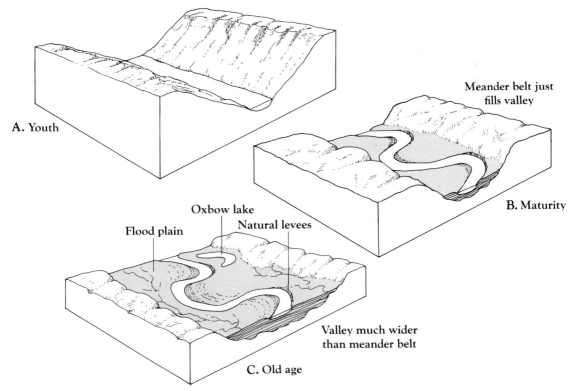

A. Youth

Meander belt just fills valley

B. Maturity

Oxbow lake
Flood plain
Natural levees

Valley much wider than meander belt

C. Old age

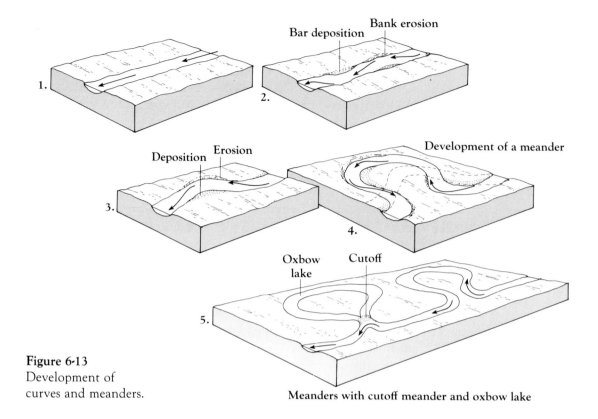

Figure 6-13
Development of
curves and meanders.

Meanders with cutoff meander and oxbow lake

An old age or broad river valley, differs from a mature valley in that the meanders occupy only a part of the valley (Figure 6-12C). The meanders are constantly changing their location, and this process widens the valley. Cutoff meanders and oxbow lakes (Figure 6-13) are features of mature and especially old-age valleys.

Broad valleys have flat bottoms that, at many places, are utilized as agricultural areas. At other places, towns and cities have been built in river valleys. The geologic term that describes these flat areas is **flood plain.** The name is apt because this is the area that is flooded at times of large flow. This is the normal behavior of a river, and flooding must be expected by anyone using a flood plain. Flooding occurs seasonally on some rivers but only rarely on others, for reasons we consider below.

When a river overflows its banks, the water is slowed abruptly as it covers the flood plain. At times of flood, the river is transporting large amounts of sediment, which is deposited in the flooded area. This renewal of silt may keep a valley fertile. Much of the river's load may be deposited on the bank of the channel where the flow slows abruptly. Such deposits over many floods build **levees** (Figure 6-12C). On some rivers—for example, the Mississippi River near New Orleans—the natural levees are much higher than the surrounding flood plain. These levees prevent many minor floods, but when rivers overtop such levees, serious floods are the result.

Many factors affect the development of a river valley, including rock type, weathering, vegetation, and stream flow. Thus, different parts of the same river are generally at different stages.

Development of Landscapes

Figure 6-14
Stages in the erosional development of an area. Very few landscapes pass through all of these stages.

Landscapes are more complicated than river valleys, and so is their erosional development. We can envision an area going through a series of stages, but probably no real landscape goes through all of the stages. The **cycle of erosion** of a landscape begins with a smooth, newly uplifted area. Rivers soon develop on this surface, and they rapidly downcut. This stage, with intrenched rivers and flat interstream divides, is called **youth** (Figure 6-14). As erosion proceeds, the river valleys widen until no flat interstream areas remain, and this stage is called **maturity** (Figure 6-14). Maturity passes into **old age** (Figure 6-14) when the interstream areas are low and rounded. At the ultimate stage, when the interstream areas are almost flat, the landscape is called a **peneplain** (almost a plain). F 6–14

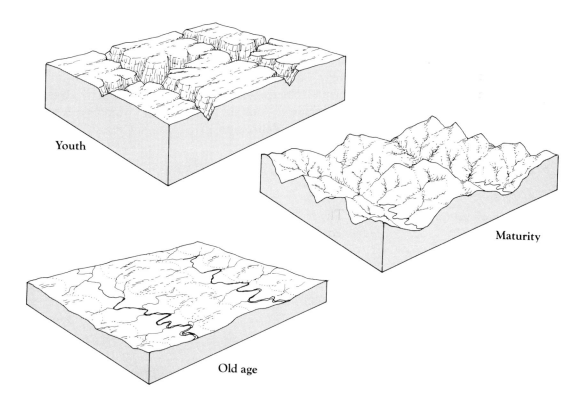

Youth

Maturity

Old age

For the cycle of erosion to go through all its stages, an area must be uplifted in a single step, and then no other changes can occur as erosion proceeds. This seems to happen very rarely because very few peneplains exist. Uplifts and climatic changes can upset the cycle, and not all uplifts are rapid. The interstream divides may not be rounded under all climatic conditions. In spite of these objections, the cycle of erosion is a means of characterizing landscapes and envisioning erosional changes.

GROUND WATER

The origin of ground water is precipitation, mainly rain, that infiltrates, or percolates, into the ground. The water moves through the ground until it reaches the **water table.** Below the water table the ground is saturated with water. The depth of the water table may change seasonally, and its shape is a subdued replica of the surface topography (Figure 6-15). Lakes, swamps, or springs may occur where F 6-15 the water table is exposed. If the downward-percolating water comes to a relatively impervious layer, such as shale, a **perched water table** may result.

Water from below the water table is recovered by drilling or digging wells. A producing well must penetrate a good reservoir rock. A **reservoir rock** is one that has much pore space to hold water, and the pores must be interconnected so that the water can flow into the well. Well-sorted sandstone is a common reservoir rock, or **aquifer,** because it has both high **porosity** and **permeability** (Figure 6-16). Ground F 6-16 water moves very slowly through rocks. The average rate of flow is estimated at about 15 meters (50 feet) per year, but it is much faster in many aquifers. If water is pumped out of a well faster than it can flow to the well, a **cone of depression** (Figure 6-17) results. Continued F 6-17

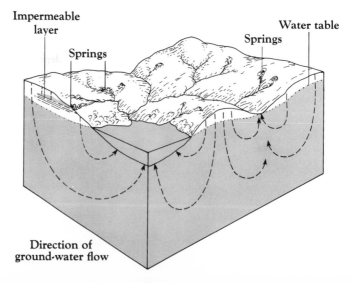

Figure 6-15
The water table is a subdued replica of the surface. Springs form where the water table is exposed.

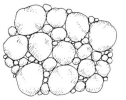

Well-sorted sandstone   **Poorly-sorted sandstone**

Figure 6-16
A well-sorted sandstone has a much higher porosity and permeability than a poorly sorted sandstone. Porosity is the amount of pore space, and permeability is a measure of how well the pore spaces are interconnected.

over-pumping can lower the water table, and many years may be required to restore the reservoir.

In some situations water may rise in or flow from a well without requiring pumping. Such wells are called **artesian wells** (Figure 6-18). F 6-18 Artesian wells form where an aquifer, such as a sandstone, lies between two impermeable beds, such as shale. The aquifer is charged with rain water where it reaches the surface at an elevation higher than the tops of the wells. The pressure that makes artesian wells flow is due to the weight of the water above the tops of the wells. The same principle allows the use of a water tank to provide water to a farm or a town.

Geologic Role of Ground Water

In areas where rocks near the surface are soluble, water may move underground instead of in rivers. This typically occurs in areas underlain by limestone. Limestone is soluble in water, especially if the water contains carbon dioxide. Water moving through cracks and pores in limestone, or any other soluble rock, dissolves the rock and ultimately creates a cave. Caves probably form in the zone of seasonal water-table fluctuation and become accessible to us only after some other event, such as an uplift, lowers the water table (Figure 6-19). F 6-19 Once caves or underground passages exist, water moves there instead of on the surface.

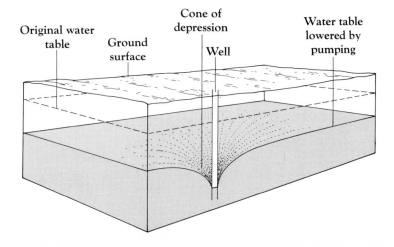

Original water table   Ground surface   Cone of depression   Well   Water table lowered by pumping

Figure 6-17
A cone of depression develops if a well is pumped too fast.

Figure 6-18
Artesian wells.

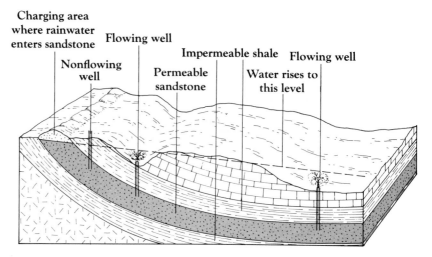

Charging area where rainwater enters sandstone
Flowing well
Impermeable shale
Flowing well
Nonflowing well
Permeable sandstone
Water rises to this level

Where caves are close to the surface, the roof may fail and a **sink-hole** develops. Water flows into caves at such sinkholes. In areas underlain by limestone, river valleys tend to be short and end at sinkholes, and the surface is pockmarked with sinkholes. Such a landscape is called **karst topography.**

Inside many caves a peculiar but beautiful type of limestone deposition occurs. Water containing much dissolved calcite may drip from the ceiling of a cave. If the drop evaporates, the limestone is precipitated on the ceiling. If this process goes on long enough, a **stalactite** forms. If some of the water falls to the floor of the cave and evaporates, **stalagmites** develop (Figure 6-20).

At some places, ground water circulation is such that the water goes deep enough to be heated. **Hot springs** form where such water reaches the surface. Most spring water contains dissolved material,

F 6-20

Water table, wet season
Water table, dry season

Figure 6-19
Caves form in the zone of seasonal water table fluctuation, and uplift of the area or lowering of the water table exposes the caves.

Lowered water table at a later time

Figure 6-20
Dripping water forms stalactites and stalagmites. (Photo of Carlsbad Caverns, New Mexico from National Park Service.)

but hot springs tend to have more because warm water is a better solvent. In some areas the heat flow from the interior of the earth is much higher than normal, and at such places hot springs and geysers are abundant. Yellowstone National Park is one such place.

Geysers are hot springs that erupt steam and water periodically. They occur where the water forms an irregular column. The water at the base of the column is heated to its boiling point, which is higher than the boiling point at the surface because of the pressure caused by the weight of the water column. Heating causes the water to expand, and so some of it flows out on the surface. This reduces the weight of the water column and so reduces pressure at the bottom and the boiling-point temperature. The bottom water is now hotter than its boiling point, so it flashes into steam and the geyser erupts (Figure 6-21).

F 6-21

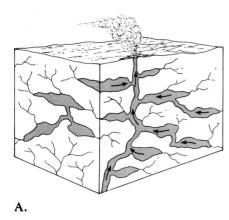

A.

B.

Figure 6-21
A. A geyser results if water is heated at depth to near its boiling point and so expands and flows out at the surface. This reduces the pressure at depth, so the water there flashes into steam, causing an eruption. **B.** Lone Star Geyser, Yellowstone National Park. (Photo by W. H. Jackson; U.S. Geological Survey.)

Areas of hot springs and geysers are of great interest as sources of *geothermal energy*. Electricity is being produced at a number of such places worldwide (see Figure 3-34, p. 87).

DESERTS AND WIND

Even in deserts, running water is the main agent of erosion. Rainfall is infrequent at any one place, but it generally comes in the form of intense cloudbursts or thunderstorms. Flash floods may develop and surprise travelers far from the area where the rain occurred. Because there is little vegetation, the rare running water carries much of the surface material to the usually dry stream bed, creating mudflows. Most deserts have little or no soil because chemical weathering is very slow in dry climates. As a result, landscapes tend to be angular, with cliffs rather than rounded hills.

The erosional stages of a desert mountain range are shown in Figure 6-22. The desert ranges of southwestern United States show examples of these stages and most of the inbetween stages. After the initial uplift of the fault-block mountains, fans are deposited in front of the range. Erosion of the uplifted block causes the slope of the

F 6-22

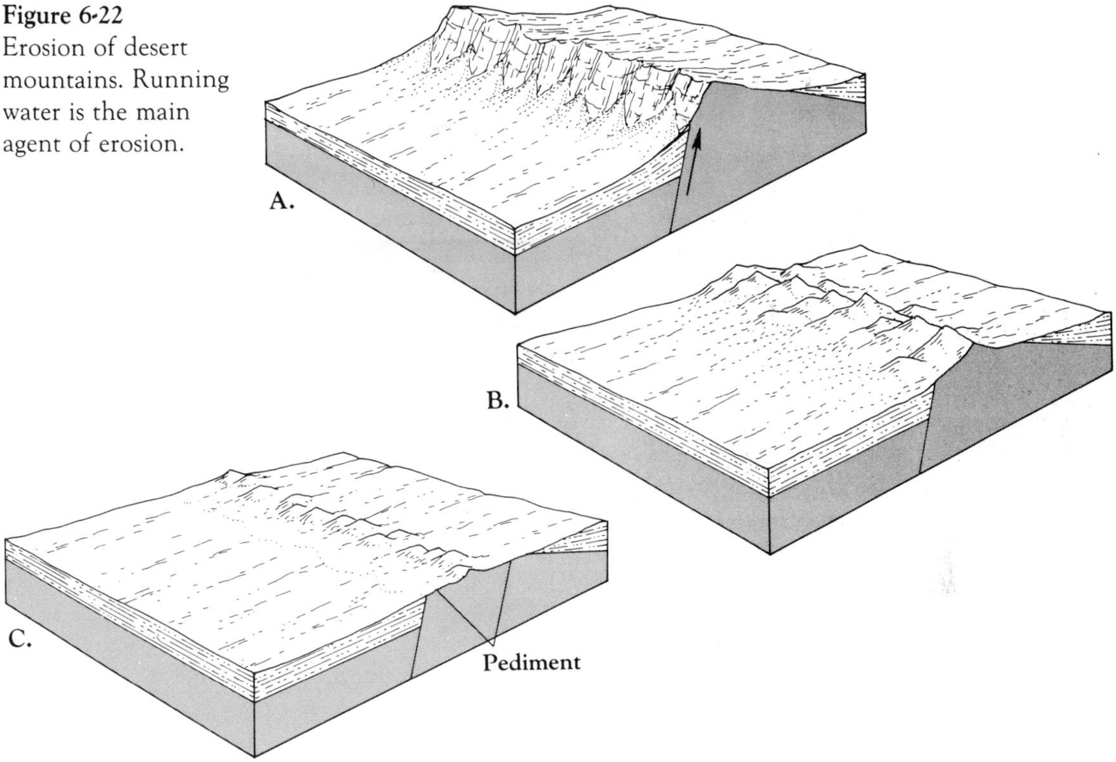

Figure 6-22
Erosion of desert
mountains. Running
water is the main
agent of erosion.

mountains to retreat. The resulting bedrock surface is continuous
with the range-front fan and is called a **pediment.** The surface of a
pediment is covered with gravel that is being transported across it,
making it very difficult to determine where the pediment ends and
the fan begins. As erosion continues, the pediments become larger
and the peaks above the pediment become smaller. The surface of a
pediment is a smooth, concave-upward surface, like the profile of a
river.

Wind

Wind only modifies the landscapes of most deserts, but locally it can
be an important agent of erosion and deposition (Figure 6-23). The
lifting and carrying away of sand or dust particles by the wind is
called **deflation.** This process is most effective in dry, unvegetated
areas and may produce hollows on the surface. The effectiveness of
deflation is shown by many deserts where the surface is covered by
pebbles too large for the wind to move. Apparently the finer materials were removed by the wind, leaving the larger pebbles to form
what is called **desert armor,** or **desert pavement** (Figure 6-24).

F 6–23

F 6–24

Figure 6-23
The effects of wind erosion and deposition in the late 1930s in Baca County, southeastern Colorado. (Photo from USDA, Soil Conservation Service.)

Figure 6-24
Desert armor, or pavement. (Photo by C. S. Denny; U.S. Geological Survey.)

Wind can carry dust to great elevations and for long distances. Thus the dust in a dust storm can be carried far from its original location. The wind can raise sand grains only a few feet, and once the grains are in motion, they bounce along the desert floor. (The movement of sand is shown in Figure 6-25.) Wind separates dust from sand and can only move a limited size range of sand grains, so wind-deposited sediments are very well sorted by size.

Sand grains bouncing along the desert can erode by a sandblasting effect. This can be seen on pebbles, called **ventifacts,** that have smooth faces caused by wind abrasion (Figure 6-26).

Wind-deposited dust is called **loess.** Loess deposits tend to be sheet-like layers of uniform grain size. Loess generally forms good agricultural soils.

Accumulations of wind-deposited sand are called **dunes.** They tend to form in the lee of obstacles where the wind velocity drops. Once formed, dunes tend to move unless they are stabilized by vege-

Figure 6-25
Movement of sediment by the wind. Dust is carried in suspension. Sand grains are lifted only a few feet and then fall back; the resulting collisions cause other sand grains to bounce up.

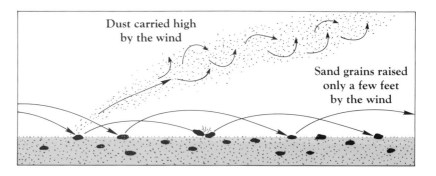

Dust carried high by the wind

Sand grains raised only a few feet by the wind

Figure 6-26
Ventifacts. Abrasion by wind-carried sand shapes and polishes pebbles.

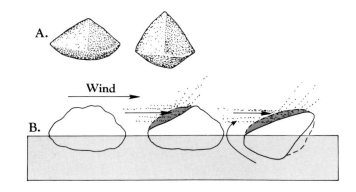

Figure 6-27
A sand dune migrates as the wind moves sand grains up the face of the dune.

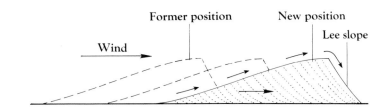

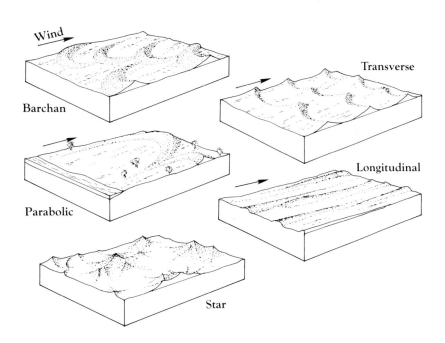

Figure 6-28
Types of sand dunes.

tation. Wind blows sand grains up the gentle face of a dune, and the grains fall off the steep face, so the dune advances (Figure 6-27). Depending on the amount of sand and the velocity and persistence of the wind, dunes can take many forms (Figure 6-28).

F 6-27

F 6-28

GLACIERS

Most erosion is done by running water, and glaciers only modify landscapes. However, the effects of glaciation commonly produce spectacular changes. *A glacier is a mass of moving ice.* Glaciers begin as snowfall, and the snow must accumulate to a thickness of a few hundred feet before a glacier is born. Generally, several seasons in which more snow falls than melts are required. As the snow accumulates, the weight of the overlying snow causes compaction and recrystallization of the deeper snow into bigger crystals of blue ice. This recrystallization is somewhat like the metamorphism of limestone into marble. When the pile of ice is thick enough, it begins to flow as a result of its own weight, much as a pile of tar or other viscous substance would flow (Figure 6-29). The flowing ice or glacier can erode, transport, and deposit sediment.   F 6-29

Glaciers can form locally in a mountain range, or they can form as ice caps covering huge areas such as Greenland and Antarctica.

Mountain Glaciers

The smaller **mountain,** or **alpine, glaciers** form in existing stream valleys in mountains because valleys are the low places where snow can accumulate. When enough ice has accumulated, the glacier begins to move down its valley. How far down the valley it goes depends upon how much snow falls in winter and how much melting occurs in sum-

Figure 6-29
Glaciers moving out of a mountain range on Ellesmere Island, Northwest Territories, Canada. (Photo from Geological Survey of Canada.)

Figure 6-30
Stakes driven into a glacier show the movement of the ice.

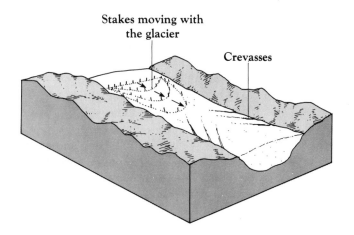

Stakes moving with the glacier

Crevasses

mer. Much of the melting occurs at the lower end, or snout, of the glacier. The snout of a glacier advances or retreats with minor changes in climate, and so glaciers record climatic changes. In most cases it takes a number of years before a climatic change affects the location of the snout.

Even in a retreating glacier, the ice at the glacier's snout is moving down the valley. The movement of the ice in a glacier can be seen if stakes are driven into the glacial ice and their positions observed at later times (Figure 6-30). The ice in the middle moves faster because it is not affected by friction at the valley sides. The ice near the bottom of the glacier behaves like a viscous fluid, but near the surface it is brittle. This and the more rapid movement in the center combine to open cracks in the glacier called **crevasses** (Figure 6-30). F 6-30

Glaciers erode their valleys by plucking and abrasion. **Plucking,** *or* **quarrying,** *is the removal of generally big fragments of bedrock by a glacier.* The blocks are loosened by frost action and become frozen to the bottom of the glacier. As the ice advances, it pulls the block out. This process requires meltwater and so is more common in temperate-climate glaciers. Abrasion results from rocks imbedded on the glacier's bottom that are rubbed over the bedrock by the moving ice. The glacier's bottom acts like a rasp.

Glacial erosion produces spectacular mountain scenery. The sides of the originally V-shaped river valley are steepened, changing the valley to a U shape, and at the same time the valley is deepened. These changes cause the tributaries to form waterfalls from **hanging valleys** (Figure 6-31). Glacially eroded valleys are straighter than F 6-31 stream valleys because the glaciers tend to remove spurs. The effects of glacial abrasion are grooves and glacial polish (Figure 6-32). At the F 6-32 head of a glacial valley, plucking deepens and steepens the valley, so that it develops a shape like a tea cup cut in half. This amphitheater-like shape is called a **cirque** (Figure 6-31). If several cirques develop

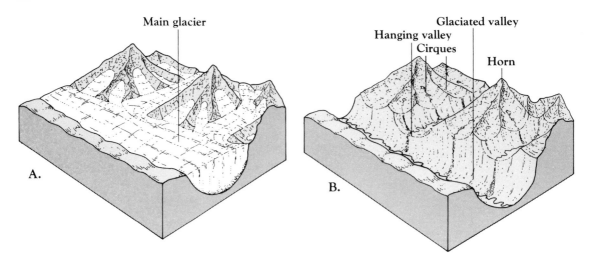

A.   Main glacier

B.   Glaciated valley
Hanging valley
Cirques
Horn

Figure 6-31
Glacial erosion changes V-shaped river valleys into U-shaped glacial valleys and turns tributary valleys into hanging valleys.

on different sides of a mountain, the peak becomes pyramid-like and is called a **horn.** Areas above the glacier are attacked by frost action and have rugged, rough surfaces that contrast with the smoothed surfaces overrun by the glacier.

Glaciers are also agents of deposition. Much of the material carried by the moving ice eventually reaches the snout of the glacier where it is deposited. The deposits are called **moraines,** and they contain a mixture of fragments of all sizes, from boulders to powdered rock. Their curved shapes, the lack of sorting, and scratches and grooves on the rocks make moraines easy to recognize. **Lateral moraines** form along the sides of a glacier. The farthest advance of a glacier is

Figure 6-32
Glacial polish and scratches near Clinton, Massachusetts. These were caused by a continental glacier. (Photo by W. C. Alden; U.S. Geological Survey.)

Figure 6-33
Glacial moraines.

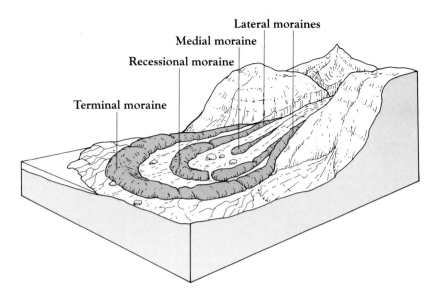

Lateral moraines
Medial moraine
Recessional moraine
Terminal moraine

marked by what is called the **terminal moraine** (Figure 6-33). Melt- F 6-33
water from the glacier may redeposit morainal material to form an
outwash plain. The general name used for all types of glacial deposits
is **drift,** and **till** is used for ice-deposited sediment.

Continental Glaciers

Continental glaciers are the result of the buildup of an ice cap over a
large area. Their erosion is less spectacular than that of mountain
glaciers and generally consists of rounding the hills and scraping off
the soil and weathered material. The deposits are widespread, with
moraines and outwash plains showing the former extent of the ice
caps. The recognition that continental glaciation had occurred came
about in part from the recognition of erratic boulders. **Erratic boulders** are boulders, in many cases quite large, that are different from
the local rock types. These foreign rocks have been carried by continental glaciers. By tracing the origins of erratic boulders, the extent
of continental glaciation was determined.

From the study of both erosional and depositional features, it is
clear that parts of North America and Europe were covered by ice
caps in the recent geologic past (Figure 6-34). The ice that caused F 6-34
continental glaciation in North America did not spill southward
from the north pole. The ice cap that was centered in northern Canada, for instance, moved radially outward from its center so that ice
moved northward in parts of northern Canada and southward in the
northern United States. The direction of movement can be determined from the study of glacial grooves and scratches because they
are smoother in the direction of movement.

Figure 6-34
Area covered by ice during the Pleistocene glaciation. Arrows show the general direction of ice movement.

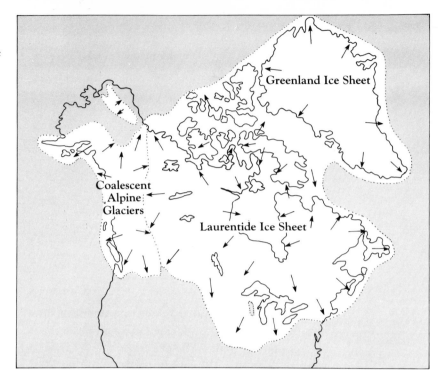

From such studies we have learned that the last glacial times consisted of four major advances of ice and many smaller ones. This sequence was determined from study of places where up to four moraines were deposited, one above the other. This Ice Age, or Pleistocene age, began in North America about 2.5 million years ago, and the last continental ice sheet melted a little over 10,000 years ago. These dates have been determined by measuring the amount of radioactive carbon-14 in organic material found in glacial deposits. (Radioactive dating is discussed in Chapter 7.)

The Pleistocene glaciation had effects far from the actual ice. During the times of glaciation, sea level was about 150 meters (500 feet) lower than at present because of the water held in the glacial ice. It is estimated that, if the present glaciers melt, sea level will rise between 30 and 40 meters (100 to 150 feet). The climate of glacial times was apparently wet because, far to the south of the glaciers, large lakes formed in the western United States (Figure 6-35). Great Salt Lake in Utah is a remnant of the Pleistocene lakes. F 6-35

The cause of the last Ice Age is not known, but many theories have been suggested to explain it and the other glaciations found in the geologic past. It is clear that temperatures must be a few degrees lower than at present to initiate glaciation. There must also be heavy precipitation so that the snow can build up to form glaciers. A cold, relatively dry climate like Siberia's is not enough to cause glaciation;

Figure 6-35
Lakes in western United States during the Pleistocene glaciation.

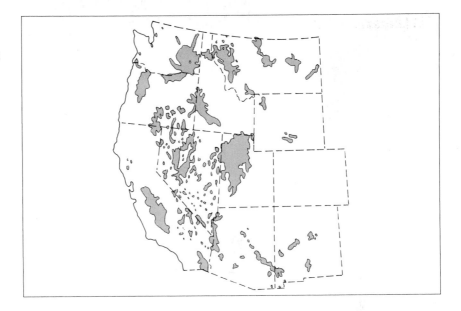

there must be much precipitation as well. Some suggested causes of glaciation are:

1. Changes in the amount of energy received from the sun, because of changes either in the sun itself or in our atmosphere. Dust in the atmosphere from extensive volcanic eruptions is one possibility.
2. Changes in the positions of the continents caused by plate tectonics and sea-floor spreading.
3. Changes in the earth's orbital motions. The direction in which the earth's axis points changes slightly, as do other orbital motions, in well-known cycles or periods that are described in Chapter 16.
4. Changes in the circulation of the oceans can have a large effect on the temperature and precipitation on the continents.

At the moment, it is difficult to choose among these and other theories.

SUMMARY

- The sun is the energy source of the *hydrologic*, or *water*, *cycle*, and this and gravity are the causes of erosion.

- *Downslope*, or *mass*, *movement* may involve only the overburden, or *regolith*, or it may also involve bedrock.

- *Creep* is slow downslope movement; with increasing amounts of water creep grades into earthflow and mudflow.

- *Slides* move by sliding. *Landslides* involve both bedrock and overburden; *rockslides* only involve bedrock.

- The accumulation of rockfall below a cliff is called *talus*.

- Rivers transport by *bed load*, *suspended load*, and *dissolved load*.

- Running water erodes by *solution*, *abrasion*, and *direct lifting*.

- The elevation of the mouth of a river is the lowest level to which it can erode and is called its *base level*.

- River deposits are *deltas* and *alluvial fans*.

- A *graded river* is one that is balanced so that its gradient and flow are able to transport its load.

- *Young* river valleys are generally V shaped and may have rapids, waterfalls, and lakes.

- *Mature* river valleys are somewhat wider so that the meander belt of the river occupies the width of the valley.

- At *old age*, the valley is much wider than the river's meander belt.

- A *youthful* landscape is one with flat interstream divides.

- At *maturity*, no flat interstream areas remain.

- In *old age*, the interstream divides are low and rounded; the ultimate stage is the *peneplain*.

- Probably very few areas ever go through the complete cycle of erosion just described.

- *Ground water* comes from the accumulation of precipitation that infiltrates. The top of the region that is saturated with ground water is called the *water table*, and, in general, it is a subdued replica of the surface topography.

- A *cone of depression* results if ground water is pumped faster than it is replaced.

- *Artesian wells* are self-flowing wells.

- *Caves* are formed by solution of soluble rocks by ground water.

- Regions of soluble rocks, such as limestone, may develop *karst topography* in which much of the water flows underground, and the surface is pockmarked with *sinkholes*.

- Even in deserts, running water is the main agent of erosion, and *pediments* are a common erosional form.

- The lifting and carrying away of fine material by wind is called *deflation*. The pebbles left behind are called *desert armor* or *pavement*.

- Wind-blown sand can erode by *sandblasting*.

- Accumulations of wind-blown dust are called *loess*; those of sand are called *dunes*.

- A *glacier* is a mass of moving ice.

- Glacial erosion changes a V-shaped river valley into a U-shaped glacial valley; it also tends to straighten the valley. The amphitheater-like head of a glacial valley is called a *cirque*.

- Glacial deposits are *moraines*, and they show the former extent of glaciers.

- Glaciers that cover large areas are called *ice-cap*, or *continental*, *glaciers*.

- Continental glaciers tend to round the topography and move the overburden.

- In the last glacial advance, the centers of ice accumulation were in Canada, and the glaciers moved radially outward in all directions from these centers.

- During such glacial times, sea level is lowered.

- Glacial climates are cool and wet; south of the ice, large lakes formed.

KEY TERMS

erosion	dissolved load	reservoir rock, or	mountain, or alpine,
downslope move-	suspended load	aquifer	glacier
ment, or mass	solution	porosity	crevasse
movement	bed load	permeability	plucking, or quarry-
bedrock	abrasion	cone of depression	ing
overburden, or	direct lifting	artesian well	hanging valley
regolith	base level	sinkhole	cirque
fall	delta	karst topography	horn
slide	alluvial fan	stalactite	moraine
flow	graded river	stalagmite	lateral moraine
creep	meanders	hot spring	terminal moraine
earthflow	flood plain	geyser	outwash plain
mudflow	levee	pediment	drift
solifluction	cycle of erosion	deflation	till
landslide	youth	desert armor, or des-	continental glacier
debris slide	maturity	ert pavement	erratic boulder
rockslide	old age	ventifact	
slump	peneplain	loess	
rockfall	water table	dune	
talus	perched water table	glacier	

REVIEW QUESTIONS

1. (a) What is downslope movement?
   (b) What force is responsible for it?
   (c) Why is downslope movement the most important process of erosion?
2. Describe each of the following and explain how they differ:
   (a) flow
   (b) slide
   (c) fall
3. (a) What is a river's load?
   (b) Contrast the following loads: dissolved, suspended, and bed.
   (c) Does the river's velocity affect its dissolved load? What effect does the velocity have on the suspended and bed loads?
4. Describe the three processes by which rivers erode their beds.
5. What is the lowest level to which a river can erode its bed?

6. What is a delta? What conditions are required for its formation?
7. Answer the following questions about graded rivers:
   (a) Is a stream's slope steeper near its headwaters or near its mouth?
   (b) Does erosion make a stream's slope steeper or gentler? What effect does deposition have on the slope?
   (c) State the relationships between erosion and deposition, and changes in water flow and load.
   (d) Explain the effects that a dam has on the slope of a previously balanced (graded) river.

8. Answer the following questions about the development of river valleys:
   (a) What is the shape of a valley during youth? Why is the youthful shape so different from the shapes that will develop later?
   (b) At what point does the mature stage begin? During maturity and old age, erosion is more horizontal than vertical. What effect does this change have on the valley?
   (c) You are canoeing a shallow, meandering stream and want to avoid getting stuck on sand spits. Is the water likely to be deeper on the inside or outside of a bend? Why?
   (d) How does a valley in old age differ from one in the mature stage?

9. (a) Describe the development of a peneplain.
   (b) Why are peneplains more an idealized model than a reality?

10. Why does well-sorted sandstone make a good aquifer?

11. Describe the development of karst topography.

12. Which causes more erosion of a desert landscape, wind or water?

13. Why are desert landscapes more angular than the landscapes of humid areas?

14. Examine Figure 6-22 and describe in words how a pediment forms.

15. Why are wind-deposited sediments well sorted?

16. Describe the process by which a dune advances.

17. Describe how a glacier forms and begins to flow.

18. (a) In what direction does a mountain glacier flow?
    (b) Which part of the glacier is moving most rapidly? Which part is moving most slowly? Why?
    (c) Which part is most brittle?

19. In what directions did the continental glaciers of North America flow? How do scientists know the directions of flow?

20. (a) Describe the two processes by which glaciers erode.
    (b) A certain valley is U shaped and has a cirque at one end. Explain how a mountain glacier created this scenery.
    (c) Describe erosion by a continental glacier.

21. (a) What is a moraine?
    (b) How do geologists recognize moraines?
    (c) What is an outwash plain?
    (d) What is an erratic boulder? Why are these boulders helpful in determining the extent of past continental glaciation?
    (e) How do geologists know that four major advances of continental glaciers occurred in North America during the Ice Age?

22. (a) When did the last continental ice sheet melt? How do geologists know?
    (b) What effect did Pleistocene glaciation have on sea level?
    (c) What was the climate like south of the glaciers?

23. (a) What two requirements are necessary to initiate glaciation?
    (b) What are some suggested causes of continental glaciation?

SUGGESTED READINGS

Baldwin, H. L., and C. L. McGuiness. *A Primer on Ground Water*. Washington, D.C.: U.S. Government Printing Office, 1963, 26 pp.

Bloom, A. L. *The Surface of the Earth*. Englewood Cliffs, N.J.: Prentice-Hall, 1969, 152 pp. (paperback).

Carter, L. J. "Soil Erosion: The Problem Persists Despite the Billions Spent on It," *Science* (April 22, 1977), Vol. 196, No. 4288, pp. 409–411.

Denton, G. H., and S. C. Potter. "Neoglaciation," *Scientific American* (June 1970), Vol. 222, No. 6, pp. 101–110.

Donn, W. L. "Causes of the Ice Ages," *Sky and Telescope* (April 1967), Vol. 33, No. 4, pp. 221–225.

Hays, J. D., John Imbrie, and N. J. Shackleton. "Variations in the Earth's Orbit: Pacemaker of the Ice Ages," *Science* (December 10, 1976), Vol. 194, No. 4270, pp. 1122–1132.

Hoyt, W. G., and W. B. Langbein. *Floods*. Princeton, N.J.: Princeton University Press, 1955, 469 pp.

Hunt, C. B. *Physiography of the United States*. San Francisco: W. H. Freeman, 1967, 480 pp.

Judson, Sheldon. "Erosion of the Land," *American Scientist* (Winter 1968), Vol. 56, No. 4, pp. 356–374.

Keefer, W. R. *The Geologic Story of Yellowstone National Park*. U.S. Geological Survey Bulletin 1347. Washington, D.C.: U.S. Government Printing Office, 1971, 92 pp.

Leopold, L. B., and W. B. Langbein. "River Meanders," *Scientific American* (June 1966), Vol. 214, No. 6, pp. 60–70.

Moore, G. W. "Origin of Limestone Caves: A Symposium," *Bulletin of the National Speleological Society* (1960), Vol. 22, Part 1, 84 pp.

Morgan, J. P. "Deltas—A Résumé," *Journal of Geological Education* (May 1970), Vol. 18, No. 3, pp. 107–117.

Rantz, S. E. *Urban Sprawl and Flooding in Southern California*. U.S. Geological Survey Circular 601-B. Washington, D.C.: U.S. Government Printing Office, 1970, pp. B1–B11.

Schumm, S. A. "The Development and Evolution of Hillslopes," *Journal of Geological Education* (June 1966), Vol. 14, No. 3, pp. 98–104.

Sheaffer, J. R., and others. *Flood-Hazard Mapping in Metropolitan Chicago*. U.S. Geological Survey Circular 601-C. Washington, D.C.: U.S. Government Printing Office, 1970, pp. C1–C14.

Thomas, H. E., and L. B. Leopold. "Groundwater in North America," *Science* (March 5, 1964), Vol. 143, No. 3610, pp. 1001–1006.

RELATIVE AGE
OF ROCKS

ABSOLUTE AGE
OF ROCKS

Radioactive Dating

CORRELATION

UNCONFORMITIES

PLATE
TECTONICS
AND LIFE—A
BRIEF EARTH
HISTORY

CHAPTER 7

EARTH HISTORY: FOSSILS, TIME, AND PLATE TECTONICS

Trilobites of Middle Devonian age on rock slab from Arkona, Ontario. Each is about 3.8 cm (1.5 in.) long. (Photo from Smithsonian Institution.)

An obvious goal of geology is to decipher the history of the earth. To determine this physical history requires both dating rocks and interpreting them. The main approach to interpretation, as we have seen, is the principle of **uniformitarianism**—which states that the present is the key to the past. In other words, we can interpret ancient rocks if we understand how similar rocks are forming today. One of the most reliable ways to date rocks is to study the fossils they contain. Thus, the study of ancient life on the earth is also an important aspect of geologic history. Recognition of the age of the earth—that is, the immensity of geologic time—is one of the great intellectual accomplishments.

RELATIVE AGE OF ROCKS

The relative ages of rocks that are near one another is easily determined. It is clear, for example, that in an undisturbed section of sedimentary rocks the oldest rocks are at the bottom and the youngest are at the top (Figure 7-1). This is the principle of **superposition**. Another means of relative dating is by cross-cutting relationships. An intrusive igneous rock is younger than the rocks that it intrudes (Figure 7-2). A fault, too, must be younger than the beds it cuts. Relationships such as these can be used to determine the relative ages of the rocks exposed in an area, as is done in Figure 7-3.

F 7-1

F 7-2

F 7-3

Figure 7-1
The principle of superposition. The oldest beds are on the bottom, and the youngest are on top.

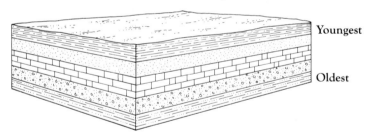

Youngest

Oldest

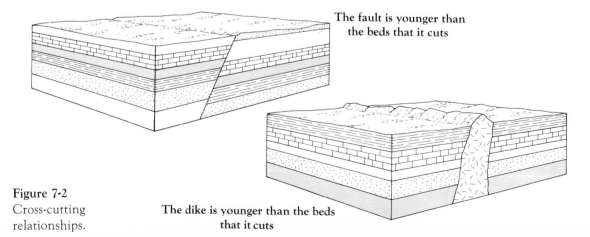

The fault is younger than the beds that it cuts

Figure 7-2
Cross-cutting relationships.

The dike is younger than the beds that it cuts

Figure 7-3
The steps in the development of an area.

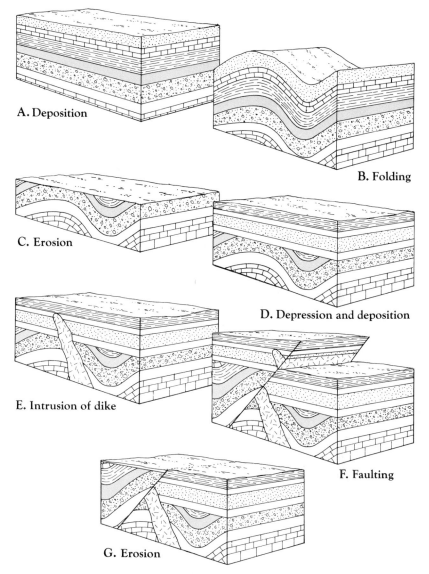

A. Deposition

B. Folding

C. Erosion

D. Depression and deposition

E. Intrusion of dike

F. Faulting

G. Erosion

The next step in dating rocks uses fossils. Around the end of the 18th and the beginning of the 19th century, it was recognized that a given bed, or group of beds, contains the same fossils every place that it is exposed. This discovery is generally credited to an English surveyor and canal builder, William Smith, but the same observation was also made in France at about the same time. Smith was concerned with the problems of canal building, and the problems encountered depended on the rocks excavated to make the canal. He could predict the ease of construction if he knew which of the rock layers was hidden under the soil. He observed that different fossils weathered out of each rock layer, and he could predict which layer lay under the soil by which fossils were on the surface. On this basis, he published the first geologic map showing the distribution of rock layers.

In the next few decades, other workers, mostly in England and western Europe, developed the geologic time scale. The **geologic time scale** (Figure 7-4) is a relative time scale based on fossils. The time scale was built by a number of individuals, each working alone in a specific area. Each one studied a group of sedimentary beds and described the rocks and their fossils. They called the rocks a **system**, and each published a book describing a system, such as the *Silurian System*. In general they chose easily recognized groups of rocks with distinct boundaries. It was possible to put the systems in the proper sequence from oldest to youngest because some of the strata in England and western Europe are both distinctive and widespread. The rocks containing coal were of economic importance and at that time were called the Coal Measures. Immediately above and below the Coal Measures were two distinct, easily traced units called the New Red Sandstone and the Old Red Sandstone. These units were soon traced through much of Great Britain, and, in the same way, so were the newly defined systems. The unit of time represented by a system is called a **period.**

The geologic time scale was established before the middle of the 19th century. After that, it was possible to date rocks in terms of the time scale—that is, to date them as Cambrian or some other of the systems—from study of the fossils in the rock. The fossils in the unknown rock were compared to those in each of the systems, and in this way their age was determined.

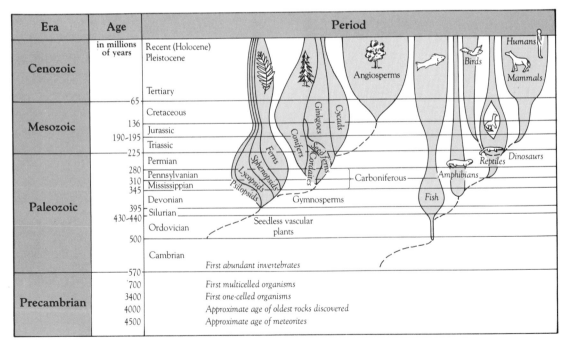

Figure 7-4   The geologic time scale.

ABSOLUTE AGE OF ROCKS

The geologic time scale enabled geologists to date rocks all over the world in terms of the systems developed in Europe, but a major problem still remained. No one knew the actual age of the systems in years. Until this was determined, the age of the earth could not be known. The age of the earth became a controversy in the 19th century, and the problem was not solved until early in the 20th century when radioactivity was understood.

The first attempts to determine the age of the earth used the rate of deposition of sedimentary rocks. Uniformitarianism implies that sedimentary beds have accumulated throughout geologic time at the same rate that they are accumulating now. Therefore, if we can measure the rate at which sediments are now accumulating and we know the total thickness of all of the sedimentary rocks of all the systems, simple division will reveal the age of the earth. Of course, the rate of sedimentation is not constant, and the thickness of each system varies from place to place. As a result, the age of the earth calculated in this way varied between a few million years and 1584 million years, with the average about 100 million years. Such results were far from satisfactory. Other attempts based on the accumulation of salt in the ocean suggested an age of about 90 million years.

The controversy heated up after 1859, the year that Charles Darwin published his book on evolution. Darwin needed a very old earth so that there would be enough time for evolution to occur. In the later part of the 19th century, Lord Kelvin, perhaps the most prominent physicist of the day, entered the fray. His interest was thermodynamics. He assumed, incorrectly as we will see, that the earth originated as a hot, molten body that has been cooling throughout geologic time. He calculated the time necessary for the earth to cool from the highest temperature at which life could exist to the present surface temperature. He continued to refine his calculations, and his best estimate was 20 to 40 million years. Many geologists and evolutionists thought that this was not long enough, but it was their estimates versus Kelvin's calculations.

Radioactivity was discovered very late in the 19th century, and by very early in the 20th century its nature was understood. When a radioactive element decays, it releases energy and changes into another element. The energy released by radioactive elements is a source of heat energy within the earth, not known to Kelvin. If the rate of radioactive decay is known, one can measure the amounts of original elements and daughter elements and calculate the age of the mineral containing the radioactive element. In this way radioactive elements can be used to date rocks, and the date is in number of years. Radioactivity has been used to date the systems of the geologic time scale, as shown in Figure 7-4.

Figure 7-5
To date a sedimentary bed by radioactive methods, one must find the youngest rock datable by radioactivity below the bed (in this case dike A) and the oldest one that cuts the bed (dike B).

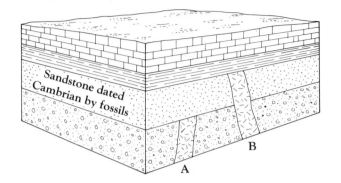

A radioactive date tells the time at which the mineral containing the radioactive element crystallized. Thus, igneous and, in some cases, metamorphic rocks are the main rock types that can be dated by radioactive means. The conditions necessary to date sedimentary rocks are shown in Figure 7-5, and such conditions are not common.   F 7-5

Radioactive Dating

Natural radioactivity occurs because the nuclei of some elements are unstable and so change spontaneously into other elements, emitting energy in the process. Radioactive dating depends upon the principle that the rate of decay of a radioactive element is constant and is unaffected by such physical conditions as temperature, pressure, and the like. The rate of decay is measured by the element's **half-life**. During each half-life, one-half of the atoms of the parent radioactive element decay, or change into the **daughter product.** Thus, if we start with one hundred atoms of a radioactive element, at the end of the first half-life, fifty will remain. During the next half-life, half of the fifty,

Figure 7-6
During each half-life, one-half of the number of remaining atoms of a radioactive element decay. At the end of one half-life, 50 per cent remain; at the end of two half-lives, 25 percent; and so forth.

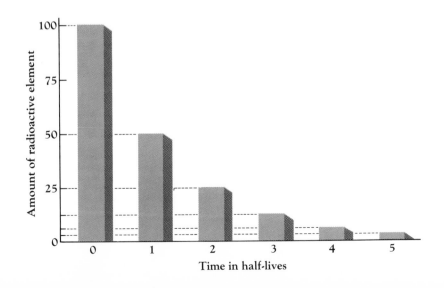

Table 7-1   RADIOACTIVE ELEMENTS USED IN GEOLOGICAL DATING

Element (Isotope)	Decay Product	Half-Life in years	Type of Decay
Carbon-14	Nitrogen-14	5,560 ± 30	single-step
Potassium-40	Argon-40	1,300 million	single-step
Rubidium-87	Strontium-87	47,000 million	single-step
Uranium-235	Lead-207	710 million	multiple-step
Uranium-238	Lead-206	4,500 million	multiple-step
Thorium-232	Lead-208	14,000 million	multiple-step

or twenty-five more, will decay, leaving twenty-five of the original atoms (Figure 7-6). F 7-6

A number of radioactive elements are used in geologic dating, as shown in Table 7-1. The isotopes of uranium and thorium are not T 7-1 abundant, but they tend to be present in the same minerals. This enables the determination of up to three different radiometric dates, and they should all be the same. Such accuracy checks are important because the radioactive decay of each of these elements goes through many steps before one of the stable isotopes of lead forms. Loss of part of any of these radioactive daughter products from weathering or metamorphism would make the age determination inaccurate. Uranium and thorium isotopes are used to date meteorites, and, because they are believed to have formed at the same time as the earth, these age determinations indicate that the earth and the rest of the solar system formed almost 5000 million years ago.

Radioactive elements that decay in a single step are less apt to have their daughter products removed. The decay of rubidium-87 to strontium-87 and potassium-40 to argon-40 are examples. Potassium-40 is widely used for dating, and loss of argon-40, a gas, has not proved to be a problem. Calcium-40, the other decay product of potassium-40, is common calcium, which is present in most rocks, so it is not used in radiometric dating.

Carbon-14 dating has been widely used in archaeology as well as in geology. Table 7-1 shows that the half-life of carbon-14 is very much shorter than the half-lives of the other isotopes used in radiometric dating. Because of this, carbon-14 generally can be used only to date materials less than 50,000 years old. Carbon-14 is formed as a result of cosmic rays coming into our atmosphere from space. The cosmic rays bombard the gases in the atmosphere, producing, among other particles, neutrons. Some of these neutrons collide with nitrogen atoms in the atmosphere, producing carbon-14 atoms. The carbon-14 atoms become part of the carbon dioxide in the atmosphere. This production of carbon-14 is a continuous process, so the amount of carbon-14 in the atmospheric carbon dioxide is constant. Carbon dioxide is used by plants in their life cycle, so carbon-14 is absorbed by plants, and from the plants it gets into all living things. Thus, a small percentage of the carbon in organisms is carbon-14, and the percentage remains constant during life. After death, no new carbon-14 is taken in, so the amount of carbon-14 in the organism decreases as the radioactive carbon-14 decays to nitrogen. The amount of carbon-14 remaining is thus a measure of the time since the organism died.

CORRELATION

In field studies, a geologist must determine whether a rock layer exposed in one area is part of a similar layer exposed somewhere else. If the bed involved is resistant to weathering, the exposures may be good enough that the bed can be traced or walked out from one place to the other. In other instances, the bed may be distinctive in some way so that it is easily recognized. Unusual minerals or fossils or some other feature may make such distinctive beds (Figure 7-7). In some F 7-7 cases it may be possible to use the sequence of beds to identify a layer (Figure 7-8). In other cases, it may be important to determine the age F 7-8 of a bed to see whether it is the same age as some other bed.

Fossils are generally used to determine geologic age, but, as we have seen, radioactivity can also be used to date rocks. At some places thin layers of volcanic ash provide what can be termed time lines because the ash fall occurred over a very short time. In most cases, however, fossils are used to date rocks. Some fossils by themselves can establish the age of the containing rocks. Such **index fossils** should be abundant, widespread, and have a short time span if they are to be truly useful. In most cases, a number of fossils must be used to determine the age of a bed. If the time spans of the fossils are known, the age of the bed can be determined (Figure 7-9).   F 7-9

The most useful fossils for dating are those of animals found in a number of different environments. Bottom-dwelling animals on muddy sea floors are different from animals found on sandy bottoms. Thus, the fossils in shale are generally different from those found in sandstone, even though the shale and the sandstone are the same age.

Figure 7-7
A distinctive bed enables correlation in this area where rock types are gradational. Volcanic-ash beds may both be distinctive and provide time lines because such beds may be deposited rapidly by a single eruption.

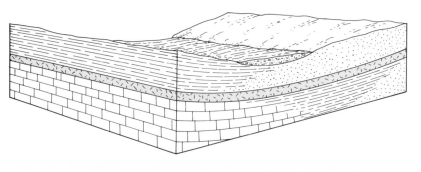

Figure 7-8
Correlation by similar sequence of beds. The beds encountered in the drill hole are similar to those exposed on the hillside.

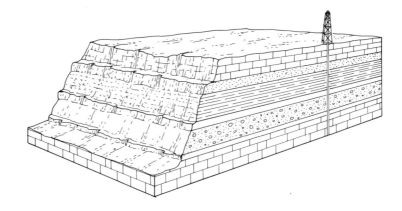

Figure 7-9
If the time spans of several fossils are known, the occurrence of more than one in a single bed can be used to date the bed more accurately.

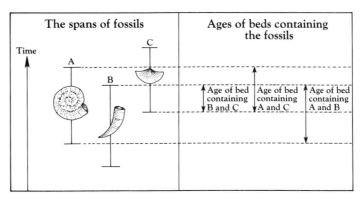

For this reason, swimming animals or those whose shells float after death make the most useful fossils for geologic dating.

UNCONFORMITIES

It should be clear that sedimentary rocks are not deposited continuously in any basin. *Breaks in the sedimentary record are called* **unconformities**. An unconformity may be the result of nondeposition, or beds may be deposited and then eroded. Many types of unconformities are possible. Sedimentation may be interrupted because the basin fills to sea level and later subsidence causes renewed deposition (Figure 7-10). An area of deposition may be uplifted, causing erosion, and later subsidence renews sedimentation (Figure 7-11). A basin may be deformed, and the folded and faulted rocks may be eroded before subsidence renews deposition. These events produce an **angular unconformity** (Figure 7-12), so called because the beds below the unconformity are not parallel to those above it. If sedimentary rocks overlie granitic or metamorphic rocks, this, too, is an unconformity. In this case, a long time of erosion was required to expose the igneous or metamorphic rocks that must have formed deep in the crust.

F 7-10

F 7-11

F 7-12

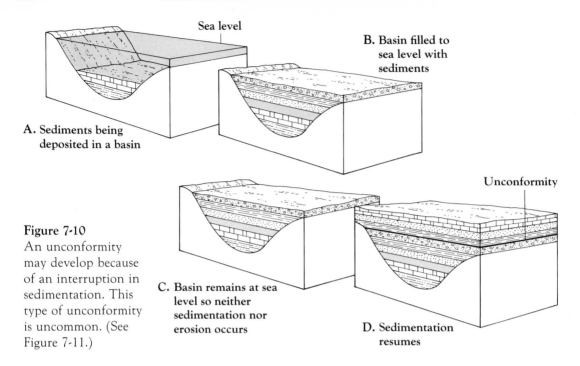

Sea level

B. Basin filled to sea level with sediments

A. Sediments being deposited in a basin

Unconformity

Figure 7-10
An unconformity may develop because of an interruption in sedimentation. This type of unconformity is uncommon. (See Figure 7-11.)

C. Basin remains at sea level so neither sedimentation nor erosion occurs

D. Sedimentation resumes

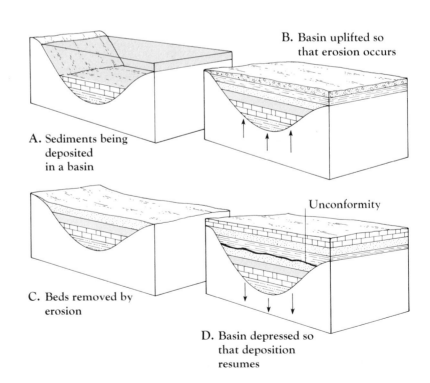

B. Basin uplifted so that erosion occurs

A. Sediments being deposited in a basin

Unconformity

Figure 7-11
More commonly, unconformities develop because of both nondeposition (as in Figure 7-10) and erosion.

C. Beds removed by erosion

D. Basin depressed so that deposition resumes

Figure 7-12
Development of an
angular
unconformity.

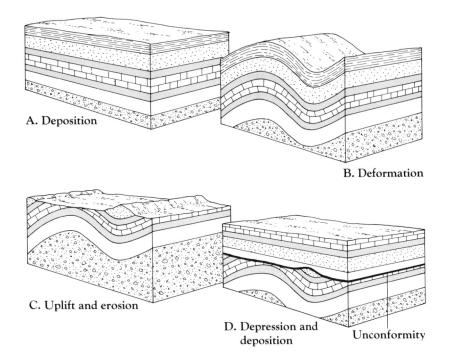

A. Deposition

B. Deformation

C. Uplift and erosion

**D. Depression and
deposition**   **Unconformity**

Another way to think of unconformities is that they are surfaces that show locally unrecorded time intervals; that is, they are local gaps in the geologic record. When the geologic time scale was established, the natural breaks in the sedimentary rocks that were used to define the original systems were, in many cases, unconformities. At other places rocks were deposited and fossils formed during the time of those unconformities. This has required revision of the time scale after it was first established.

PLATE
TECTONICS
AND LIFE—A
BRIEF EARTH
HISTORY

The rocks in which the earth's history, especially its early history, is recorded are, of course, incomplete because of later deformation, metamorphism, and erosion. Even well-preserved rocks can be very difficult to interpret. Because of these problems, many details remain obscure; but the main outlines are clear, at least for the last 1000 million years.

The earth apparently formed at the same time as the sun and the rest of the solar system, and its earliest history is in the realm of astronomy. The oldest rocks so far found are almost 4000 million (4 billion) years old, and there the story begins. The oldest life yet found is about 3400 million (3.4 billion) years old and consists of bacteria and blue-green algae. These very primitive organisms have continued to exist to the present, so their life style is well understood.

They differ from more advanced organisms in that their reproductive system does not involve genes from two parents. For that reason their evolution, at least compared with that of more advanced organisms, is almost nonexistent or very slow. The first organisms capable of real evolution that have been found are green algae about 1000 million (1 billion) years old. From this point on, the story moves much faster.

The geologic time scale and its periods are based on fossils, so it is not surprising that geologic history is known in much more detail after life became fairly abundant. Without fossils, geologic dating is generally difficult. Abundant life began in the Cambrian Period. Between the appearance about 1000 million (1 billion) years ago of organisms capable of evolution and the earliest Cambrian rocks about 570 million years ago, almost all of the present groups of **invertebrate** animals evolved. The fossil life forms found in Cambrian rocks, while very different from present-day animals, are not primitive like the Precambrian life. Very little record of this evolution has been preserved, probably because the organisms had no shells or skeletons capable of preservation. Indeed, the appearance of abundant fossils in Cambrian rocks may not mean that life suddenly became more abundant but only that hard parts capable of being preserved as fossils had evolved. It seems reasonable to assume that, as life became more abundant, more predators evolved and also became more abundant. The evolution of hard parts may have provided a better chance for survival.

According to paleomagnetic data, the continents came together about 800 million years ago, in the late Precambrian. Our knowledge of their positions before that time is very sketchy. Even for most of the Paleozoic Era, our knowledge is only somewhat better, and revision will probably be necessary. Similar late Precambrian fossils of multicellular animals have been found at at least ten places worldwide, also suggesting that the continents were together at that time (Figure 7-13).   F 7-13

Just prior to the start of the Cambrian Period, North America and Europe separated from the southern continents. For the most part, the southern continents remained more or less together throughout the Paleozoic Era. The separation created an ocean with geosynclines on both shores (Figure 7-14). The present Pacific and Arctic margins   F 7-14 of North America were also areas of sediment accumulation. The life in the early Cambrian seas consisted mainly of trilobites and brachiopods, and they were similar throughout much of the world during early Cambrian time. Later the life became more diverse, probably as a result of the separation of the continents. By the early Ordovician Period, all the invertebrate animal groups were present.

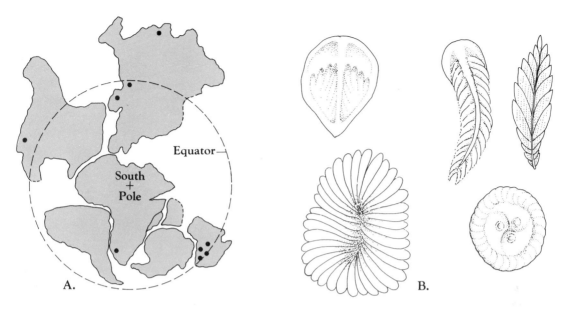

Figure 7-13 Late Precambrian positions of continents, and the places where late Precambrian multicellular animals have been found. The worldwide distribution of these animals suggests that the continents were together at this time.

Figure 7-14
Early Paleozoic continental positions. Seas in western Europe and northeastern North America had similar trilobites, and the rest of North America and most of the rest of the world had a different fauna. During the Paleozoic, several collisions occurred between North America and Europe and Africa, forming mountains.

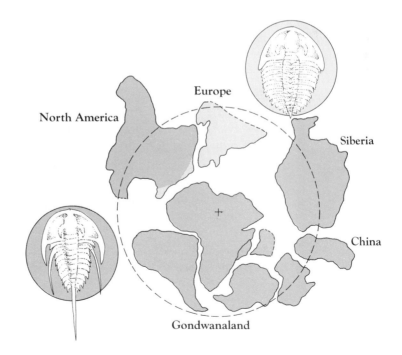

Figure 7·15 Reconstruction of an Ordovician sea floor. The conical-shelled animals with tentacles are nautiloid cephalopods and were early predators. Also shown are trilobites, corals, and bryozoans. This scene is based on fossils found near Cincinnati, Ohio. (Photo from Smithsonian Institution.)

During the Paleozoic Era, the geosyncline on the east coast of North America was deformed several times by collisions with Europe and Africa. The deformation culminated near the end of the Paleozoic Era in the collision of North America with the other continents, forming the Appalachian Mountains. Mountain building also occurred in the west in the mid-Paleozoic, probably because a convergent plate boundary formed along the west coast of North America. These deformations were only a part of the continental movements. At the beginning of the Paleozoic Era, the south pole was in northern Africa. As the continents moved, the pole was first in southern Africa and, finally, by the end of the Paleozoic Era, in its present location in Antarctica. Another effect of this movement was that the climate of much of North America became drier.

Life, too, changed during the Paleozoic Era. At first, there were only invertebrates, and they lived on the sea floor. In the Ordovician and Silurian Periods, some of the invertebrates became mobile preda-

Figure 7-16
Dunkleosteus, a Devonian armored fish. Some of these fish were 10 meters (30 feet) long. (Photo from Smithsonian Institution, No. 490.)

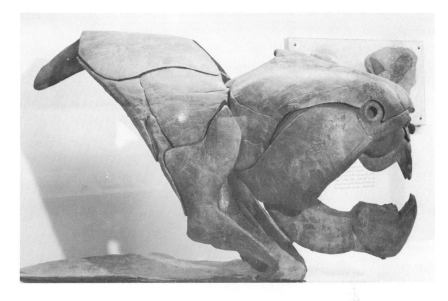

tors (Figure 7-15). The first **vertebrates** were fish that first appeared in the late Cambrian, and by the Devonian Period (Figure 7-16) they were the main predators in the seas. In middle and late Paleozoic time, some of the invertebrates on which the predators fed developed spines and other defenses (Figure 7-17). The mountain building and the changing climate caused by continental movements created new environments into which life evolved. Before animals could move onto the continents, plants had to form. The first land plants appear in Middle Ordovician rocks, and they must have developed rapidly because from the Devonian Period onward every period has coal deposits (Figure 7-18). The present-day lungfish that can crawl from puddle to puddle was probably similar to the ancestor of the first **amphibians** that are found in Devonian rocks. By the Pennsylvanian Period, the amphibians had evolved into the **reptiles** that could live their whole lives on the land.

At or near the end of the Paleozoic Era, the life on the earth changed markedly. This change was obvious to early geologists, and this is why they subdivided geologic time into the Paleozoic (old life), Mesozoic (middle life), and Cenozoic (young life) Eras. Many of the marine invertebrates became extinct at this time, perhaps because the collisions that created a single continent reduced the area of shallow seas in which they lived (Figure 7-19). The single large continent would also have caused more climatic extremes than smaller land masses. The extinctions did not extend to the land-dwelling vertebrates nor to plant life.

In the Mesozoic Era, the single large continent of the late Paleozoic broke up, and the present-day continents moved apart. The breakup

F 7-15
F 7-16

F 7-17

F 7-18

F 7-19

Figure 7-17
Spiny brachiopods of Permian age from Glass Mountains, in Texas. The spines were protection against predators. (Photo from Smithsonian Institution.)

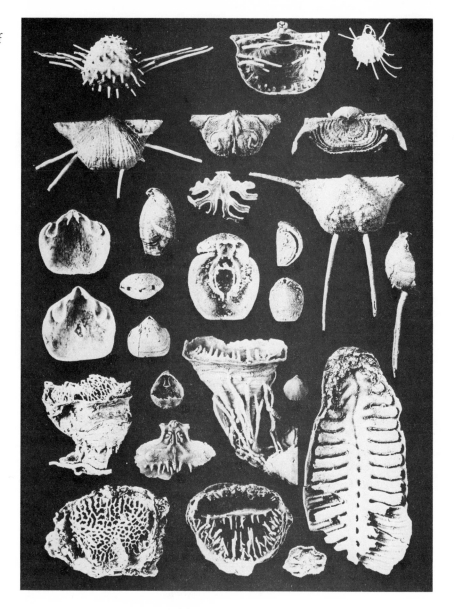

began in the Triassic Period (Figure 7-20). The late Paleozoic reptiles F 7-20 found on all continents are quite similar, as would be expected. This similarity persisted into the Mesozoic Era, as the continents diverged. In the Mesozoic Era, the dominant reptiles were the **dinosaurs** (Figure 7-21). The Mesozoic seas had many varieties of **cephalopods**, swim- F 7-21 ming predators whose shells evolved rapidly, making them nearly ideal index fossils (Figure 7-22). The first birds appeared in Jurassic F 7-22 rocks. The first **mammals** are of Triassic age, but they did not become dominant until the dinosaurs disappeared. Modern plants, the an-

Figure 7-18 Devonian forest in western New York. (Painting by C. R. Knight, Field Museum of Natural History.)

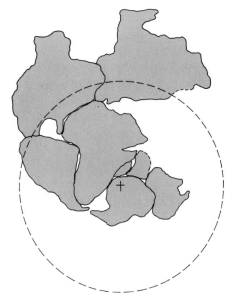

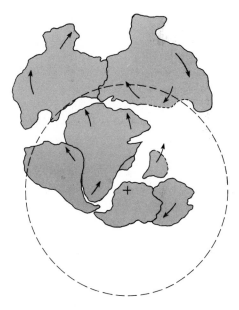

Figure 7-19
End of the Paleozoic. From the late Paleozoic through the Mesozoic, the land dwellers were similar on all continents. The reduction in shallow seas and climatic change that resulted from joining the continents caused many extinctions.

Figure 7-20
The breakup of the single continent began in the early Mesozoic, but the continents remained close enough that animal migrations could occur.

giosperms, became abundant during the Cretaceous Period. Mountain building occurred in western North America at times during the Mesozoic Era.

Figure 7-21
Dinosaurs. *Top:* Carnivorous dinosaur, *Ceratosaurus,* Colorado. About 7 meters (22 feet) long. *Middle:* Plant-eating dinosaur, *Diplodocus,* Utah. About 26 meters (85 feet) long. *Bottom left:* Plated dinosaur, *Stegosaurus:* About 6 meters (20 feet) long. *Bottom right:* Horned dinosaur, *Triceratops.* About 2 meters (6 feet) long. (Photos from Smithsonian Institution. *Top,* No. 43497; *Middle,* No. 43492; *Bottom left,* No. 43494; *Bottom right,* No. 43493.)

Figure 7-22 Cretaceous sea bottom, showing numerous straight and coiled cephalopods, as well as many clams and snails. This scene is based on fossils found in western Tennessee. (Photo from Smithsonian Institution.)

The boundary between the Mesozoic and the Cenozoic is marked by a profound change in life. On the lands the dinosaurs disappeared, and in the oceans so did the cephalopods. Many theories have been proposed to explain these and other extinctions, but none is totally satisfactory. The invertebrate fossils from the Cenozoic oceans are largely clams, snails, and oysters. During the Cenozoic the continents continued to move apart as they had in the Mesozoic, and as a result the mammals diversified and flourished. Mountains formed in western North America, and, of course, all of the present-day topographic features of the earth formed during the Cenozoic.

SUMMARY

- *Relative ages* of rocks can be determined from *superposition;* that is, in undisturbed sedimentary rocks, the oldest rocks are at the bottom and the youngest at the top. *Cross-cutting* features such as intrusive bodies and faults are also used; such features are younger than the rocks they cross-cut.

- *Fossils* are also used to date the rocks that contain them. The *geologic time scale* was developed from study of fossils, and this time scale enables worldwide dating of fossiliferous rocks.

- *Absolute dating* of rocks is done by *radioactive* methods. Natural radioactivity occurs because the nuclei of some elements are unstable and spontaneously change into other elements, emitting energy in the process. The rate of radioactive decay is measured by the *half-life,* and this rate of change is not affected by temperature, pressure, or other physical conditions.

- In one half-life, one-half of the atoms of a radioactive element change into the *daughter* element. In the next half-life, one-half of the remaining atoms change, and so forth. The natural radioactive elements used in geologic dating and their half-lives are:
uranium-238 to lead-206
    7.1×10^8 or 710 million years
uranium-235 to lead-207
    4.5×10^9 or 4500 million years
thorium-232 to lead-208
    1.4×10^{10} or 14,000 million years
rubidium-87 to strontium-87
    4.7×10^{10} or 47,000 million years
potassium-40 to argon-40
    1.3×10^9 or 1300 million years
carbon-14
    5660 ± 30 years

- *Correlation* is the tracing or recognition of a rock unit or layer at different places. This may be accomplished by actually walking out the layer from one location to the other, by recognition of distinctive properties of the layer such as unusual minerals or fossils, or by determination of sequence of beds. Correlation may also be done by dating the beds using either fossils or *time lines,* such as thin layers of volcanic ash.

- *Index fossils* should be abundant, widespread in many environments, and have a narrow time range to be really useful in dating rocks.

- *Unconformities* are breaks in the sedimentary record where geologic time is not represented locally by rocks. Unconformities may be formed by nondeposition, or deposition and later erosion, or by a combination of both. If the older beds are deformed and tilted before deposition resumes, an *angular unconformity* results.

- The oldest rocks found so far on the earth are about 4000 million (4 billion) years old. The earth is probably nearly 5000 million (5 billion) years old.

- The oldest life yet found is about 3400 million (3.4 billion) years old and is bacteria and blue-green algae. The first organisms capable of true evolution are green algae, which are about 1000 million (1 billion) years old. The first really abundant life represented by fossils begins in the Cambrian Period, about 570 million years ago.

- In late Precambrian time the continents came together; and, just prior to the start of the Cambrian, North America and Europe separated from the southern continents.

- During the Paleozoic Era, the geosyncline on the east coast of North America was deformed several times by collisions. Near the end of the Paleozoic, the final collision formed the Appalachian Mountains and brought all of the continents together into one supercontinent.

- The life of the early Paleozoic was invertebrates such as brachiopods and trilobites. By mid-Paleozoic, fish, the first vertebrates, had developed. Amphibians, the first land-dwelling vertebrates, soon evolved from the fish. Plant life, too, developed during the Paleozoic, making terrestrial animals possible.

- The single continent that formed at the end of the Paleozoic destroyed many shallow-sea areas and caused climatic changes. As a result many marine invertebrates died out, and the life of the Mesozoic Era was much different.

- During the Mesozoic, the single large continent broke up and the present Atlantic Ocean formed.

- The Mesozoic seas contained much invertebrate life. Cephalopods are nearly ideal index fossils. On the land the dinosaurs were dominant. The first birds and the first mammals are found in Mesozoic rocks.

- A profound change in life marks the end of the Mesozoic and the beginning of the Cenozoic Era. Many of the cephalopods died as did all of the dinosaurs. Mammals became the dominant vertebrates. The reason for these changes is not known, and many theories have been proposed.

- During the Cenozoic, the present topography of the earth developed.

KEY TERMS

uniformitarianism	daughter product	amphibian
superposition	correlation	reptile
geologic time scale	index fossil	dinosaur
system	unconformity	cephalopod
period	angular unconformity	mammal
radioactivity	invertebrate	angiosperm
half-life	vertebrate	

REVIEW QUESTIONS

1. State the principle of superposition.
2. A vertical igneous rock body intrudes two horizontal beds of sedimentary rock. Which rock is oldest: the bottom sedimentary bed, the top sedimentary bed, or the igneous intrusive? Which rock is youngest?
3. What relationship between sedimentary beds and fossils did William Smith notice?
4. (a) On what is the geologic time scale based?
   (b) At the time the scale was developed, neither radiometric dating nor Darwin's concept of evolution was known. How were geologists able to put the systems into chronological order?
5. Kelvin calculated that life had existed on earth for 20 to 40 million years. Why did evolutionists and many geologists feel that this was much too short a time span? What source of heat did Kelvin leave out of his calculations?
6. (a) During radioactive decay, an unstable isotope of an element releases _____ and changes into an isotope of another _____.
   (b) Define half-life.
   (c) If the half-life of a radioactive isotope is 100,000 years, how much of the element will be left in 300,000 years? $66^2/3\%$, 50%, $33^1/3\%$, 25%, or $12^1/2\%$
7. (a) Why is most radiometric dating done on igneous rocks?
   (b) Under what conditions can sedimentary rocks be radiometrically dated?
   (c) Why can weathering make the date inaccurate?
8. What is the age of the earth and the rest of the solar system believed to be? How was this date obtained?
9. (a) Would carbon-14 be used to date an igneous rock? Why or why not?
   (b) Would carbon-14 be used to date fossils from the Cretaceous System? Why or why not?
   (c) Explain why the amount of carbon-14 in organic remains dates an organism's death.

10. What characteristics does a good index fossil have?
11. (a) Define an unconformity.
    (b) Describe a situation in which an unconformity is created by a period of non-deposition.
    (c) Describe a situation in which an unconformity is created by a cycle of deposition, erosion, and deposition.
    (d) What causes an angular unconformity? How is it identified?
    (e) In a particular area, sedimentary rocks are found directly overlying granite. Why is this an unconformity?
12. Make your own chart of the earth's history to help yourself collate and review the information. Start by making three columns. Above the left column write "Eras and Periods"; above the middle column write "Geologic Events"; above the right column, "Biologic Events." As you reread the section on "Plate Tectonics and Life," fill in your chart. Next, answer the following questions:
    (a) By the beginning of the Cambrian Period, most invertebrates had evolved. Why is there so little fossil evidence of their evolution?
    (b) Outline the history of continental movement from the late Precambrian to the early Cenozoic. Be sure to include formation of the Appalachian Mountains.
    (c) Late Paleozoic fossils of marine invertebrates and of reptiles support the geologic evidence that one giant continent existed at that time. Explain why these fossils support the geologic evidence.
    (d) What do the terms *Paleozoic*, *Mesozoic*, and *Cenozoic* mean? Describe the major forms of life that existed during each of these eras.

SUGGESTED READINGS

Alvarez, L. W., and others. "Extraterrestrial Cause for the Cretaceous-Tertiary Extinction," *Science* (June 6, 1980), Vol. 208, No. 4448, pp. 1095–1108.

Barghoorn, E. S. "The Oldest Fossils," *Scientific American* (May 1971), Vol. 224, No. 5, pp. 30–54.

Colbert, E. H. *Wandering Lands and Animals.* New York: E. P. Dutton, 1973, 323 pp.

Deevey, E. S., Jr. "Radiocarbon Dating," *Scientific American* (February 1952), Vol. 186, No. 2, pp. 24–33. Reprint 811, W. H. Freeman, San Francisco.

Eicher, D. L. *Geologic Time*, 2nd ed. Englewood Cliffs, N.J.: Prentice-Hall, 1976, 152 pp. (paperback).

Eisley, L. C. "Charles Lyell," *Scientific American* (August 1959), Vol. 201, No. 2, pp. 98–106. Reprint 846, W. H. Freeman, San Francisco.

Engel, A. E. J. "Geologic Evolution of North America," *Science* (April 12, 1963), Vol. 140, No. 3563, pp. 143–152.

Hallam, A. "Continental Drift and the Fossil Record," *Scientific American* (November 1972), Vol. 227, No. 5, pp. 56–66.

Hay, E. A. "Uniformitarianism Reconsidered," *Journal of Geological Education* (February 1967), Vol. 15, No. 1, pp. 11–12.

Kurtén, Björn. "Continental Drift and Evolution," *Scientific American* (March 1969), Vol. 220, No. 3, pp. 54–64.

Langston, Wann, Jr. "Pterosaurs," *Scientific American* (February 1981), Vol. 244, No. 2, pp. 122–136.

McAlester, A. L. *The History of Life*, 2nd ed. Englewood Cliffs, N.J.: Prentice-Hall, 1977, 160 pp. (paperback).

Nichols, R. L. "The Comprehension of Geologic Time," *Journal of Geological Education* (March 1974), Vol. 22, No. 2, pp. 65–68.

Ralph, E. K., and H. N. Michael. "Twenty-five Years of Radiocarbon Dating," *American Scientist* (September 1974), Vol. 62, No. 5, pp. 553–560.

Sclater, J. G., and Christopher Tapscott. "The History of the Atlantic," *Scientific American* (June 1979), Vol. 240, No. 6, pp. 156–174.

Stone, Irving. *The Origin: A Biographical Novel of Charles Darwin.* Garden City, N.Y.: Doubleday, 1980, 744 pp.

PART II

OCEANOGRAPHY

Perhaps nowhere is the earth's dynamism more striking than in the oceans. The tides run daily, swallowing the lower beaches and pushing great quantities of water up the rivers. Then the tides retreat back to sea, exposing the mud flats on which shore birds search for food. Waves and currents remove beaches and fill in harbors—to the distress of homeowners along the shore. And, finally, deep in the basins of the oceans lies the evidence for a much greater change—the rifting and colliding of the earth's giant plates.

Chapter 8 provides a tour of the ocean floors. The tour starts with the continental shelves, which are shallow areas that extend from many coasts. At the seaward edge of the shelves, the bottom plunges down the steep continental slopes to the plains of the ocean basins. Rising out of these plains is a ridge that branches into every ocean of the world. This mid-ocean ridge is where new oceanic crust is being made. The chapter explains the theory of sea-floor spreading and describes the great trenches where the old ocean floor disappears down into the melting pot of the mantle. The chapter concludes with a look at the sediment that coats the bottom of the sea.

Chapter 9 describes the water that fills the ocean basins. The composition of sea water—what makes it salty and what gases are dissolved in it—and the temperatures at different depths and different latitudes are all considered. Next the chapter charts the great wind-driven surface currents, such as the Gulf Stream, and then it explains how the earth's rotation causes these currents to spin in giant circles called *gyres*. The chapter concludes with a description of the deep ocean currents.

Chapter 10 explores the coastal environment. It explains how tides occur; discusses what causes waves, rip currents, and longshore currents; and describes the remodeling of shorelines by the ocean's constantly moving waters.

OCEANOGRAPHY

CONTINENTAL SHELF

CONTINENTAL SLOPE

THE OCEAN BASIN FLOORS

Abyssal Plains

Submarine Mountains

MID-OCEAN RIDGES

VOLCANIC ISLAND ARCS AND DEEP TRENCHES

SEA-FLOOR SPREADING

SEA-FLOOR SEDIMENTS

Lithogenous Sediment

Biogenous Sediment

Hydrogenous Sediment

Drilling operations on the Glomar Challenger. *(Photo courtesy of Deep Sea Drilling Project.)*

THE OCEAN FLOORS AND THEIR SEDIMENTS

OCEAN-OGRAPHY

The oceans cover about 70 per cent of the earth's surface, so their water and its movements and the architecture of the ocean bottom are important in understanding our planet. Unlike the separate continents, the oceans are a single interconnected body of water. Earth is the water planet, but for many years the oceans were mysterious, somewhat frightening places. The first understanding of the ocean surface developed because winds and currents were used by sailing ships. Then, much later, instruments were developed that revealed information about the deep, unseen ocean and its floor.

Oceanography is the study of the oceans; but this is really too broad, so oceanographers are divided into **physical oceanographers,** who study the motions of sea water; **chemical oceanographers,** who study the chemistry of sea water; **biological oceanographers,** who study the life in the ocean; and **geological oceanographers,** who study the shape of the ocean bottoms and their rocks.

Oceanography began as exploration of the earth. The great voyages of exploration started in 1492 with Columbus, and the culmination of these early voyages was Ferdinand Magellan's circumnavigation of the earth from 1519 to 1522. Voyages of exploration and mapping continued for at least the next 250 years. During much of this period, navigation methods were somewhat crude. Latitude can be determined fairly easily in the northern hemisphere by measuring the angle between the horizon and the north star (Figure 8-1). Longitude is more difficult to measure, and accurate determination of lon-   F 8–1

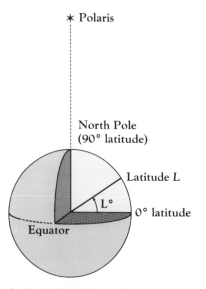

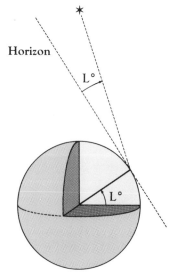
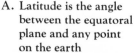

Figure 8-1
Latitude is the angle between the horizon and Polaris, the north star.

A. Latitude is the angle between the equatoral plane and any point on the earth

B. Because of the distance to Polaris, an accurate diagram cannot be drawn

A.

Figure 8-2
A. H.M.S. *Challenger,*
the first
oceanographic
research vessel, off
Cape Challenger, the
southern extremity of
Kerguelen Island,
1874. **B.** *Glomar
Challenger,* a modern
research vessel that
can recover drill cores
from the deep sea
floor. (Photo **A,** from
National Maritime
Museum, London;
and **B,** from Deep Sea
Drilling Project.)

B.

gitude was difficult until a clock, or **chronometer,** that would function on a wave-tossed ship was invented in 1761. Accurate pendulum
clocks were in use on land, but seagoing chronometers were not possible until the spring-driven escapement mechanism, still in use in
many watches, was invented. The last important voyages of exploration were made by James Cook between 1768 and 1779. His work was
much more scientific than that of the earlier voyages, and he charted

much of the Pacific Ocean and looked for the southern continent, Antarctica.

American involvement in oceanography began much later, although Benjamin Franklin did produce a chart showing the location of the Gulf Stream in 1770. Matthew Fontaine Maury was a U.S. Navy officer who, using records and log books, produced charts showing winds and currents that were very valuable to seamen. In 1855 he published *The Physical Geography of the Sea*, summarizing his work.

Scientific oceanography began with the voyage of the *Challenger*. H.M.S. *Challenger* (Figure 8-2A) sailed in 1872 with the mission of F 8-2 exploring the depths of the ocean. Her scientists spent three and a half years determining the depths of the oceans, dredging bottom samples, trawling and netting marine life at all depths, and measuring salinity and temperature at all depths. Among other things, they discovered 4717 new species of marine organisms and measured the depth of the Marianas Trench in the western Pacific, the deepest place in the oceans at 10,867 meters (35,630 feet). Little more was done for 50 years until the German *Meteor* voyages of 1925–1927. This expedition charted the bottom topography of the south Atlantic.

Currently, most oceanographic research is conducted by universities or research institutes; and, because of the expense, much of the research is government sponsored. An example is the Deep Sea Drilling Project, carried out by a group of academic and private research institutions under federal grants. This group built the *Glomar Challenger* (Figure 8-2B), a drilling ship that can position itself accurately enough to be able to reenter a drill hole in deep water. Without this ability drill bits cannot be changed, and so deep drilling is not possible.

CONTINENTAL SHELF

At many places the **continental shelf** is the border between the oceans and the continents (Figure 8-3). The shelf is an area of shallow F 8-3 water that separates the continents from the deep ocean. The average water depth on the shelves is about 60 meters (200 feet). The shelves slope very gently from the coast to their edges, which are generally at depths of 120 to 150 meters (400 to 500 feet). At the edge of the shelf, the slope becomes much steeper as the *continental slope* begins. Continental shelves average about 70 kilometers (43 miles) wide, although the range is from none at all to 1300 kilometers (800 miles).

Most continental shelves would appear to be flat, monotonous surfaces if they could be viewed. They are covered with terrestrial sediment, which in many cases was deposited during the Ice Age, when sea level was much lower because the water was held in the glacial ice. At some places, river-cut valleys can be seen on the continental

Figure 8-3
A. Continental shelf and slope form the boundary between ocean and continent at many places. Vertical exaggeration is generally necessary to show details. **B.** Actual profiles with no vertical exaggeration. (**B** adapted from U.S. Geological Survey.)

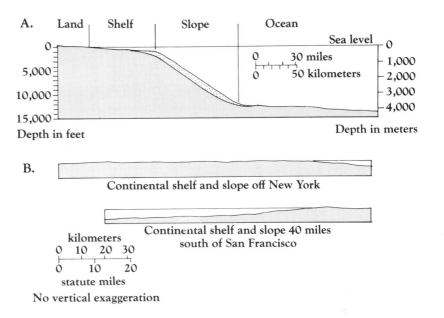

shelves. On the Pacific coast of North America, the shelves are only a few tens of kilometers wide, and at many places they are divided into ridges and basins.

The continental shelves are important economically because of commercial fishing and because at many places the sedimentary rocks contain petroleum.

The deep rocks that underlie the continental shelves are continental; that is, they are granitic rocks. The rocks that underlie the sediments in the deep ocean are basaltic. Geologically, the continental shelves are part of the continents, and the geologic transition between continent and ocean occurs on the continental slope.

CONTINENTAL SLOPE

The **continental slope** is the rather abrupt transition between the continental shelf and the ocean floor. It begins at the edge of the shelf, where in a distance of a few tens of kilometers the incline steepens from almost flat on the shelf to several degrees on the continental slope. The continental slopes generally slope between 2 and 6 degrees, but some are much steeper. Although this may not seem very steep, few mountain ranges are so abrupt. In a distance of between 50 and 100 kilometers (30 and 60 miles), the continental slope deepens from 150 meters (500 feet) at the edge of the continental shelf to 3700 meters (12,000 feet) on the ocean floor. Near the bottom, the continental slope becomes much less steep, making the transition between the slope and the ocean floor gradual. This transition area is called the *continental rise*. Most continental rises are composed of sediments eroded from the continents.

Most continental slopes are composed of sedimentary rocks. Their structures appear to be diverse, and some of the possible structures are shown in Figure 8-4. The present surfaces of continental slopes are sculptured by both deposition and erosion. Deposition is suggested because the sediment carried to the ocean by rivers must move across the slopes to reach the ocean basins. Erosion is shown by canyons and landslide scars.

Submarine canyons are deep, commonly V-shaped valleys that are cut into the continental slopes (Figure 8-5). They are generally big features with tributary valleys and curved paths. Some are clearly associated with rivers on the nearby continents—for example, the Hud-

F 8-4

F 8-5

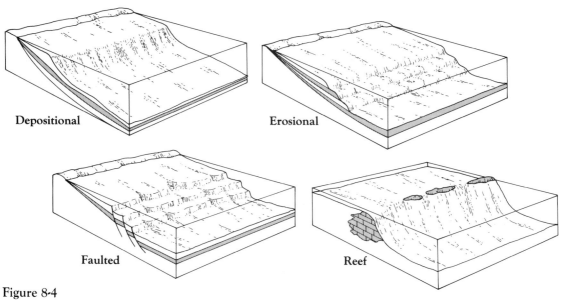

Figure 8-4
Types of continental slopes.

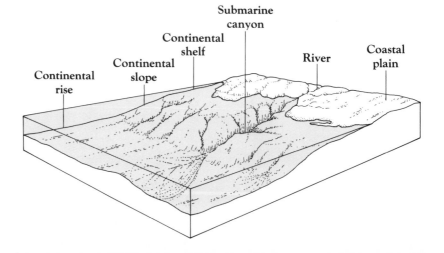

Figure 8-5
Submarine canyons are cut into continental slopes.

son and Congo Rivers. The Hudson canyon is connected with the Hudson River by a valley cut into the continental shelf. In cases like this, the valley on the continental shelf could have been eroded during the Ice Age, when the ocean level was lower than at present. The canyons cut into the continental slope could not have been eroded by rivers, because sea level would have to be lowered by more than 3 kilometers (2 miles) for this to be possible. For this reason, and because many submarine canyons are not associated with rivers, another process is suggested.

Turbidity currents are the erosional agents that cut submarine canyons and move sediments down the continental slope. A **turbidity current** *is a mixture of sediment and water that, because it is heavier or denser than the sea water, moves downslope along the bottom.* A turbidity current moving downslope is capable of erosion, and the muddy water can transport even large boulders. The discovery of turbidity currents came about when earthquakes caused submarine slumps or landslides, which, in turn, caused turbidity currents. As the turbidity currents moved along the sea floor, they broke a number of submarine telephone cables. Careful analysis of the times and places where the breaks occurred enabled calculations of the speed of the turbidity currents. They move at 10 to 50 kilometers per hour (6 to 30 miles per hour).

At the foot of the continental slope is the **continental rise.** This is the area where the surface gradually changes angle between the continental slope and the flat sea floor. The continental rise is a generally smooth area with, in some places, hills like those on the sea floor. This suggests that the continental rise is a depositional feature, perhaps caused by the deposits of turbidity currents and other sediment movement down the slope. The small channels that cross the rise may be the route of sediment to the ocean floor.

THE OCEAN BASIN FLOORS	**Abyssal Plains**

The ocean basin floors are the areas between the continental rises and the mid-ocean ridges. Defined in this way, the ocean floors form about 30 per cent of the earth's surface, or about the same percentage as the continents. Depth soundings have shown that these surfaces are at places the smoothest and flattest on earth, but at other places they are interrupted by submarine hills and mountains. The smooth, flat surfaces are generally close to the continental rises, which probably are the source of their surface sediments. These areas are called **abyssal plains.**

The water depth over the ocean floors is about 4.4 kilometers (2.75 miles). Beneath the water, seismic studies and some drilling reveal a uniform structure that begins with a surface layer of unconsolidated sediment about 300 meters (1000 feet) thick. This layer has apparently formed from particles falling from the surface. These particles consist of dust settling on the ocean and biologic material sinking from the surface waters. They sink to the bottom like exceedingly slow snowflakes and smooth and obscure the bottom topography much as a snowfall obscures a land surface. Below is a layer of older consolidated sediments and basalt, averaging about 2 kilometers (1.2 miles) thick. Below that is the oceanic crust of basaltic composition, which is everywhere close to 4.8 kilometers (3 miles) thick. These layers are shown in Figure 4-8, p. 102.

Submarine Mountains

The ocean floors are interrupted at many places by hills and mountains. Those that rise less than 300 meters (1000 feet) above the ocean floor are called **abyssal hills,** and the higher ones are **submarine mountains,** or **seamounts.** At places, the submarine mountains reach above sea level and become islands, such as the Hawaiian Islands. These islands rising above the floor of the oceans are basalt volcanoes, and samples dredged from submarine mountains are also basalt (Figure 8-6). This, together with the shape of the submarine mountains, indicates that they are also basalt volcanoes. The basaltic composition is not surprising because the crust beneath the oceans is basalt, but the reason for the volcanoes and the fact that in many cases the seamounts and islands are in lines are problems that will be addressed later in this chapter.

The areal extent and number of submarine mountains were recognized during World War II when depth sounders were used extensively to map the oceans. One of the discoveries was that many of the submarine mountains have flat tops. Dredge samples from these **flat-topped seamounts,** or **guyots,** showed that they, too, are basalt volcanoes. Along with basalt pebbles, the dredge samples also had shal-

F 8-6

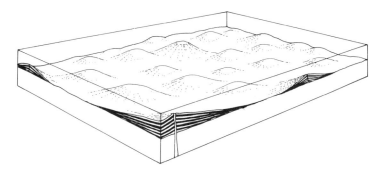

Figure 8-6
The submarine mountains are mainly basalt volcanoes.

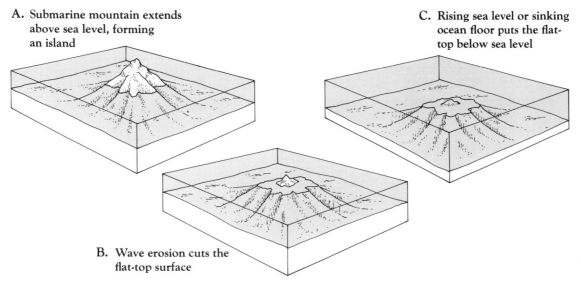

A. Submarine mountain extends above sea level, forming an island

B. Wave erosion cuts the flat-top surface

C. Rising sea level or sinking ocean floor puts the flat-top below sea level

Figure 8-7
The steps in the formation of flat-topped seamounts, or guyots.

Figure 8-8
Flat-topped seamount, or guyot, with several wave-cut benches. Each bench records a different height at which wave erosion occurred.

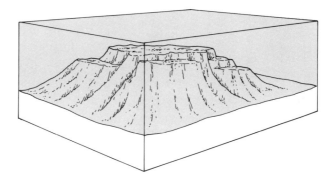

low-water fossils. The obvious interpretation is that at one time the guyots were at sea level, and erosion by waves produced the flat tops (Figure 8-7). This would account for both their shape and the presence on them of shallow-water fossils. There is almost no other way that volcanoes could be flattened, and wave erosion is further suggested by wave-cut benches on some of the guyots (Figure 8-8). (Similar wave-cut benches are preserved on some coasts, and wave erosion is described in Chapter 10.)

The flat-topped seamounts record some of the history of the sea floors. The wave-cut flat tops show that either the sea floor was much higher when the wave erosion occurred, or sea level was much lower, or some combination of both. Dating the fossils tells when the ero-

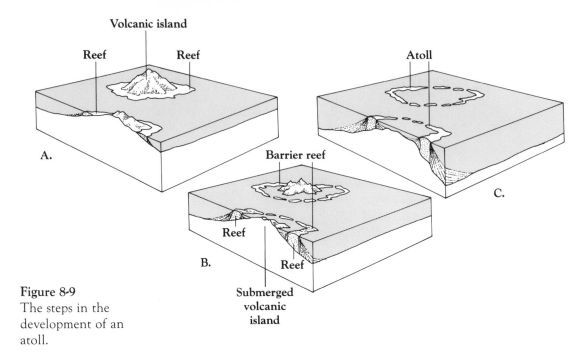

Figure 8-9
The steps in the development of an atoll.

sion occurred. The flat tops are up to about 1500 meters (5000 feet) below sea level, and the fossils are of Late Cretaceous and Early Cenozoic age (about 100 million to 60 million years old). So far it has only been possible to show that such changes in elevation or sea level did take place in part of the Pacific Ocean, but details are obscure.

Other evidence of changes in sea level is recorded by coral **atolls** in the Pacific Ocean. The significance of atolls was recognized by Charles Darwin in 1842. He observed that many islands in warm waters are ringed by coral reefs and that at places only the coral remains. Coral reefs are built by a number of organisms, mainly coral and algae, that live near sea level where the breaking waves provide food and oxygen. The coral reef is composed of both living organisms and skeletal remains. The living part of the reef is generally near the low-tide level, and, if sea level changes, the organisms build new structures so that they remain near this level. The changes in sea level caused by the Ice Age were not known in Darwin's time. Darwin reasoned that, if an island with a fringing reef subsided slowly enough that the organisms could build up the reef as the island sank, the fringing reef would first become a barrier reef with a lagoon between the reef and the island; and, finally, if the island completely sank, only an atoll would remain (Figure 8-9). It was not until 1952, when F 8-9 drilling on Eniwetok Atoll penetrated through the thick reef into volcanic rock, that Darwin's theory was proved.

MID-OCEAN RIDGES

One might think that the deepest parts of the oceans would be near their centers, far from the shores, but this is not true. Near the middle of all of the oceans except the Pacific is a ridge, or rise, where the ocean bottom rises up to about 3000 meters (10,000 feet) above the abyssal plains. In the Pacific Ocean, the ridge is closer to the Americas (Figure 8-10). The mid-ocean ridges are connected and so form F 8-10 the longest continuous ridge, 64,000 kilometers (40,000 miles), on the earth. The extent of the mid-ocean ridges was not realized until the 1950s, when the continuous echo sounder became available, making it possible to make continuous depth profiles while a ship is underway. At places, the mid-ocean ridge rises above sea level, forming islands such as the Azores, Ascension Island, St. Helena, Tristan da Cunha, and Iceland. These are all volcanic islands, and some have had volcanic activity in the very recent past. This volcanic activity, together with the many shallow earthquakes that occur along the mid-ocean ridges, shows that they are very active areas.

In cross-section a typical mid-ocean ridge is 500 to 5000 kilometers (300 to 3000 miles) wide (Figure 8-11). The ridge rises gently from the F 8-11 abyssal plain, and the surface becomes rougher as the center is approached. Near the center, the ridge is made of volcanoes in rows. At the center a **rift valley** up to 50 kilometers (30 miles) wide generally

Figure 8-10
The mid-ocean ridges are the longest continuous ridge on the earth.

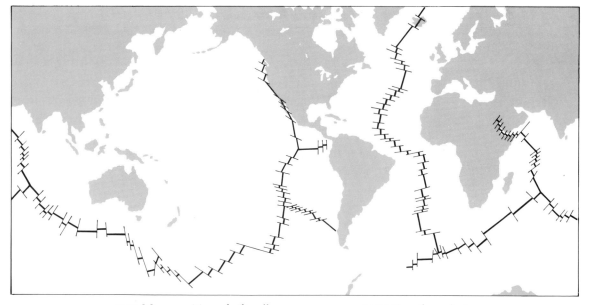

—— **Mean position of rift valley**   === **Transform faults**

Figure 8-11
Topographic map of a portion of the Atlantic Ocean. (From a painting by Heinrich Berann; courtesy of Aluminum Company of America.)

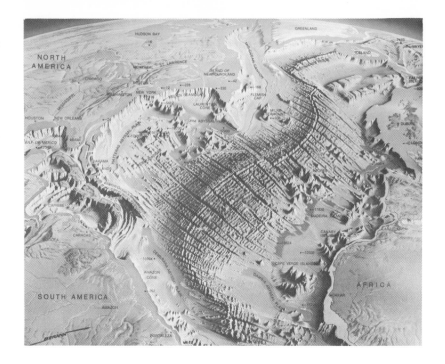

occurs. All of these features are offset by faults that are at near right angles to the axis of the ridge. These faults are the **transform faults** described in Chapter 4. Where they extend into the ocean basin floor, the topography is irregular, and they have been called **fracture zones** (see Figure 8-10).

VOLCANIC ISLAND ARCS AND DEEP TRENCHES

The borders between continents and ocean are either the continental shelf and slope or trenches and volcanic arcs. The volcanic arcs, in contrast to continental slopes, are very active areas; they are the sites of volcanic activity more violent than that of the mid-ocean ridges, and the deepest earthquakes occur in these regions. At a volcanic island arc, the ocean basin ends at a deep, narrow trench, and land-ward of the trench at some places are volcanic islands (see Figure 8-12). At other places, the volcanoes are on the continent, rather than forming islands.

The trenches are the deepest places in the ocean. The deepest is the Marianas trench, which is 10,867 meters (35,630 feet) below sea level. The trenches are narrow and deep, and must be young features, because, if they were old, they would long since have been filled with debris from the nearby volcanic islands. In the 1920s and 1930s, the Dutch geophysicist F. A. Vening-Meinesz discovered that gravitational attraction is lower than normal in the trenches. This implies that the materials there are less massive than elsewhere. His method

Figure 8-12
Sea-floor spreading.

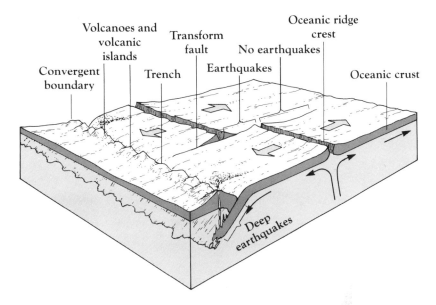

of measuring gravity was to time accurately the period of swing of a pendulum. He worked in U.S. and Dutch Navy submarines in order to avoid the effects of waves.

Sloping downward from the trenches are zones of earthquake activity. The deepest earthquakes come from these zones, and they are as deep as 700 kilometers (435 miles). Because of the depth and slope, the deep quakes occur under the continent. The earthquakes, both deep and shallow, outline the sloping zone shown in Figure 8-12. The source of the volcanic rocks that form the volcanic islands is apparently in this sloping zone.

The volcanic island arcs are so named because the islands are formed by volcanoes, and these volcanic islands appear as curves or arcs on a map. The volcanoes are formed of the rock called *andesite*, so the volcanic island arcs differ in composition from the rest of the ocean, where the rock type is basalt. The andesite volcanoes of the island arcs have more violent and explosive eruptions than the basalt volcanoes of the oceanic islands. Because island arcs form much of the margin of the Pacific Ocean, they have been called the "rim of fire."

SEA-FLOOR SPREADING

The explanation of these features of the oceans has led to the theory of **sea-floor spreading** (Figure 8-12). According to this theory, the oceans are the youngest parts of the earth's surface. New ocean floor is created at the mid-ocean ridges by basaltic magma rising from within the earth. As new ocean floor is formed at the ridges, the older floor is pushed away from the ridges. Thus, the ocean floor is progressively older at greater distances from the ridges. This explains the

Figure 8-13
Movement of plates
over permanent hot
spots that cause
volcanoes may
account for some
volcanic islands in the
Pacific Ocean.

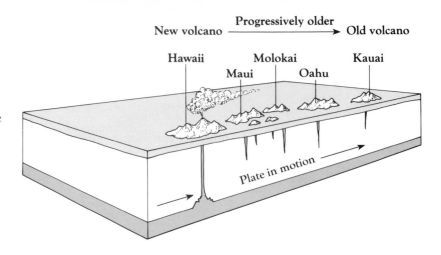

volcanic activity and earthquakes at the mid-ocean ridges and why
the sediments on the ocean basin floor become older and thicker
away from the ridges. At the trenches, the ocean floor is believed to
move down under the continents. Friction of the downward-moving
slab could account for the earthquakes, and increased heat and pressure at depth could cause melting and so be the origin of the volcanic
rocks.

Recognition of sea-floor spreading on a global scale has led to the
concept of a number of crustal plates (see Figure 4-15, p. 107). The
plates are thin slabs, and most consist of both ocean and continent.
The plates are created at the mid-ocean ridges, are consumed at the
trenches, and move past each other along transform faults. As they
move, they carry the continents with them. (The evidence for sea-
floor spreading, plate tectonics, and continental drift is presented in
more detail in Chapter 4.)

The theory of sea-floor spreading explains submarine mountains
and oceanic islands that form in lines. As the oceanic plate passes
over an area in the mantle where conditions are right for the formation of basaltic magma, the basalt moves through the oceanic plate
and forms a seamount or an island. Such places have been called **hot
spots.** That some process like this has occurred at some places is
shown by islands in lines, with progressively older rocks in the direction in which the plate is moving (Figure 8-13).

F 8-13

SEA-FLOOR SEDIMENTS

Oceanic sediments are generally discussed in terms of their origin.
Lithogenous sediments *come primarily from weathering and erosion of
continental rocks and also from volcanic eruptions.* **Biogenous sediments**
are the remains of marine organisms. **Hydrogenous sediments** *are precipitated directly from sea water by chemical reactions.* A few particles found

in marine sediments are of extraterrestrial origin, such as tiny meteorites, and these are termed **cosmogenous.**

Lithogenous Sediment

Lithogenous sediment is carried to the seas by several agents. Most is carried by rivers, which are the source of the sediments of deltas, estuaries, and much of the continental shelves. Some of the fine material that eventually sinks to the floors of the ocean basins comes from river sediments and is carried far from shore by ocean currents. In many parts of the world, currently active or ancient glaciers deposited many of the oceanic nearshore sediments. Many lithogenous sediments are transported far from the continents by wind. Wind-borne dust originates from wind erosion and from volcanic eruption.

Most of the nearshore sediments are lithogenous and in time form the ordinary sedimentary rocks described in Chapter 3, such as shale, mudstone, sandstone, and conglomerate. The continental shelf and slope are composed of such rocks at most places. Turbidity currents on the continental slopes may carry nearshore coarse sediments into deeper water. Such sediments are recognized by the mixture of coarse and fine sediment and of shallow-water and deep-water fossils. In addition, such rocks generally have **graded bedding** (Figure 8-14), in which the coarsest, densest particles settle before the finer material. F 8-14

The sediment on the ocean floors far from land falls to the bottom particle by particle. The settling rates are very slow. A silt-size particle may require six months, and a clay or dust-size particle may require hundreds of years to reach the ocean bottom. The rate of accumulation of deep-sea sediment is, therefore, very slow—centimeters or fractions of centimeters per 1000 years. Because these particles sink so slowly and oxygen is dissolved in sea water, they are oxidized by the time they reach the bottom. The iron in the sediments is oxidized to the ferric state (iron rust), giving the sediments a red or brown color. Such sediments are called **red clays** because of their color and particle size. Clay is a mineral name, and the "red clays" are not composed

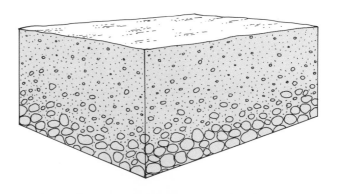

Figure 8-14
Graded bedding forms where the larger fragments are at the bottom of a bed.

completely of clay. The name originated from the *Challenger* expedition before it was possible to identify clay minerals. X-ray techniques are now used to identify clay minerals.

Biogenous Sediment

Biogenous sediment is composed of the remains of organisms. Many of the dissolved materials brought into the sea are used by marine plants and animals to build bones, shells, and skeletons. The materials most commonly used for this purpose are calcium carbonate ($CaCO_3$), silica (SiO_2), and calcium phosphate. Most of the organisms live near the surface, and only some of the hard parts survive the fall to the bottom. The resulting *sediments that contain at least 30 per cent biogenous material are* **oozes.** They are generally named for the type of material, such as foraminiferal ooze or diatom ooze.

The most common biogenous sediment is made of calcium carbonate. *Foraminifera* are the most common of the larger shells, *coccoliths* are the common smaller remains, and *pteropods* are relatively rare. If recognizable biogenous sediment is to accumulate on the bottom, the surface waters must be quite productive, and the dilution by lithogenous material cannot be too high. As the calcium carbonate shells sink slowly toward the bottom, they pass through deeper, cooler waters that contain dissolved carbon dioxide. The warm surface waters are saturated with calcium carbonate, but calcium carbonate will dissolve in the cool water containing carbon dioxide. This is the reason that calcium carbonate rocks are common at shallow depths but relatively rare in deep water.

The silica oozes are composed mainly of the remains of *diatoms* and *radiolaria,* with some *silicoflagellates* and rare sponge spicules. Biogenous sediments composed of calcium phosphate or bone are generally made of shark's teeth and whale earbones. Both silica and calcium phosphate are somewhat soluble at all depths in the oceans.

Hydrogenous Sediment

Hydrogenous sediments are those that are precipitated directly from sea water. They are of great interest because of their potential value, but we know little about how they form. The deposits are **manganese nodules** (Figure 8-15). The nodules are composed of about 23 per cent manganese, 6 per cent iron, and 1 or 2 per cent copper, cobalt, and nickel. These nodules are up to about 10 centimeters (4 inches) in diameter and cover 10 to 20 per cent of the ocean floors. They grow very slowly at rates estimated at 1 millimeter per million years in some areas and up to one hundred times faster in other regions. In one case, growth on a naval artillery shell dredged up off southern

F 8-15

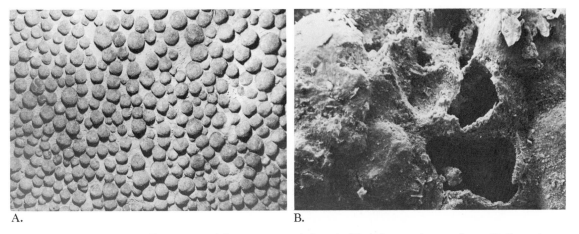

A.   B.

Figure 8-15 Manganese nodules. **A.** Nodules on the sea floor. **B.** Scanning electron microscope photo of the surface of a nodule. (Photo **A** from Lamont-Doherty Geological Observatory; **B** from Scripps Institution of Oceanography, University of California, San Diego.)

California was at the rate of 10 centimeters per century. Because they grow so slowly, they can form only on surfaces where little sedimentation is occurring, although nodule layers have been discovered by drilling bottom sediments.

Oceanographers devise ingenious instruments for their explorations. Figures 8-16 and 8-17 show a few of them.

F 8-16

F 8-17

Figure 8-16
A. Lowering a Peterson grab sampler.
B. Unattended bottom photographer.
(From Scripps Institution of Oceanography, University of California, San Diego.)

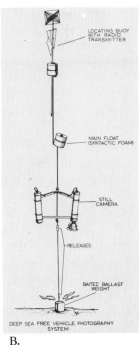

A.   B.

Figure 8-17
A. Deep Tow vehicle operation. **B.** Deep Tow vehicle (From Scripps Institution of Oceanography, University of California, San Diego.)

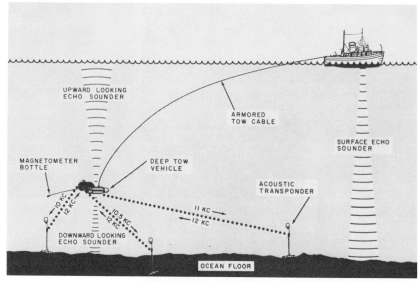

A.

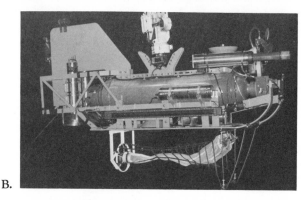

B.

SUMMARY

- The oceans cover about 70 per cent of the earth's surface.

- The *continental shelf* is the border between the ocean and the continent. Continental shelves have average water depths of 60 meters (200 feet), and at their outer edges the depth is generally between 120 and 150 meters (400 to 500 feet). Their width averages about 70 kilometers (42 miles), but the range is great.

- The *continental slope* is the transition between the continental shelf and the ocean floor. Continental slopes are between 50 and 100 kilometers (30 to 60 miles) wide on the average and slope between about 150 meters (500 feet) and 3700 meters (12,000 feet).

- Between the ocean floor and the continental slope is the *continental rise*.

- All of these features are composed of sedimentary rocks.

- *Submarine canyons* are deep V-shaped valleys cut into continental slopes. Some were eroded by rivers when the oceans were lower.

- *Turbidity currents* may also erode continental slopes and cut submarine canyons. A turbidity current is a mixture of sediment and water that, because it is heavier or denser than sea water, moves downslope along the bottom.

- The ocean basin floors lie between the continental rises and the mid-ocean ridges.

- *Abyssal plains* are smooth, flat surfaces, generally near continental rises, that are probably formed by deposition of material moving down continental slopes.

- The ocean basin floors are covered with about 300 meters (1000 feet) of unconsolidated sediments; below them is the oceanic crust, about 4.8 kilometers (3 miles) thick, which is composed of older consolidated sediments and basalt.

- The ocean floors are interrupted by *abyssal hills* and *submarine mountains*. These features are generally basaltic volcanoes.

- *Guyots* or *flat-topped seamounts* are submarine mountains with smooth, flat surfaces cut by wave action. They imply either change in elevation of the seamount or changes in sea level.

- *Coral atolls* are built by organisms that live near sea level, and they imply a slow submergence of the original island or a rise in sea level.

- *Mid-ocean ridges* are places where the bottom rises up to 3000 meters (10,000 feet) above the abyssal plains. They are between 500 and 5000 kilometers (300 to 3000 miles) wide and are interconnected to form a single ridge 64,000 kilometers (40,000 miles) long.

- Mid-ocean ridges are active volcanic areas and have many small earthquakes. They are offset at places by *transform faults.*

- At some borders between ocean and continent, deep trenches and volcanic arcs occur. At the *volcanic island arcs,* the volcanoes form arcuate chains of islands, and at other places the volcanoes are on the continent.

- The *trenches* associated with volcanic arcs are the deepest places in the oceans and are areas of negative gravity anomalies.

- The deepest earthquakes on earth, up to 700 kilometers (420 miles) deep, are also associated with volcanic arcs.

- *Sea-floor spreading* explains many of the features of the oceans. New basaltic crust is created at the mid-ocean ridges, and the newly created sea floor moves away from the ridges. The sea floor is consumed at the trench-volcanic arcs where the ocean floor moves down, causing the trench and the deep earthquakes. The melting ocean floor is the source of the volcanic rocks.

- *Lithogenous sediments* come from erosion of continental rocks and from volcanic eruptions. Turbidity currents deposit them on the continental slopes. Far from land much is dust-size and is carried by the wind. The material sinks slowly, falling on the bottom like snow. These particles are oxidized during their slow descent and so are called *red clays,* although most are not truly clay minerals.

- *Biogenous sediment* is composed of the remains of organisms. Sediments containing 30 or more per cent biogenous material are called *oozes.*

- *Hydrogenous sediments* are precipitated directly from sea water. *Manganese nodules* are the most common type of hydrogenous sediment.

KEY TERMS

oceanography
physical
  oceanography
chemical
  oceanography
biological
  oceanography
geological
  oceanography
chronometer
continental shelf

continental slope
submarine canyon
turbidity current
continental rise
abyssal plain
abyssal hill
submarine mountain,
  or seamount
flat-topped seamount,
  or guyot

atoll
mid-ocean ridge
rift valley
transform fault
fracture zone
volcanic arc
volcanic island arc
trench
sea-floor spreading
hot spot

lithogenous sediment
biogenous sediment
hydrogenous
  sediment
cosmogenous
  sediment
graded bedding
red clay
ooze
manganese nodule

REVIEW QUESTIONS

1. What kind of information can the *Glomar Challenger* give us about the oceans?

2. Where are the continental shelves? Describe some effects that the Ice Age had on them.

3. Geologically speaking, are the continental shelves part of the continents or part of the oceans? Why?

4. State where the continental slopes are found and describe the characteristics that give them their name.

5. Are turbidity currents a threat to surface ships? Explain your answer by describing a turbidity current.

6. What mystery appears to have been solved by the discovery of oceanic turbidity currents?

7. Are the continental rises believed to be erosional or depositional features? What evidence can you cite?

8. The abyssal plains are constructed of three layers. The top layer gives the plains their smooth appearance; how has this layer been created? The middle and deep layers are composed of what materials? From your reading of the entire chapter, what do you think might be the origins of the middle and deep layers?

9. What features of seamounts indicate that either sea level has changed or parts of the ocean have sunk? Briefly explain why each of these features indicates such change.

10. Describe how the coral atolls of the Pacific are believed to have formed.

11. Draw a cross section of a mid-ocean ridge. Label the area of active volcanoes and the rift valley.

12. As one moves away from a mid-ocean ridge, does the ocean floor become progressively younger or older? What kind of magma is extruded from the volcanoes of the mid-ocean ridges?

13. How does the theory of sea-floor spreading explain your answers to Question 12?

14. What is believed to happen to the ocean floor at the trenches?

15. The volcanic island arcs are composed of what kind of rock? Use the theory of sea-floor spreading to explain the possible origin of these islands.

16. Explain how lines of basaltic oceanic islands may have been formed.

17. How does the lithogenous sediment of the ocean basin floors differ from that of the continental shelves?

18. Three conditions must be met before significant amounts of the biogenous sediment calcium carbonate can form. What are those three conditions?

19. Name the two other most common biogenous sediments and state one source of each.

20. How are manganese nodules formed?

SUGGESTED READINGS

Bonatti, E. "The Origin of Metal Deposits in the Oceanic Lithosphere," *Scientific American* (February 1978), Vol. 238, No. 2, pp. 54–61.

Burke, K. C., and J. T. Wilson. "Hot Spots on the Earth's Surface," *Scientific American* (August 1976), Vol. 235, No. 2, pp. 46–59.

Dugolinsky, B. K. "Mystery of Manganese Nodules," *Sea Frontiers* (1979), Vol. 25, No. 6, pp. 364–369.

Emery, K. O. *A Coastal Pond Studied by Oceanographic Methods.* New York: Elsevier, 1969, 82 pp.

Emery, K. O. "The Continental Shelves," *Scientific American* (September 1969), Vol. 221, No. 3, pp. 106–125.

Fine, J. C. "Exploring the Ocean Bottom in Manned Submersibles," *Sea Frontiers* (1978), Vol. 24, No. 6, pp. 327–334.

Hallam, A. "Alfred Wegener and Continental Drift," *Scientific American* (February 1975), Vol. 232, No. 2, pp. 88–97.

Heirtzer, J. R., and W. B. Bryan. "The Floor of the Mid-Atlantic Rift," *Scientific American* (August 1975), Vol. 233, No. 2, pp. 78–91.

Mark, K. "Coral Reefs, Seamounts, and Guyots," *Sea Frontiers* (1976), Vol. 22, No. 3, pp. 143–149.

Rice, A. L. "H.M.S. *Challenger*—Midwife to Oceanography," *Sea Frontiers* (1972), Vol. 18, No. 5, pp. 291–305.

Rona, P. A. "Plate Tectonics and Mineral Resources," *Scientific American* (July 1973), Vol. 229, No. 1, pp. 86–95.

Schlee, S. "The Curious Controversy over Coral Reefs," *Sea Frontiers* (1971), Vol. 17, No. 4, pp. 214–223.

Sclater, J. G., and Christopher Tapscott. "The History of the Atlantic," *Scientific American* (June 1979), Vol. 240, No. 6, pp. 156–175.

Shepard, F. P. "Submarine Canyons of the Pacific," *Sea Frontiers* (1975), Vol. 21, No. 1, pp. 2–13.

Shepard, F. P. "Coral Reefs of Moorea," *Sea Frontiers* (1976), Vol. 22, No. 6, pp. 360–366.

Smith, R. V., and D. W. Kinsey. "Calcium Carbonate Production, Coral Reef Growth, and Sea Level Change," *Science* (November 26, 1976), Vol. 194, No. 4268, pp. 937–939.

SEA WATER

LAYERING IN THE OPEN OCEAN

Surface Zone

Transition Zone

Deep Zone

DEVELOPMENT OF THE THERMOCLINE

DEVELOPMENT OF THE HALOCLINE

SURFACE CURRENTS

Coriolis Effect

Gyres

Ekman Spiral

Mapping Ocean Currents

VERTICAL CURRENTS

CIRCULATION OF THE DEEP WATERS

CHAPTER 9

OCEAN WATER AND ITS CIRCULATION

Lowering a temperature probe to measure heat flow on the ocean floor. (Photo from Scripps Institution of Oceanography, University of California, San Diego.)

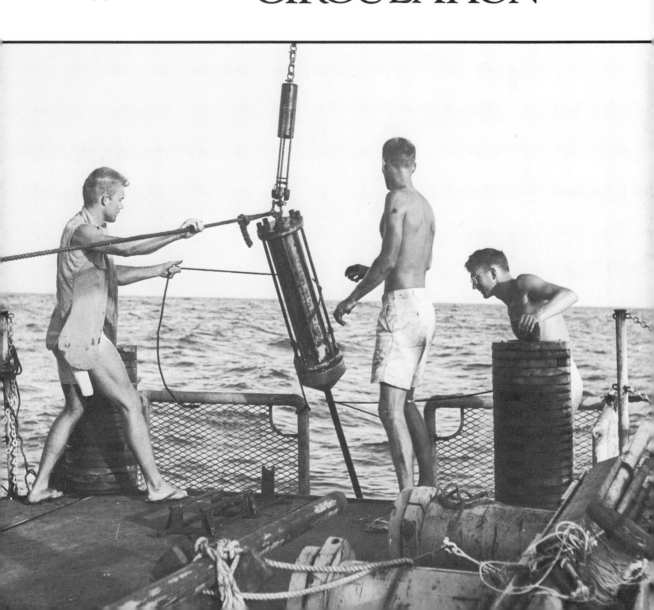

SEA WATER

The most obvious attribute of sea water is that it is salty. It is a complicated mixture of many ions, but six ions make up over 99 per cent of the dissolved elements (Table 9-1). (**Ions** are atoms that have gained or lost one or more electrons.) Some of the minor constituents, however, are necessary for some of the organisms that live in the oceans. The average sea water from all over the world contains 3.47 per cent by weight dissolved material. This is the **salinity** of sea water, which is usually expressed as 34.7 parts per thousand (0/00) rather than 3.47 per cent.

T 9-1

The ultimate source of the water in the oceans is believed to be the interior of the earth. Water in the form of steam is one of the major gases emitted by volcanoes, and some of this water comes from deep within the earth. The chlorine and sulfur in sea water are also believed to come from volcanoes. The other four major elements in sea water—sodium, magnesium, calcium, and potassium—are common in the rocks of the earth's continents. Weathering processes release these elements from the rocks, and rivers carry them into the oceans.

The salinity of surface water does vary. In polar regions, where melting ice and runoff from the land is high, the salinity of surface water may be as low as 33 parts per thousand. In tropical areas of high evaporation and little rainfall, such as near 30 degrees north and south latitude, the salinity may be as high as 37 parts per thousand. The Red Sea is in an area of high evaporation and little rainfall or river runoff, and here the salinity is 41 parts per thousand.

The salinity of sea water is difficult to measure directly because of chemical reactions that occur if the water is boiled away in order to weigh the salts in a given volume. Many measurements of salinity have shown that the oceans are very well mixed, and so the ratios of the elements in sea water are nearly constant, even though the total

Table 9-1 AVERAGE COMPOSITION OF DISSOLVED MATERIAL IN SEA WATER

Ion	Weight Per Cent
Chlorine	1.92
Sodium	1.07
Sulfate (SO_4^{2-})	0.25
Magnesium	0.13
Calcium	0.04
Potassium	0.04
Others	0.02
Total	3.47% = 34.7 parts per thousand

Figure 9-1
The temperature at which freezing of sea water begins depends on salinity.

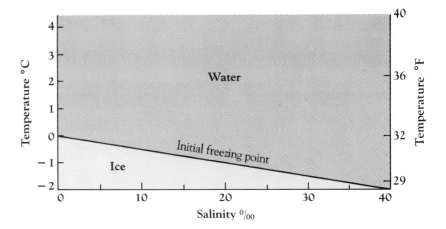

amount of all of the salts—the salinity—varies. This means that, if the amount of any element in sea water can be measured, the salinity can be easily calculated because the proportions are almost always the same. A common way to measure the salinity is to measure the chlorinity of a water sample. Salinity is most commonly determined from measurement of electrical conductivity and temperature.

Because sea water contains salts, it does not have a fixed freezing point; that is, freezing occurs over a range of temperatures. Pure water, of course, freezes at 0°C (32°F). Salts lower the initial freezing temperature of sea water to about −1.9°C (28.6°F) for average salinity (Figure 9-1). The temperature of maximum density also depends on salinity.

F 9-1

In addition to having salts dissolved in it, the ocean also has dissolved gases. Cool water can hold more gas than can warm water. Nitrogen makes up about 78 per cent of atmospheric air, and, not surprisingly, it is also the major gas in the ocean, making up 64 per cent of the dissolved gas. As in the atmosphere, nitrogen is inert in the oceans. Many sea organisms require nitrogen, but they obtain it from other plants or animals in their food chain.

Oxygen is the next most abundant gas in the oceans, at 34 per cent of the dissolved gas. (Oxygen is only about 21 per cent of the atmosphere.) The oxygen in the oceans comes from surface mixing with the atmosphere and from production by marine photosynthetic plants. Most of the oxygen in the oceans is near the surface, where it supports animal life.

Carbon dioxide is important to the green plant life in the oceans. Green plants use large amounts of this gas in their life cycles. Carbon dioxide in the oceans comes from surface mixing, animal breathing, and decaying of organic material. It makes up 1.6 per cent of the ocean's gases, about 50 times more than its proportion in the atmosphere. In sea water, carbon dioxide gas forms the carbonate ion

(CO_3), which, especially in warm, shallow waters, is precipitated as calcium carbonate, $CaCO_3$. Calcium carbonate is the mineral calcite, and the rock it forms is limestone. Much of the earth's carbon dioxide is dissolved in sea water.

Hydrogen sulfide is the only other important gas in the oceans. It originates in the decay of organic material and has the very disagreeable smell of rotten eggs. Hydrogen sulfide makes up only about a half per cent of the gases in normal sea water, and, if its concentration becomes much higher, it is lethal to much of the life in the oceans. Higher concentrations can develop if the oxygen in sea water becomes depleted, as can occur if the circulation is impaired.

A few other gases, such as argon, are dissolved in sea water, but their amount is very small, and they are of little consequence.

LAYERING IN THE OPEN OCEAN

In the open ocean—that is, away from the influence of land—the ocean has distinct layers that are caused by differences in the density of sea water. The principal causes of these density differences are salinity and temperature. Of the two, temperature is more important.

Surface Zone

The upper layer is called, not too surprisingly, the **surface zone** or sometimes the **mixing zone** (Figure 9-2). The water here is the least dense and the warmest; hence, it floats on top. The thickness of the surface layer is determined largely by mixing, and the chief agent of mixing is wind. Because temperature is most important in determining the density, the extent of the surface layer changes seasonally. F 9-2

As shown in Figure 9-2, the surface zone does not extend poleward beyond about 50 degrees north and south latitude, because the cold surface water near the poles is similar to the deep water. The surface zone comprises about 2 per cent of the volume of the oceans.

The sun's radiation is the main agent that heats the surface zone, and the wind mixes the heated water. Because the oceans cover about 70 per cent of the earth, much of the sun's radiation is absorbed by them. It should be clear that the oceans over a year must lose as much heat as they absorb; otherwise, they would be getting hotter. The oceans lose about 50 per cent of this heat energy by evaporation. The evaporation process requires heat, which is released to the atmosphere when the evaporated water vapor condenses to form clouds. On the average, about one meter of water is evaporated from the ocean each year. About 10 per cent of the ocean's heat is removed by the air in contact with it. The rest of the ocean's heat is reradiated. Thus, the ocean is very important in the earth's weather.

Figure 9-2
The layered structure of the oceans. The deep zone extends to the surface at high latitudes.

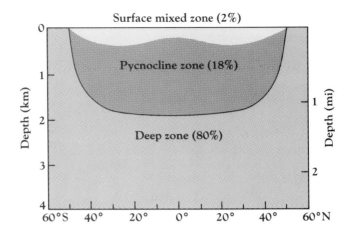

The oceans are able to absorb much heat from the sun for several reasons. Water is able to contain more heat than can the rocks of the continents. The mixing by wind spreads the heat energy over a considerable depth and thus prevents the heat from being reradiated at night, as happens on the continents. For these reasons, the temperature of the oceans is more constant than that of the continents, and so the climates near oceans have fewer extremes than inland areas.

Transition Zone

Below the surface zone is the **transition zone,** sometimes called the **pycnocline zone.** In this layer the density of the water changes rapidly with depth (Figure 9-3). Because the water in this zone is denser toward the bottom of the zone, it forms a very stable buffer that prevents mixing of surface and deep waters. This transition zone is absent near the poles, where the deep waters and the surface waters have similar characteristics. The waters of the transition zone are about 18 per cent of the oceans by volume. This zone forms because the temperature and salinity of sea water change rapidly with depth; these changes will be discussed in the next section. The temperature and salinity also vary with latitude, and this is why the upper surface of the transition zone is not of uniform depth (see Figure 9-2).

F 9-3

Deep Zone

The **deep zone** contains 80 per cent of the ocean's water. The temperature and salinity of the deep waters are quite uniform. The salinity is in the very narrow range of 34.6 to 34.9 parts per thousand. The temperature is generally just a few degrees Celsius, ranging from about 5°C (40°F) just below the transition zone in tropical areas to

Figure 9-3
The pycnocline, or transition zone, is the layer in which the density of sea water changes rapidly with depth. The change in density occurs because both temperature (thermocline) and salinity (halocline) change rapidly with depth.

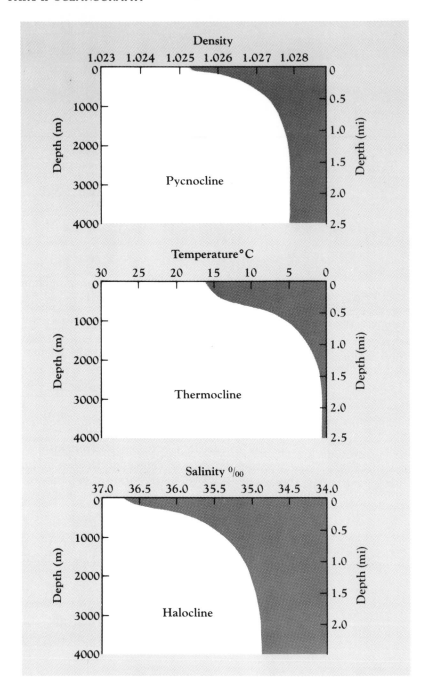

about 1°C (34°F) near the bottom. In the polar regions, the surface waters are near 0°C (32°F) and so differ very little from the deep-zone waters. Near the poles, the deep water comes into contact with the atmosphere, and wind mixing causes solution of oxygen and the loss of carbon dioxide. The surface water near the poles is at times colder and therefore more dense than the rest of the deep water, so it sinks, carrying the oxygen into the deep zone. This deep circulation will be further considered later in this chapter. The process is important in providing oxygen for deep-dwelling organisms. Because sunlight cannot penetrate into most of the deep zone, all life there depends for food on material falling from the surface. Turbidity currents may also carry food into the deep zone.

DEVELOPMENT OF THE THERMOCLINE

Over most of the oceans the location of the transition zone is determined mainly by the temperature of the water and not so much by the salinity. *The zone in which the temperature changes rapidly with depth is called the* **thermocline** (Figure 9-3). Temperature is important in F 9-3 controlling the density of the oceans because of its large range, from about −1.7°C to 30°C (29° to 86°F). The waters of the surface zone are well mixed by the wind and so are of uniform temperature. The deep waters also have a fairly uniform temperature and are much colder than the surface waters. The thermocline is the zone separating the surface and deep temperatures, and at many places is at the same location as the transition zone.

DEVELOPMENT OF THE HALOCLINE

The zone in which the salinity changes rapidly with depth is called the **halocline** (Figure 9-3). The salinity of the surface waters is generally F 9-3 higher than that of the deep water. The reason for this is that the surface waters are generally warm, and so evaporation occurs, making the surface waters saltier. The salinity of the surface zone varies between 33 and 37 parts per thousand, and in the deep zone the variation is only between 34.6 and 34.9 parts per thousand. Other factors that affect surface salinity are precipitation and runoff from rivers, which, of course, reduce the salinity, and the effects of polar ice. When ice forms on the polar seas, the salt is excluded from the sea ice. Thus the ice on the polar seas is fresh-water ice, generally including some pockets of brine. When sea ice is forming, the excluded salt increases the salinity of the surrounding waters. Conversely, when the polar ice melts, the fresh water reduces the salinity.

SURFACE CURRENTS

The surface currents in the open ocean are driven by the winds. This circulation can be thought of as part of the transporting toward the poles of heat received from the sun in tropical areas. Most of the heat is transferred by the winds, but about one-fifth is transported by the oceans. The generalized wind pattern over the oceans is shown in Figure 9-4A, and the major ocean surface currents in Figure 9-4B. To F 9-4 see the correspondence better, we must consider how winds induce surface currents. First, however, we must consider the effect of the earth's daily rotation.

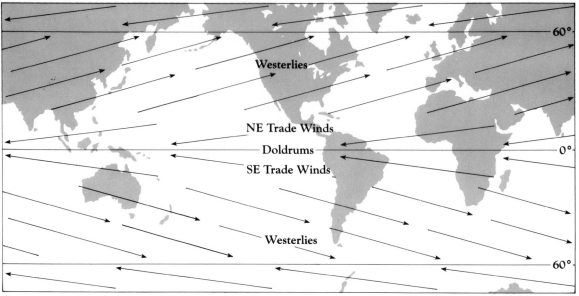

A.   Planetary wind pattern

B.   Ocean current pattern

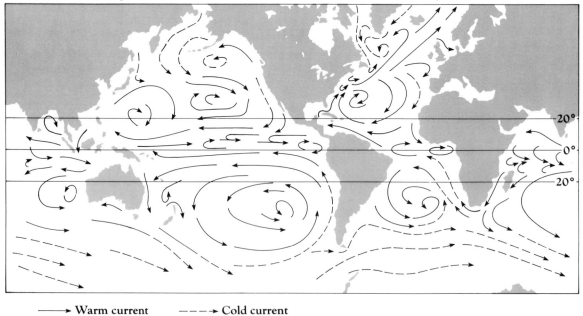

——▶ **Warm current**   ———▶ **Cold current**

Figure 9-4 A. Surface winds. **B.** Surface water currents. Winds cause the main surface currents in the oceans.

Coriolis Effect

As winds and water currents—or rockets and artillery shells—move, the earth turns beneath them so that their paths over the earth curve. As an example, assume that a rocket is shot northward from the equator. While the rocket is in flight, the earth turns on its axis so that the rocket's path relative to the earth is deflected to the right. (Notice that, to an observer in space above the north pole, the rocket would be seen to travel due north in a straight line.) This apparent change in direction is real to us on earth, and is called the **Coriolis effect.** In the northern hemisphere the deflection is to the right of the direction of movement, and in the southern hemisphere it is to the left. The Coriolis effect is maximum at the poles and diminishes toward the equator, where it goes through zero and changes direction. (The Coriolis effect is discussed further in Chapter 13).

Because of the Coriolis effect, a wind-induced surface current should move at about 45 degrees to the direction of the wind. In the ocean, the currents are actually at 45 degrees or less from the wind direction, and the angle is lower where shores affect the currents. Looking back at Figure 9-4, we can now see that the correspondence between the winds and the surface currents is more apparent when the influence of the Coriolis effect is taken into account.

Gyres

The major surface currents form loops on the map. *The name given to these nearly closed loops is* **gyre.** Notice that the main gyres are centered (Figure 9-4) over the tropical areas at approximately 30 degrees north and south latitude. The effect of the continents in forming the gyres is obvious on the map. Just north of Antarctica, where there are no continents to interrupt the currents, the West Wind Drift circles the earth between 50 and 60 degrees latitude.

Each of the major gyres in the Atlantic and Pacific Oceans has a westward-moving current north or south of the equator. Between these two equatorial currents is the eastward-moving Equatorial Countercurrent. This current returns some of the water moved by the North and South Equatorial Currents in this region of very weak winds. The equatorial currents move between 3 and 6 kilometers per day (2 and 4 miles per day), and so the water becomes warm because it remains so long in the tropics. When the warm equatorial currents near the continents, they are deflected northward and southward by the winds and the Coriolis effect. This is the source of the Gulf Stream current that moves along the east coast of North America. The currents along the east coasts (the western sides of the oceans) move rapidly, and so the water remains warm. The Gulf Stream travels between 40 and 120 kilometers per day (25 and 75 miles per day). When the waters get north and south of about 40 degrees latitude, the westerly winds and the Coriolis effect turn the currents eastward. As the waters move eastward, they finally cool, although the Gulf Stream is still warm when it reaches Europe. As the currents near the eastern shores of the oceans, the continents and the Coriolis effect turn them southward, completing the gyre.

Ekman Spiral

If we look at the surface currents a little more closely, we will learn more about ocean currents. The discussion of Coriolis effect led us to understand that the wind tries to move the surface water in the same direction that it is blowing, but the Coriolis effect diverts the water to about 45 degrees from the wind direction. In the northern hemisphere, the diversion is to the right. The wind causes the surface water to move, and this surface water drags the next lower layer of water along with it. However, the Coriolis effect also acts on this next lower layer and deflects it farther to the right. In turn, that layer drags the next layer, and it too is diverted even farther to the right. This process, in theory, will continue all the way to the bottom. This is called the **Ekman spiral,** and it describes the water motion in an ocean current (Figure 9-5).

F 9-5

A steady wind blowing across an expanse of ocean will cause the surface water to move at a speed about 2 per cent of the wind speed. According to the Ekman spiral, at some depth the water will be moving in the opposite direction to the surface water. This reversing

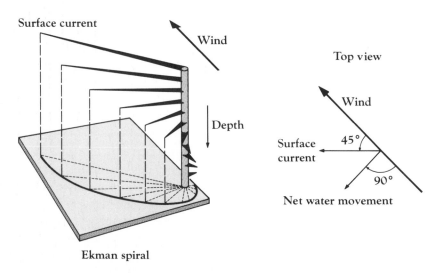

Figure 9-5
The Ekman spiral. The surface current is about 45 degrees from the wind direction, and the net water movement at all depths is at a right angle to the wind. The figure shows the deflection in the southern hemisphere. The deflection is to the right in the northern hemisphere.

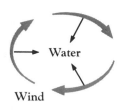

Figure 9-6
The net water movement is toward the center of the gyre because of the Ekman spiral.

depth occurs at about 100 meters (330 feet), and it is generally considered the bottom of the wind-driven currents. The speed at this depth is about 4 per cent of the speed of the surface water. A very important effect of the Ekman spiral is that, if all levels of water movement are considered, the net water movement is at 90 degrees to the wind direction (Figure 9-5). This means that, in the northern hemisphere, water is moving 90 degrees to the right of the wind; in the southern hemisphere, the movement is to the left. The result is that water is moved toward the centers of the gyres (Figure 9-6).   F 9-6

As predicted by the Ekman spiral, the water surface of the oceans is higher in the centers of the gyres, because water is transported there. The total difference in elevation of the sea surface is only about 2 meters (6 feet), so it is difficult to measure. The water near the top of one of these "hills" will begin to flow downhill under the influence of gravity. As soon as it begins to move, the Coriolis effect acts on it and, in the northern hemisphere, deflects it to the right. As the water changes direction, the Coriolis effect acts on the new direction, diverting it farther to the right. This process continues until the water is moving almost parallel to the hill, and the gravity force that started the water moving is balanced by the Coriolis effect (Figure 9-7). Because of friction, the water flows around the hill and downward. This   F 9-7
flow is called a **geostrophic current,** and some of the major ocean currents are of this type.

Mapping Ocean Currents

The earliest attempts at mapping the location as well as measuring direction and speed of currents were from recording the drift of ships. Other approaches used drift bottles and similar devices thrown into the ocean with instructions to the finder to return the enclosed card with information on where and when the bottle was found. Such devices, however, only tell the starting and finishing positions and the total time elapsed, and they do not reveal the actual route. Some of these floating devices are weighted, so that they drift at a desired depth and record deeper currents (Figure 9-8). Some devices drag a   F 9-8

Figure 9-7
Geostrophic currents flow around a hill of water. They form if the Coriolis effect is balanced by the gravity force.

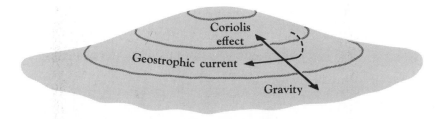

Figure 9-8
A drogue for measuring currents below the surface.

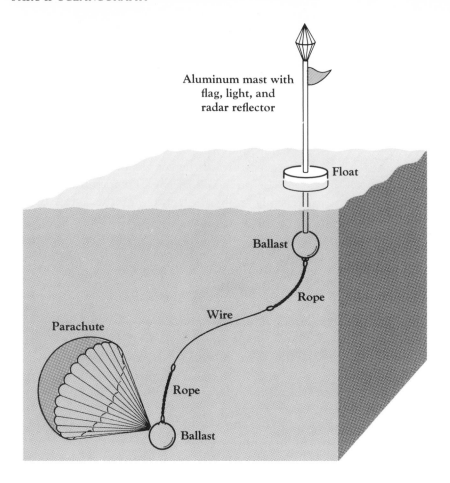

Aluminum mast with flag, light, and radar reflector

Float

Ballast

Rope

Wire

Parachute

Rope

Ballast

float along the surface, and others reveal their locations by sending sound waves to a nearby research ship. Other instruments, such as current meters, are lowered on wires and measure the speed and direction of the current at any desired depth.

VERTICAL CURRENTS

Winds can also cause vertical movements of the surface waters, especially along coastlines. If the wind is blowing toward the southeast along the west coast of a continent, as it does along the west coast of North America, it causes a net movement of water in the offshore direction. This causes an *upwelling* of water from the deeper part of the surface zone (Figure 9-9A). Upwelling water commonly carries nutrients with it, so such areas are commonly excellent for fishing. The opposite situation, where the water flow is landward, results in sinking (Figure 9-9B).

Near the equator in open water the winds cause westward-moving surface equatorial currents. North of the equator the net flow of wa-

F 9-9

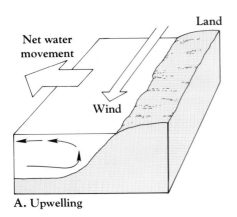

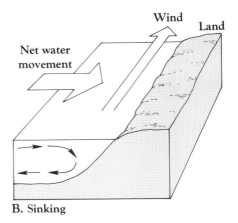

A. Upwelling

B. Sinking

Figure 9·9
Wind can cause vertical currents, especially near coastlines.

ter is northward away from the equator, and south of the equator the water flows southward, as predicted by the Ekman spiral. As a result of these movements away from the equator, deeper waters rise, resulting in upwelling in the open ocean.

CIRCULATION OF THE DEEP WATERS

The circulation of deep water in the oceans is controlled almost entirely by density differences. *The principal factors controlling density are temperature and salinity, so the deep circulation is called* **thermohaline circulation.** The sources of most of the deep water are at high latitudes where the deep waters are in contact with the surface. The principal source areas are near Antarctica in the south and off Greenland in the north. The Arctic Ocean basin is isolated from other oceans by submarine ridges that prevent outflow of deep water, and so it is not a source of deep water for the other oceans. The Mediterranean Sea is a source of heavy, warm, saline water.

The densest water in the oceans comes from the Antarctic region. The waters near the surface there are about −1.9°C (28.6°F), the temperature at which sea water begins to freeze, and have a salinity of 34.6 parts per thousand. This water sinks as it moves around Antarctica as part of the West Wind Drift. Then, as it reaches the bottom, it begins its journey northward along the bottom. The trip carries it north of the equator to near the Aleutian Islands in the Pacific and to near Nova Scotia in the Atlantic (Figure 9-10).   F 9-10

A bit farther north of Antarctica, near 50 degrees south latitude, is another area of sinking surface waters. The water here is not so dense as that farther south, so it sinks to depths of 1 or 2 kilometers. It, too, can be traced north of the equator to about 20 degrees north latitude.

The cold water that sinks near Greenland flows along the bottom of the Atlantic Ocean until it meets the denser bottom water from the Antarctic. The North Atlantic water flows over the Antarctic

Figure 9-10
Deep-water
circulation in the
Atlantic Ocean.
(After Davis, 1977)

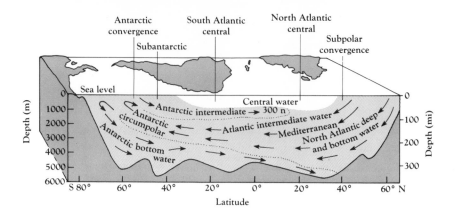

bottom water and eventually reaches the West Wind Drift just north of the Antarctic continent.

All of these bottom waters are traced by their temperature and salinity. Bottom currents move very slowly, about 1 kilometer per day (0.6 mile per day), although along the west sides of the ocean basins the flow may be ten times faster. At these slow speeds the water requires between 500 years and a few thousand years to complete the cycle and return to the surface. The bottom waters eventually mix with other water and so lose their identities. Eventually, the deep waters must return to the surface by moving through the transition zone waters. Such movements must be very slow, probably less than 10 meters (33 feet) per year. The movements of the bottom water are part of the general movement of equatorial heat toward the poles. The surface gyres do not tend to cross the equator, but the deep circulation moves water between the northern and the southern hemispheres.

Figure 9-11
Heavy saline waters
from the
Mediterranean Sea
flow into the Atlantic
Ocean.

Another source of deep water is the Mediterranean Sea. The connection between the Mediterranean and the Atlantic is the Strait of Gibraltar, where a sill restricts the circulation to shallow depths. The

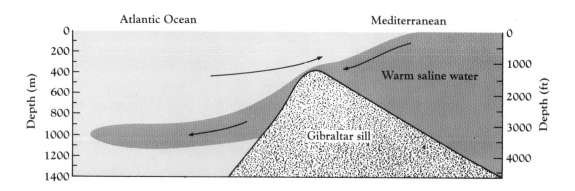

Mediterranean is in a warm area, and evaporation is high, making the waters quite saline. The heavy saline waters from the Mediterranean spill over the sill and flow into the Atlantic Ocean, forming a current at a depth of about 1 kilometer (0.6 mile) (Figure 9-11).

F 9-11

The level at which any of the deep waters flow is determined by their density. A mass of water will go to the depth at which the water below is denser and the overlying water is less dense.

SUMMARY

- The *salinity* of average sea water is 34.7 parts per thousand, and the relative amounts of the dissolved ions in sea water are nearly constant.

- The main gases dissolved in sea water are nitrogen and oxygen, and there are lesser amounts of carbon dioxide and hydrogen sulfide.

- In the open ocean there are three layers (Figure 9-2): *surface*, or *mixing*, *zone*; *transition*, or *pycnocline*, *zone*; and *deep zone*.

- The *thermocline* is the zone in which the temperature changes rapidly with depth.

- The *halocline* is the zone in which the salinity decreases rapidly with depth.

- The surface currents are caused by the wind.

- The *Coriolis effect* deflects the currents to the right in the northern hemisphere and to the left in the southern hemisphere.

- The major surface currents form loops that are called *gyres*.

- The Coriolis effect causes each layer to be deflected more so that at some depth the water is moving in the direction opposite to the surface water. This motion is called the *Ekman spiral*. The net water movement is 90 degrees to the wind direction.

- The resulting very low hills of water cause *geostrophic currents*. In a geostrophic current, the gravity force is balanced by the Coriolis effect, and the current flows around the hill.

- *Upwelling* may be caused by wind currents near a coast (Figure 9-9A).

- The deep circulation is called *thermohaline circulation* because the causes are temperature and salinity. Cold, heavy water sinking near the poles and moving toward the equator is an example.

KEY TERMS

ion
salinity
surface zone, or
  mixing zone

transition zone, or
  pycnocline zone
deep zone
thermocline

halocline
Coriolis effect
gyre
Ekman spiral

geostrophic current
thermohaline
  circulation

REVIEW QUESTIONS

1. What is believed to be the original source of the water in the oceans?
2. What six ions account for most of the salinity of the sea? What are their probable sources?
3. Does the salinity of the surface waters vary from place to place? Why or why not? Does the ratio of chlorine to magnesium vary from place to place?
4. What effect does salinity have on the freezing point of water?
5. What are the two principal causes of density differences in the ocean waters? Of these two, which has more effect on the layering of the open ocean?
6. Why is the surface zone of the open ocean sometimes called the mixing zone? Why does the extent of the surface zone vary with the seasons?
7. Why do the surface and transition zones disappear at about 50 degrees north and south?
8. Describe the deep zone. Is there much variation in its temperature and salinity?
9. Why does the thermocline coincide with the transition zone in most places?
10. What drives the surface currents of the open ocean?
11. If the wind is blowing toward the northeast, and the surface of the sea is moving due east, is this area in the northern or southern hemisphere? Why? What is causing the water to deviate from the direction in which it is being pushed?
12. Describe the Ekman spiral.
13. Why does the Ekman spiral predict that sea level is higher in the centers of the gyres?
14. Describe the flow of a geostrophic current.
15. Explain how upwelling occurs:
    (a) along the coast of California.
    (b) out in the open oceans near the equator.
16. Why is the movement of the deep waters of the oceans called thermohaline circulation?
17. Name the two areas in which most of the deep waters originate. Which is the primary factor causing the waters from these two areas to sink: temperature or salinity? From where does the densest water in the oceans come?
18. Water flowing out of the Strait of Gibraltar sinks to the deep waters of the Atlantic, yet the Mediterranean is a warm sea. Why does this water sink?

SUGGESTED READINGS

Baker, D. J., Jr. "Models of Ocean Circulation," *Scientific American* (January 1970), Vol. 222, No. 1, pp. 114–121.

Baker, R. D. "Dangerous Shore Currents," *Sea Frontiers* (1972), Vol. 18, No. 3, pp. 138–143.

Bascom, Willard. "Ocean Waves," *Scientific American* (August 1959), Vol. 201, No. 2, pp. 74–84.

Gregg, M. "Microstructure of the Ocean," *Scientific American* (February 1973), Vol. 228, No. 2, pp. 64–77.

Kort, V. G. "The Antarctic Ocean," *Scientific American* (September 1962), Vol. 207, No. 3, pp. 113–128. Reprint 860, W. H. Freeman, San Francisco.

MacIntyre, F. "Why the Sea Is Salt," *Scientific American* (November 1970), Vol. 223, No. 5, pp. 104–115.

MacIntyre, F. "The Top Millimeter of the Ocean," *Scientific American* (May 1974), Vol. 230, No. 5, pp. 62–77.

Mason, P. "The Changeable Ocean River," *Sea Frontiers* (1975), Vol. 21, No. 3, pp. 171–177.

Munk, Walter. "The Circulation of the Oceans," *Scientific American* (September 1955), Vol. 193, No. 3, pp. 96–104. Reprint 813, W. H. Freeman, San Francisco.

Rubey, W. W. "Geologic History of Sea Water— An Attempt to State the Problem," *Bulletin of the Geological Society of America* (1951), Vol. 62, pp. 1110–1119.

Smith, F. G. W. "The Simple Wave," *Sea Frontiers* (1970), Vol. 16, No. 4, pp. 234–245.

Smith, F. G. W. "The Real Sea," *Sea Frontiers* (1971), Vol. 17, No. 5, pp. 298–311.

Smith, F. G. W. "Measuring Ocean Movements," *Sea Frontiers* (1972), Vol. 18, No. 3, pp. 166–174.

Stewart, R. W. "The Atmosphere and the Ocean," *Scientific American* (September 1969), Vol. 221, No. 3, pp. 76–105.

TIDES

WAVES

Characteristics

Breaking Waves

Refraction and Longshore Currents

WAVE EROSION

WAVE TRANSPORTATION

EVOLUTION OF SHORELINES

ESTUARIES

CHAPTER 10

THE COASTAL ENVIRONMENT

Wave erosion near Depoe Bay on the Oregon coast. (Photo courtesy Oregon Travel Information Section.)

TIDES

At most coastlines, the constant rise and fall of the tides is perhaps the most obvious daily change. From classical times onward, people have realized that the moon is somehow involved with the tides, because they are later each day by about the same length of time as is the moon's rising. The relationship is not simple between the times of the tides and the moon's passage overhead. If it were, prediction of high tide as so many hours before or after the moon's passage would be easy. Before scientific studies of tides, tidal prediction was a family secret of local seafarers.

The scientific understanding of tides began with Newton and his law of gravity, which states that there is a force of attraction between any two masses. The gravity, or force of attraction, depends on the size of the masses and their distance apart. On this basis, it is easy to see that there is a tidal bulge beneath the moon as the ocean water is pulled toward the moon by gravity. The problem is in understanding why there is also a high tide on the opposite side of the earth at the same time.

The cause of the second tidal bulge is the centrifugal force caused by the revolution of the earth-moon system. The moon revolves around the earth, and the axis of revolution is their mutual center of mass. In such a system, the gravitational attraction of the moon is greater on the side of the earth nearer the moon than it is on the opposite side of the earth. The gravitational force and the centrifugal force are exactly balanced at the centers of both earth and moon. The centrifugal force is the same all over the earth, but, as we have seen, the gravitational force varies with the distance from the moon. On the side of the earth near the moon, the gravitational force is greater than the centrifugal force; on the opposite side, the centrifugal force is greater than the gravitational force (Figure 10-1). In this F 10-1 manner, tidal bulges develop on opposite sides of the earth.

In the same way, the earth-sun system also creates a tide on earth. In this case, the tidal force is only about 45 per cent as much as the

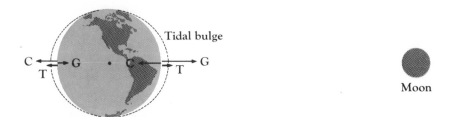

Figure 10-1 Tides result from centrifugal force and gravitational force. The centrifugal force is the same everywhere on earth, but gravitational force differs because the distance to the moon is different on opposite sides of the earth.

Figure 10-2
Neap and spring tides.

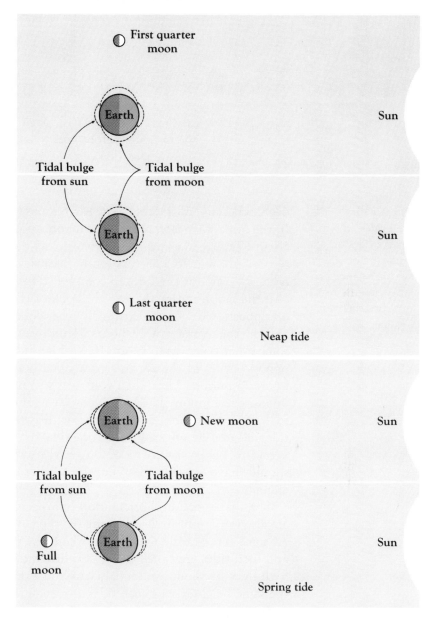

moon's. At new moon and full moon, the tidal forces are additive, producing the highest tides, which are called **spring tides** (Figure 10-2). At the first-quarter and third-quarter phases of the moon, the tidal attraction of the sun and moon are at right angles, so the tidal range is least. Such tides are called **neap tides** (Figure 10-2).

F 10-2

There are other astronomical complications to the tide-generating process. The moon's orbit around the earth is elliptical, so that its distance from earth changes. The same is true of the earth's orbit around the sun. Such changes affect the amount of gravitational attraction. In the course of a year, the sun's position in the sky varies 47

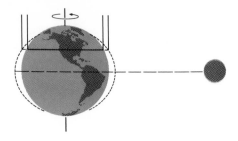

A. Moon on equator

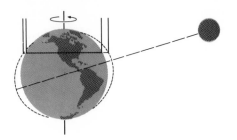

B. Moon above equator

Figure 10-3
Mixed tides, in which alternate tides on a latitude are of different heights, can occur when the moon is not directly over the equator.

degrees because of the tilt of the earth's axis (see Figure 11-18, p. 285). In the same way, each month the moon's position in the sky changes about 57 degrees as it moves between 28.5 degrees north of the celestial equator and 28.5 degrees south. Because of these complications, it takes 19 years for the tides to complete a full cycle.

If the moon is directly over the equator (Figure 10-3A), two points on opposite sides of the earth will experience the same high tide now and on the average of 12 hours 25 minutes later. If the moon is at some other point, as in Figure 10-3B, high tide will be experienced on opposite sides of the earth, as described before. But in this case, 12 hours 25 minutes later, the two points will not be directly below and opposite the moon. The later high tide at those points will not be so high as the first one. Such a place, with two high tides and two low tides, all of different heights, in one **tidal day** of 24 hours 50 minutes, is said to have a **mixed tide.** Our analysis suggests that most places should have mixed tides, but two other types of tides are also common.

The shape, size, and orientation of an ocean basin, bay, or estuary are also important in determining the type of tide as well as its height. At some places, only one high and one low tide occur each tidal day; and such tides are called **daily,** or **diurnal, tides.** At other places, two high tides of about the same height and two low tides, also of about the same height, occur each tidal day; such tides are called **semidaily,** or **semidiurnal, tides.** Figure 10-4 shows examples of the three tide types. Mixed tides occur over most of the west coast of North America, and semidaily tides are common on the east coast of the United States. Daily tides are found in parts of the Gulf of Mexico.

The movement of water as the tide ebbs and flows is generally not a very important agent of erosion. The main exception is at those places where the tidal currents must move through narrow inlets between the ocean and a bay or estuary. At such places, the tidal currents are concentrated in a small area, and they can be effective agents of erosion and transportation. Tidal currents of this type keep some harbors open without the need for dredging of the inlet. *Tidal*

F 10-3

F 10-4

Figure 10-4
Types of tides. **A.**
Semidaily, or
semidiurnal. **B.**
Mixed. **C.** Daily, or
diurnal.

current should not be confused with *tidal wave*, a term sometimes used for *tsunami*. A **tsunami** is an earthquake-generated wave that can cause much damage.

Tides are not limited to bodies of water; they also affect the atmosphere. The rock surface of the earth is also deformed a few inches by tidal forces.

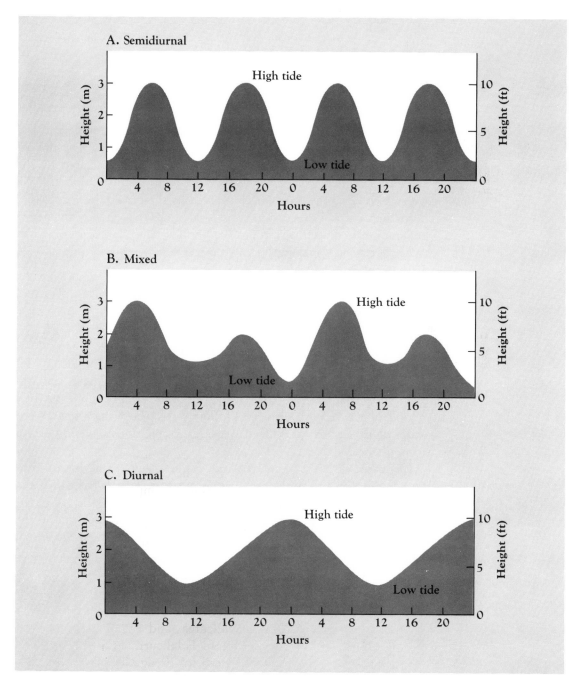

WAVES

One cannot help but be impressed with the seemingly boundless energy of waves breaking on a shore. A careful observer will note not only that they change somewhat with wind, tide, and season, but also that at times their height and energy change for no apparent reason. Watching waves will also convince one that they are agents of erosion, wearing away the shore and carrying away the debris. At sea, waves are cursed by many for causing seasickness.

Characteristics

The characteristics of water waves that are easiest for an observer to measure are their height (H) and their **period** (T), the time between successive crests. Waves are generally described by their height and wavelength (Figure 10-5), but wavelength is difficult to measure directly. The period of a wave and the **wavelength** are related by the velocity at which the wave moves. The relationship is that the velocity is equal to the wavelength divided by the period. In deep water, both theory and observation show that there the velocity in meters per second is equal to 1.56 times the period. This enables an observer to calculate the wavelength (L) (in meters): F 10-5

$$V = \frac{L}{T} = 1.56T$$

$$L = 1.56T^2$$

In shallow water, the velocity in meters per second is equal to 3.1 times the square root of the water depth in meters. These relationships show that the velocity of waves in deep water depends on their period or wavelength, and in shallow water the waves slow as the water depth decreases. As we will see, these relations have important consequences.

Ocean waves are caused by wind. When the wind begins to blow on a body of water, tiny ripples form on the surface. If the wind continues, the ripples turn into waves. The wavelength and height of the waves depend on the wind speed, how long the wind blows, and the **fetch,** which is the distance over which the wind blows. Storms or series of storms are major wave producers. Winds, especially during

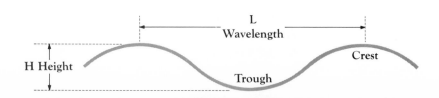

Figure 10-5
The parts of a wave.

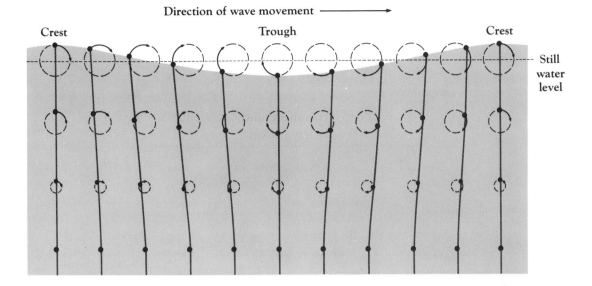

Figure 10-6 As a wave passes a point, a water particle travels in a circle. At the surface, the diameter of the circle is the height of the wave. The circular motion dies out with depth.

storms, are not constant in either speed or direction, and the resulting waves are choppy. The speed at which the waves travel can be faster than the winds that produce them. Long after the wind or storm that caused the waves has died out, the waves continue their journey, and they may travel thousands of kilometers before reaching a shore. The waves leaving their source area tend to recombine into larger, long-period waves called **swells**. The swell tends to fan out, and the longer-period, longer-wavelength waves travel faster than the shorter-wavelength waves. At sea, most waves are less than 2 meters (6 feet) high, and only 10 to 15 per cent are over 15 meters (50 feet). The highest wave ever measured at sea was 34 meters (112 feet).

Ocean waves transport energy with very little loss, and it is the wave form, not the water, that moves long distances in the open ocean. In shallow water the water moves with the wave, as any surfer knows, but in deep water only the wave form moves. In deep water, the water moves in a circle as the wave passes, as shown in Figure 10-6. At the surface the diameter of the circle is the height of the F 10-6 wave. The size of the circle becomes smaller with depth and finally dies out at a depth of one-half the wavelength. The movement of the water as a wave passes is much like the movement of a field of tall grass or wheat as the wind forms a wave that moves across the field. Actually, there is some slight forward movement of the water because the advance at the top of the circle is slightly more than the return motion (Figure 10-7). As the energy in a wave of a given wavelength F 10-7 increases, its height increases. If the height is doubled, the amount of

Figure 10-7
The actual movement of a water particle as a wave passes results in a slight forward motion.

energy increases four times. As the height of a wave increases, its crest changes from rounded to more pointed.

Breaking Waves

When a wave travels into an area where the depth of the water is less than one-half of the wavelength, the wave is affected by the bottom. The motion of the water as the wave passes changes from circles to ellipses that become flatter with depth until at the bottom the water moves horizontally back and forth. The friction on the bottom slows the wave, causing its height to increase and its speed and wavelength both to decrease. This effect, plus the fact that the front of the wave is slowed more than the rear, causes the wave to break. Generally, a wave breaks when its height is about one-seventh of its wavelength, and this commonly occurs when it reaches a depth of about one and one-half of the wave height. Most wave erosion occurs at depths of less than 9 meters (30 feet). Wave heights over 6 meters (20 feet) are rare. However, severe wave damage as much as 30 meters (100 feet) above sea level has been reported, which indicates that in storms the water from breaking waves can be driven very high. At the other extreme, some waves have wavelengths over 600 meters (2000 feet), so they can cause some bottom erosion in water as deep as 300 meters (1000 feet).

When waves break, the water runs up on the beach. Some of it percolates into the beach, and some flows back toward the ocean. Generally, the return flow is very gentle, but at times it may be concentrated in a narrow channel called a **rip current.** A swimmer caught in a rip may be carried to where the current spreads out—up to 300 meters (1000 feet) from the shore (Figure 10-8). This is a fright- F 10- ening experience, but by swimming a short distance parallel to the shore, one can generally get away from the rip current and swim back to shore. Trying to swim back through a rip current is very difficult. Some beaches have constant rip currents, but may have them only when the wind or waves are from certain directions. Rip currents can, in many cases, be seen as darker, deeper places where there is less surf.

Breaking waves are of two main types. **Spilling waves** break when the oversteepened front of the wave spills forward (Figure 10-9). The F 10- wave continues to advance, but its height is reduced as the wave spills. Such waves are good for surfing. **Plunging waves** are more spectacular. The top of the wave curls over, trapping an air pocket, and the escape of the compressed air causes much foam and splashing (Figure 10-10). Plunging waves tend to form from long, gentle waves F 10- on gently sloping shores.

Figure 10-8
Development of rip
currents.

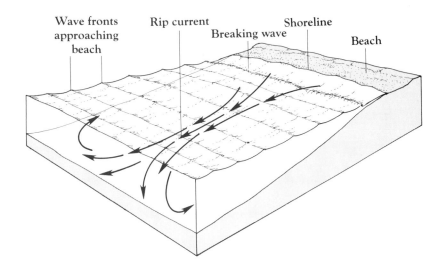

Wave fronts
approaching
beach

Rip current

Breaking wave

Shoreline

Beach

Refraction and Longshore Currents

Figure 10-9
Spilling wave. (Photo
from NOAA.)

As we noted, waves are slowed by the ocean bottom when they reach
an area where the water depth is one-half their wavelength. If a wave
approaches the shore obliquely, the shoreward part is slowed first. As
the wave continues toward the coast, more and more of it is slowed
(Figure 10-11). This slowing causes a change in the wave's direction. F 10-11

Figure 10-10
Plunging wave.
(Photo from Hawaii
Visitors Bureau.)

Figure 10-11
As a wave moves
onshore, it is slowed
when the water
motion extends to the
bottom.

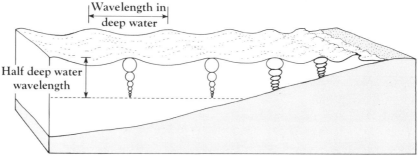

This change in direction is called **refraction.** The effect of refraction on waves approaching a smooth shore obliquely (as in Figure 10-12) is F 10-12 to turn the waves so that they approach the shore more nearly parallel to it than their original direction.

Although most waves approach a shore more or less straight because of refraction, they generally come at a slight angle to the shore. Because of this, the breaking waves run up on the beach in the direction that they come onto the shore. They return to the ocean by flowing straight down the slope of the beach. This means that the water and any material being moved by the waves moves along the beach. Such movement is termed a **longshore current** (Figure 10-13), F 10-13 and longshore currents can transport large quantities of sand or other material along a beach. Because the sand on most beaches is constantly being moved by longshore currents, beaches have been called "rivers of sand."

The sources of the material carried by longshore currents are wave erosion and the debris carried by rivers. Many tons of sand move past

Figure 10-12
Wave refraction. **A.**
Slowing of waves in
shallow water causes a
change in direction.
B. Waves off Point
Año Nuevo,
California. (Photo
from USDA-ASCS.)

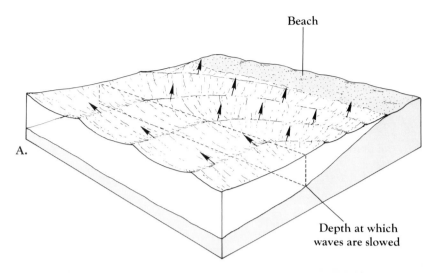

a beach each year, and, barring storm and seasonal changes, the appearance of the beach generally does not change appreciably.

Eventually, however, the beach gradually begins to get smaller. Worried property owners erect barriers to prevent erosion of the beach by the longshore current (Figure 10-14). The barriers only slow the process, however, because erosion occurs on one side of each barrier, and deposition occurs on the other. A common cause of the loss of beach as described is the damming of rivers so that their water can be used for many purposes. The erosional debris carried by the river is deposited in the lake behind the dam and never gets to the ocean.

Figure 10·13
Longshore currents result from waves that approach the coast at an angle. The breaking waves run up on the beach at an angle and run back down the slope.

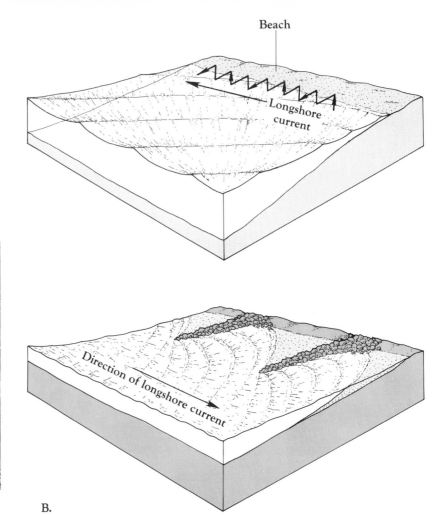

A.

B.

Figure 10·14 A. Jetties at Fenwick Island, Maryland. (Photo by R. Dolan, U.S. Geological Survey.) **B.** The effect of a barrier, or jetty, is to cause accumulation of sand on one side and erosion on the other.

Then the longshore currents use energy to erode and transport, so they begin to erode the beach.

Refraction also focuses the energy of waves on headlands and so helps to make coastlines smoother. As the waves approach the shore, as shown in Figure 10-15, they are slowed by the bottom in the area off the headland but not in the bay, so they curve toward the headland. At sea, the waves have about the same amount of energy everywhere along the wave. In Figure 10-15, lines are drawn at right angles to the waves as they refract and curve into the shore, and these lines show that the energy of the waves is concentrated on the headlands

F 10–1

Figure 10-15
Refraction of waves concentrates waves on headlands.

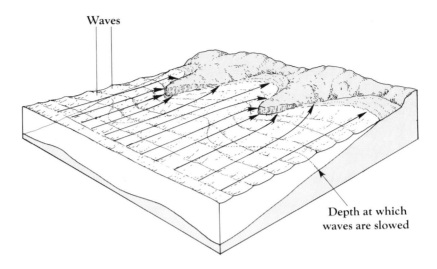

Waves

Depth at which
waves are slowed

and is reduced in the bays. Therefore, wave erosion is faster on the headlands than in the bays.

WAVE EROSION

The energy in waves comes from the wind, and this energy is transported by the waves with little loss until a shore is met. At the shore, the breaking waves transform the energy into heat and into work in the form of erosion and transportation. The agents of wave erosion are solution, abrasion, and the impact and pressure of the breaking waves. Only a few rocks are soluble, so solution is the least important form of wave erosion.

Abrasion is the grinding done by sand and gravel moving back and forth as waves run up on the shore and then flow back. The back-and-forth movement of sand is easily observed at most beaches and is the reason that most beach sands are rounded and frosted. Much wave abrasion occurs at depths of 9 meters (30 feet) or less, where the back-and-forth movement of sands on the bottom abrades flat surfaces called **surf-cut,** or **wave-cut, benches.** At many places these benches can be seen at low tide (Figure 10-16). At other places F 10-16 later uplift of the area or retreat of the sea leaves them above sea level.

The most spectacular form of wave erosion is impact and pressure caused by breaking waves, which can split large blocks out of sea cliffs. Breaking waves may undercut cliffs, causing them to collapse and retreat. The big waves that break off large blocks are, at most shores, relatively rare and generally come from storms far at sea. The rate of erosion and cliff retreat is generally slow, but at some places can be as rapid as 9 meters (30 feet) per year. The erosion generally occurs during periods of high storm waves (Figure 10-17). F 10-17

Figure 10-16
Wave-cut bench near Arecibo, Puerto Rico. (Photo by C. A. Kaye; U.S. Geological Survey.)

Figure 10-17
Wave erosion has created the spectacular coastline of Oregon. (Photo from Oregon State Highway Department.)

WAVE TRANSPOR- TATION

Most of the sediment transported by waves is carried along the coast by longshore currents, as already described. Along most coasts, a dominant wind and wave direction determines the direction of the longshore currents. On the west coast of North America, the direction is generally southward. Waves are also involved in the eventual movement of sediment toward the ocean basin floor. The longshore flow of sediment is interrupted at places where submarine canyons cut into the continental slope and shelf. At these places, the sediment moves down the canyon toward the sea. On a more local scale, rip currents that return the water moved onshore by waves carry sediments toward the ocean. Storm waves and winds may also stir up bottom sediments, which then move downslope toward the ocean.

Waves also cause seasonal changes to many beaches. During the summer, the waves generally move sand shoreward, and beaches are wide and have gentle slopes. In the winter the waves are generally higher and have larger wavelengths. These waves erode the beach, making it narrower and steeper. The sand that formerly was on the beach is moved to an offshore bar. In summer the gentler waves move the sand from the bar back to the beach. This pattern is embroidered onto the general longshore movement of the sand.

EVOLUTION OF SHORELINES

Coastlines are areas of constant change because the great energy of waves is concentrated there. In the last 10,000 years, sea level has risen about 100 meters (330 feet) as the glaciers of the last Ice Age melted. The coasts, then, are geologically young features, and this helps to account for their diversity and beauty.

Describing coastlines is difficult because some are undergoing erosion and others are being built up by deposition. Although melting glaciers have caused a worldwide rise in sea level, at some places the continents are moving either up or down. The rock type and the initial shape of a coast are also important in determining its eventual form.

Coastal erosion is generally easy to recognize. The undercutting of *sea cliffs* by breaking waves is an obvious example of erosion. At such locations the waves generally form *wave-cut benches*, or *terraces*, in front of the cliffs, and the longshore current removes the erosional debris. If the areas has headlands, as erosion continues, *stacks*, *sea caves*, and *arches* may develop (Figure 10-18). On more gentle coast- F 10-18

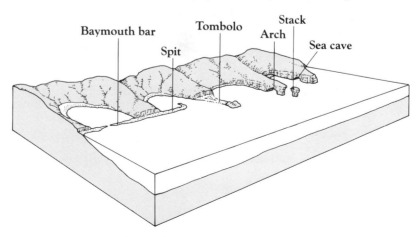

Figure 10-18
Shoreline features.

Figure 10-19
The Texas Gulf coast is a barrier island coast.

lines where cliffs do not form, erosion is more difficult to recognize. Gentle coasts without beaches may be areas of erosion.

Coastlines dominated by deltas are clearly areas of deposition. Reefs and marshes are also generally areas where the coast is building outward toward the ocean. Coasts with **barrier islands** (Figure 10-19) are very common and generally are formed by waves. Some are formed by waves and wind moving sediments landward, and others may be built by longshore currents. Still other barrier islands may be the result of exposure of offshore bars by rising land. Longshore currents form a number of depositional features. **Spits** are deposits, generally of sand, extending from a headland or similar feature (Figure 10-18). If the bar extends across a bay, it is called a **baymouth bar** (Figure 10-18). A bar extending between an island and the shore is a **tombolo.**

Changes in sea level may be caused by either uplift or depression of the continent or by rising or falling of the ocean water. Changes in the level of the ocean would, of course, be worldwide, and changes in the elevation of the land are localized. Melting glaciers have raised sea level worldwide about 100 meters (330) feet, but not all coasts show evidence of rising sea level. The areas covered with glacial ice have also risen as the weight of the overlying ice was removed when it melted. At most places it is not possible to tell whether the continents moved or sea level moved or both moved, but only that relative movement has occurred.

If the continent was uplifted or sea level went down, the area is said to be **emergent.** At an emergent coast, the newly uplifted area is now subjected to breaking waves, and sea cliffs generally develop (see Figure 10-16). Of course, one can think of other ways that sea cliffs might be formed. The clearest evidence for emergence is found at those places where successive uplifts of the land or falling of sea level has produced one or more wave-cut terraces above the present sea level (Figure 10-20). At some places the successive terraces look like a giant stairway above the sea, but the oldest and highest terraces are generally rounded by weathering and erosion. F 10-20

Submergent coasts *develop if the continent is depressed or sea level rises.* The effect is that coastal valleys are drowned or flooded, and they become **estuaries** (Figure 10-21). In some cases the estuary shape is that of the main stream and its tributaries. Chesapeake Bay and San Francisco Bay are examples of submergent coasts. On smoother, flat- F 10-21

Figure 10-20
Development of wave-cut terraces.

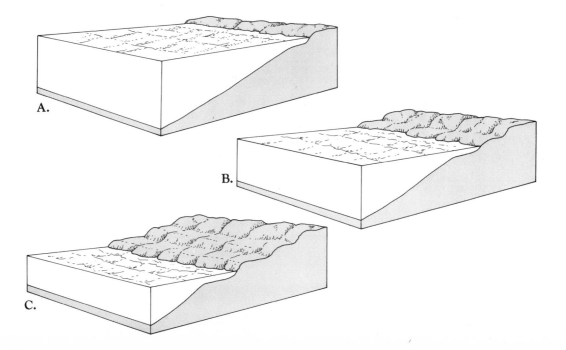

Figure 10-21
Submergent coastline.

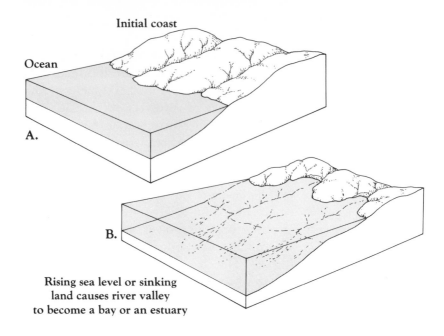

Rising sea level or sinking
land causes river valley
to become a bay or an estuary

ter, plains-type coastlines, the evidence for submergence is not so clear. *Barrier island* coasts are commonly on such plains coasts, and they have developed on about one-third of the world's coasts. Because barrier island coasts are so common, and sea level has risen as the glaciers melted, there may be a relationship between them. With slow submergence, winds and waves may build barrier islands.

At any particular coast, erosion or deposition may be occurring, and the coast may also be emerging or submerging. The changes that occur to the coast will depend on the energy of the waves, the rock types, and the initial shape of the coast. In general, if the coast is stable—that is, no major changes in these factors occur—the end product will be a smooth coastline. If the coast is irregular, with many bays, the changes in it will be most noticeable. Wave refraction will concentrate the wave energy on the headlands. They will be eroded, and spits and baymouth bars will develop. Deposition of river-carried debris may occur in the bays. Eventually, a smooth coast will result (Figure 10-22). On initially smooth coastlines the results are harder to F 10–22 predict, but barrier islands may develop, with sedimentation occurring in the resulting lagoons.

ESTUARIES

Estuaries form at river mouths and in bays into which rivers flow. They are places where river water mixes with salt water, forming **brackish water.** The circulation in an estuary depends on the size and shape of the bay, the amount of river discharge, and the tides. The fresh river water is less dense than the ocean water and therefore

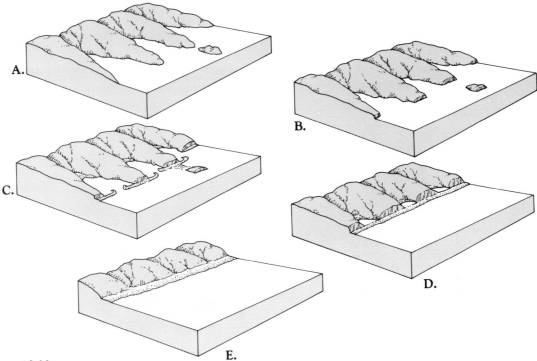

Figure 10-22
Wave erosion changes an initially irregular coastline to a smooth coastline.

flows on top of the heavier salt water. In an estuary, the bottom of the outward-flowing river water mixes with the top of the ocean water, and the resulting brackish water is carried into the ocean by the out-ward-flowing river water (Figure 10-23). The amount of mixing may F 10-23 vary seasonally and depends on the factors listed above. In most cases, the inflow of salt water on the bottom is many times the amount of river water flowing out. If river discharge is small and the tides are active, much mixing results, and the inflow of salt water at the bottom may be more than 20 times the river discharge. At the

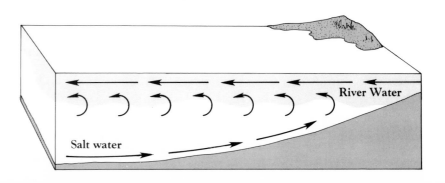

Figure 10-23
Light river water flowing above heavier salt water causes mixing in an estuary.

other extreme is the Amazon River, whose volume is so great that all of the mixing occurs on the continental shelf, far at sea.

Estuaries and bays are important parts of the oceans to us because of their commercial and recreational uses. They provide seafood and commercial and recreational harbors. Industrial development around seaports and increasing population have led to the discharging of large amounts of industrial and municipal wastes into bays and estuaries (Figure 10-24). If fishing and recreational uses are to be preserved, the ability of tide and river discharge to flush the waste to sea must not be exceeded on either a daily or seasonal basis. The effect of the wastes on the open ocean must also be considered (Figure 10-25). F 10-24

F 10-25

Figure 10-24
Shoreline development. **A.** Fenwick Island, and **B.** Ocean City, Maryland. (Photos by R. Dolan; U.S. Geological Survey.)

A.

B.

Figure 10-25
The breakup of th tanker *Argo Merchant* off the Massachusetts coast, December, 1976. (Photo from NOAA.)

SUMMARY

- Tides are caused by the moon's gravity and the centrifugal force caused by the revolution of the earth-moon system. As a result, each point on the earth should experience two high and two low tides in a tidal day of 24 hours 50 minutes.

- The earth-sun system also creates tides in the same way, and the effects of the moon and sun are additive at full and new moon, causing the highest high tides called *spring tides*. The lowest high tides occur at first- and last-quarter moon and are called *neap tides*.

- Waves are characterized by their heights and their periods or wavelengths.

- Waves are caused by wind.

- In deep water only the wave form moves; the water travels in circles as the wave passes.

- In shallow water the waves are slowed by friction on the bottom, causing the height to increase; eventually this causes the wave to break.

- The slowing of waves in shallow water causes the waves to bend or *refract*.

- Refraction causes waves to erode headlands more than bays.

- Waves approaching a shore at any angle cause a net longshore movement of sediment.

- Wave erosion causes *surf-cut*, or *wave-cut*, *benches* at depths of less than 9 meters (30 feet).

- Coasts undergoing rapid erosion are characterized at many places by sea cliffs.

- At places where the land has risen or sea level has fallen, wave-cut benches may be far above the present sea level.

- An *emergent coast* is one where the continent was uplifted or sea level went down.

- A *submergent coast* is one where the continent was depressed or sea level rose.

- One type of submergent coast is a deeply embayed coast where river valleys have been drowned by rising sea or sinking continent.

- *Estuaries* form at river mouths where sea water and river water mix, forming *brackish water*.

KEY TERMS

spring tide	tsunami	refraction	baymouth bar
neap tide	period of a wave	longshore current	tombolo
tidal day	wavelength	surf-cut, or wave-cut,	emergent coast
mixed tide	fetch	bench	submergent coast
daily tide, or diurnal	swell	barrier island	estuary
tide	rip current	spit	brackish water
semidaily tide, or	spilling wave		
semidiurnal tide	plunging wave		

REVIEW QUESTIONS

1. Tides are caused by what two forces?
2. During a spring tide, where are the sun and moon positioned relative to the earth? Where are the sun and moon positioned during a neap tide?
3. If you were to take into account only the movement of the moon, would you expect to find mixed tides everywhere? Why or why not?
4. Sketch a couple of waves and indicate what is measured to determine wave height and wavelength. Define the period.
5. Why is a wave undisturbed until it reaches an area where the water depth is only half the wavelength?
6. What causes a wave to slow in shallow water?
7. Examine the illustration of wave refraction (Figure 10-12) and explain in your own words why the wave is turning parallel to the shore.
8. Draw a line to represent the shore and a line at a right angle to it to represent a jetty. (A jetty is a long, narrow barrier, usually of rocks or cement, that extends out into the water.) Add an arrow to show the direction from which the waves usually come (you choose a direction). Now label the side of the jetty on which the longshore current is depositing material and label the side on which erosion is occurring.

9. What is an important source of the sediments carried by the longshore currents?
10. Normally sediments are carried along the coast. Describe three ways in which the sediments can be transported out to sea, causing loss of beach material.
11. What do people who live along sandy shores mean when they talk of the "summer beach" and "winter beach"?
12. Some areas of the coast are scalloped with headlands (points) and coves or bays. Where is erosion greater? Why?
13. List three ways in which waves erode. Which of these generally creates the least erosion? Which is most effective during storms? Which creates wave-cut benches?
14. You are spending the summer bicycling along the coast. What features might you see that would indicate that the shore is being eroded? Later in the summer you come to an area where a great deal of deposition is occurring; describe the scenery.
15. How can one tell whether a coast is emerging or submerging?
16. Describe a barrier island. What are some ways in which scientists think it may have been formed? Along what kind of coast are barrier islands usually found?
17. Describe circulation in an estuary.

SUGGESTED READINGS

Bascom, Willard. "Beaches," *Scientific American* (August 1960), Vol. 203, No. 2, pp. 80–94.

Bascom, Willard. *Waves and Beaches.* Garden City, N.Y.: Doubleday, 1964, 267 pp. (paperback).

Carr, A. P. "The Ever-Changing Sea Level," *Sea Frontiers* (1974), Vol. 20, No. 2, pp. 77–83.

Carson, Rachel. *The Sea Around Us,* rev. ed. New York: Oxford University Press, 1961, 230 pp. (Also available in paperback.)

Emiliani, C. "The Great Flood," *Sea Frontiers* (1976), Vol. 22, No. 5, pp. 256–270.

Fairbridge, R. W. "The Changing Level of the Sea," *Scientific American* (May 1960), Vol. 202, No. 5, pp. 70–79.

Grasso, A. "Capitola Beach," *Sea Frontiers* (1974), Vol. 20, No. 3, pp. 146–151.

Higgins, C. G. "Causes of Relative Sea-Level Changes," *American Scientist* (December 1965), Vol. 53, No. 4, pp. 464–476.

Hoyt, J. H. *Field Guide to Beaches.* ESCP Pamphlet Series, PS-7. Boston: Houghton Mifflin, 1971, 45 pp.

Hoyt, J. H. "Shoreline Processes," *Journal of Geological Education* (January 1972), Vol. 20, No. 1, pp. 16–22.

Inman, D. L., and B. M. Brush. "The Coastal Challenge," *Science* (July 6, 1973), Vol. 180, No. 4094, pp. 20–32.

King, C. A. M. *Beaches and Coasts.* London: Edward Arnold, 1959, 403 pp.

Mahoney, H. R. "Imperiled Sea Frontier—Barrier Beaches of the East Coast," *Sea Frontiers* (1979), Vol. 25, No. 6, pp. 329–337.

Officer, C. B. "Physical Oceanography of Estuaries," *Oceanus* (1976), Vol. 19, No. 5, pp. 3–9.

Schubel, J. R., and D. W. Pritchard. "The Estuarine Environment," *Journal of Geological Education.* Part I (March 1972), Vol. 20, No. 2, pp. 60–68. Part II (September 1972), Vol. 20, No. 4, pp. 179–188.

Schumberth, C. J. "Long Island's Ocean Beaches," *Sea Frontiers* (1971), Vol. 17, No. 6, pp. 350–362.

PART III
METEOROLOGY

Long before there were satellite pictures on television, we learned to make fairly reliable short-term weather predictions by observing cloud patterns, changes in wind directions, visibility, and temperature. But the science of meteorology consists of much more than merely forecasting the weather. In the next five chapters we will learn something of our planet's atmosphere and what causes weather patterns and climates.

Chapter 11 examines the structure and composition of the atmosphere. We will see how the gases that compose the atmosphere protect life from ultraviolet radiation and keep the earth warm. Then we will consider the seasonal heating and cooling of the earth that results from the changing amount of sunlight received at each latitude as the earth orbits the sun.

Chapter 12 examines atmospheric water vapor and moisture. We will study the meaning of humidity, the importance of evaporation and condensation, and the various types of clouds, fog, and precipitation.

The subjects of Chapter 13 are air pressure and wind. Wind is the result of air moving from areas of high pressure to areas of low pressure. The chapter describes the "local winds," such as sea and land breezes, and the circulation of air on the planetary scale. Winds are caused by the heat from the sun and the earth's rotation.

Chapter 14 describes both common and violent storms. Air masses, weather fronts, and the other features pictured on the weather maps are described. The changing weather as a storm passes is discussed, and finally the violent storms, thunderstorms, tornadoes, and hurricanes are described.

Chapter 15 focuses on climate. This chapter explores the geologic and astronomical causes of climatic change and illustrates how human activities can affect climate. The chapter details the numerous causes and detrimental effects of air pollution and also discusses our deliberate attempts at climate modification.

ATMOSPHERE
Composition

Structure

HEAT AND TEMPERATURE
Temperature Measurement

Heat Transfer

HEATING OF THE ATMOSPHERE
Heat Balance

Sun's Illumination

Heat Transport

Temperature Lag

CHAPTER 11

THE ATMOSPHERE AND ITS ENERGY

Storms in the Pacific Ocean. (Satellite image from NOAA Satellite Field Service Station, Redwood City, California.)

To most people, meteorology means weather forecasting. They think of either the TV weather report, with its elaborate displays, or a short paragraph in the newspaper, perhaps with an accompanying map, or a sentence or two at the end of the radio news. Although these forecasts are accurate most of the time, their failures are more often remembered. To a meteorologist, **meteorology** *means the study of the atmosphere and its changes and movements,* and, of course, weather forecasting is only a part of this science. Our goal in this unit will be to understand in a general way how the atmosphere works and what causes characteristic weather patterns and climates. **Weather** *is the short-term changes in the atmosphere,* and it includes such elements as temperature, humidity, pressure, clouds, precipitation, and direction and speed of wind. The **climate** *of an area is the typical combination of these weather elements that has characterized that area over a long term and takes into account seasonal changes as well as ranges and extremes of weather elements.*

ATMOSPHERE   Composition

Air is a mixture of gases. The mixture is uniform—that is, it is well mixed—up to a height of about 100 kilometers (60 miles). Dry air is composed by volume of about 78 per cent nitrogen, 21 per cent oxygen, 1 per cent argon, and very small amounts of carbon dioxide and other gases, as shown in Table 11-1. In addition, it can contain up to   T 11-1
3 per cent water vapor by volume. Of these gases, the carbon dioxide and the water vapor are the most important in the study of meteorology. The atmosphere also contains variable amounts of tiny liquid and solid particles, some of which are natural but many of which, unfortunately, come from cities and industry. (The effects of such pollutants are described in Chapter 15.)

Table 11·1   COMPOSITION OF AIR

Gases	Percentage by Volume	
Nitrogen	78.084	
Oxygen	20.946	Dry air
Argon	0.934	
Other gases	0.003	
Main Variable Gases		
Carbon dioxide	0.033	(Considered variable because the amount is increasing slowly)
Water vapor	0–3	

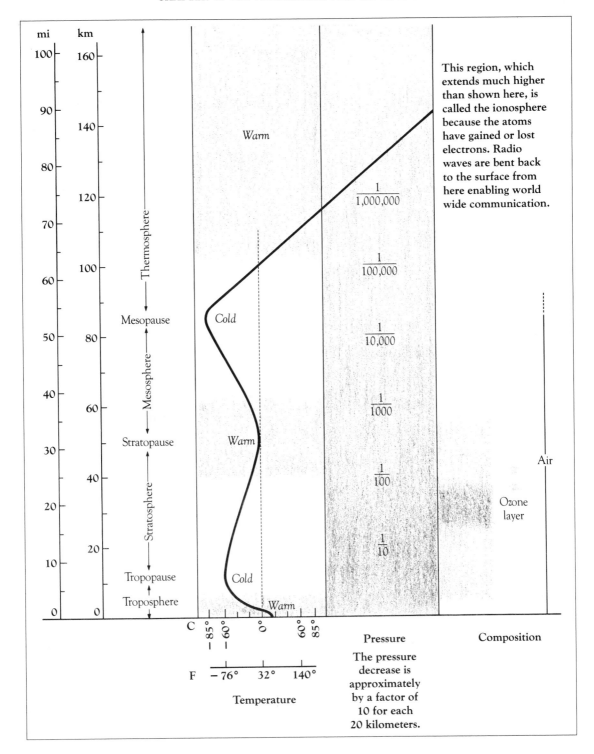

This region, which extends much higher than shown here, is called the ionosphere because the atoms have gained or lost electrons. Radio waves are bent back to the surface from here enabling world wide communication.

The pressure decrease is approximately by a factor of 10 for each 20 kilometers.

Figure 11-1
Structure of the atmosphere.

Most of the gases are concentrated in the lower atmosphere. The density of the atmosphere is reduced as one rises, as shown by the reduction of pressure illustrated in Figure 11-1.

F 11-1

A.

Figure 11-2
Upper-atmosphere measurements. **A.** Closeup of the instrument package called a radiosonde. **B.** Releasing instrument package attached to balloon. (Photo **A** from The Bendix Corporation; **B** from NOAA.)

B.

Structure

Measurement of temperature in the atmosphere, by means of instruments carried aloft by balloons (Figure 11-2), has revealed the structure of the atmosphere. The lowest part is where almost all weather occurs and is called the *troposphere*. In the **troposphere** *the temperature falls on the average about 0.65°C per 100 meters (3.5°F per 1000 feet) as one rises. The top of the troposphere is called the* **tropopause,** and its elevation changes with latitude and season. It is lower in winter than in summer, and typical heights are about 15 to 16 kilometers (9 or 10 miles) at the equator, 5 to 6 kilometers (3 to 4 miles) at the poles, and 10 kilometers (6 miles) in temperate regions. In the troposphere, weather systems keep the air in motion.

The **stratosphere** *extends from the tropopause to about 50 kilometers (30 miles).* In the stratosphere the temperature is constant up to about 20 kilometers (12 miles) and then increases to the **stratopause,** *the boundary of the stratosphere.* The temperature at the stratopause is about 0°C (32°F), but this temperature and the others shown in Figure 11-1 vary over a small range. The temperature rise in the stratosphere is caused by activity in the ozone layer.

The **ozone layer,** or **ozonosphere,** *a part of the stratosphere, extends from about 10 to 50 kilometers (6 to 30 miles), and the maximum concentra-*

F 11-2

F 11-1

tion is about 5 parts per million (ppm) at about 30 kilometers (18 miles). That is, there are about 5 **ozone** (O_3) molecules for each million molecules of air. This tiny amount of ozone, which is itself lethal in higher concentrations, shields us from lethal ultraviolet radiation from the sun. The sun's ultraviolet radiation causes the ordinary oxygen (O_2) molecules to separate into single oxygen (O) atoms. The sun's ultraviolet energy is absorbed by this reaction. Most of the oxygen atoms recombine to the stable O_2 molecules, but a few form the O_3, or ozone, molecule. The ozone molecules are also unstable and eventually break down to ordinary oxygen. The ozone also absorbs ultraviolet radiation from the sun, and it is this absorption that heats the stratosphere. Obviously, the ozone layer is very important to life at the earth's surface. Supersonic aircraft that fly in the stratosphere could upset these processes, and this is one reason that such aircraft are controversial.

Above the stratopause is the **mesosphere,** *another zone of falling temperature that extends to about 85 kilometers (50 miles).* The temperature at the **mesopause,** *or top of the mesosphere,* is about $-85°C$ ($-120°F$), but it varies about $25°C$ above and below this value. Above the mesopause is the **thermosphere,** *a zone of increasing temperature that extends several hundred kilometers.* Temperatures reach very high values in the thermosphere, but, as we will see in the next section, temperature has little meaning in this zone of very thin atmosphere.

HEAT AND TEMPERATURE

Heat *is a form of energy* and so is of fundamental importance. The sun's radiant energy is changed to heat in the atmosphere, and the heat is what causes motion of the atmosphere and, in turn, causes weather. **Temperature** *is a measure of the intensity of heat.* Better definitions can be framed in terms of kinetic theory. **Kinetic theory** *assumes that all matter is composed of molecules that are in motion. The heat of an object is proportional to the total motion, or kinetic energy, of all of its molecules. The temperature is proportional to the average energy of each molecule.*

Now we can understand the meaning of temperature in the thermosphere, as mentioned at the end of the last section. In the thermosphere there are very few molecules, but they move at a high velocity. This means that the temperature is very high, but, because there are so few molecules, very little heat is involved, so a person in the thermosphere would not sense this high temperature.

Temperature Measurement

Many thermometers depend on differential expansion to indicate the temperature. Differential expansion simply means two substances that expand at different rates when the temperature is increased. In the familiar red alcohol or mercury thermometers, the alcohol or the

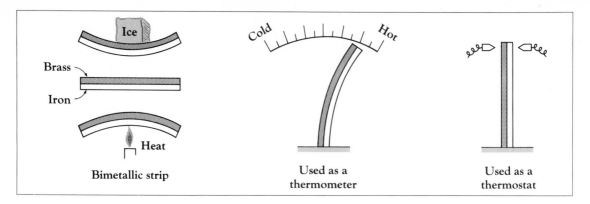

Figure 11-3 Bimetallic strip can be used as a thermometer or a thermostat.

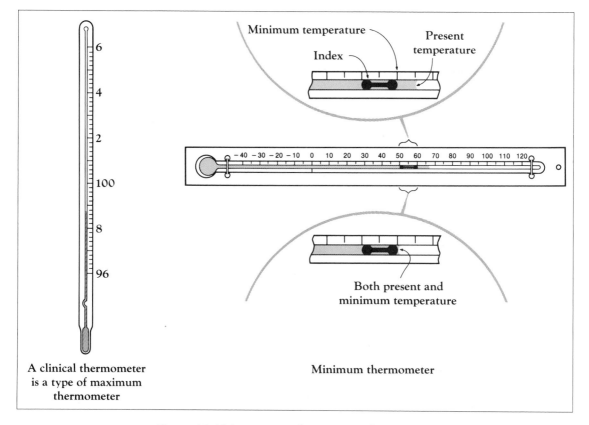

Figure 11-4 Maximum and minimum thermometers. The maximum thermometer has a construction that allows fluid to rise in the column but prevents its return to the bulb. It is reset by spinning or shaking the fluid into the bulb. In the minimum thermometer the index is pulled down by surface tension when the alcohol column shrinks; but if the alcohol expands, it moves past the index, which records the lowest temperature reached. It is reset by tilting so that the index moves to the top of the alcohol column.

Figure 11-5
Meteorologist reading maximum and minimum thermometers that are in a standard instrument shelter. (Photo from NOAA.)

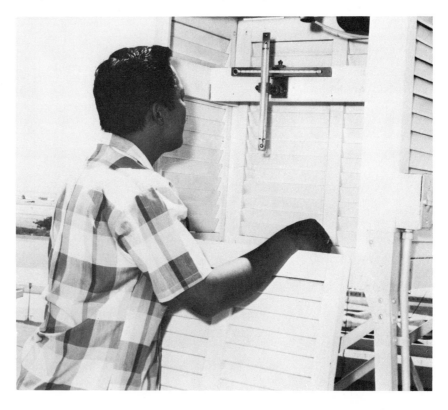

mercury expands or contracts much more than the glass tube, so that the fluid level rises or falls in the tube and so indicates the temperature. The bimetallic strip is another type of thermometer. Here, one metal expands more than the other, and so the strip bends (Figure 11-3).   F 11-3

Recording thermometers and maximum-minimum thermometers (Figure 11-4) are in common use. Such devices greatly reduce the   F 11-4 number of temperature readings that should be taken.

To standardize temperature readings, a standard thermometer shelter (Figure 11-5) is used. Its height is about eye level (1.5 meters, or 5   F 11-5 feet), it is painted white, and it has louvered sides to allow air circulation. The importance of standardized temperature readings can be seen by comparing the readings of, for instance, a thermometer in the sun, one in the shade, and one mounted on a dark wall.

A comparison of the Celsius and Fahrenheit temperature scales is given in Figure 11-6. These scales are in common use.   F 11-6

Heat Transfer

The movement of heat in the atmosphere is the cause of most weather. Heat moves from high temperatures to low temperatures, and movement stops when temperatures are the same.

Figure 11-6
Comparison of
Celsius and
Fahrenheit
temperature scales.

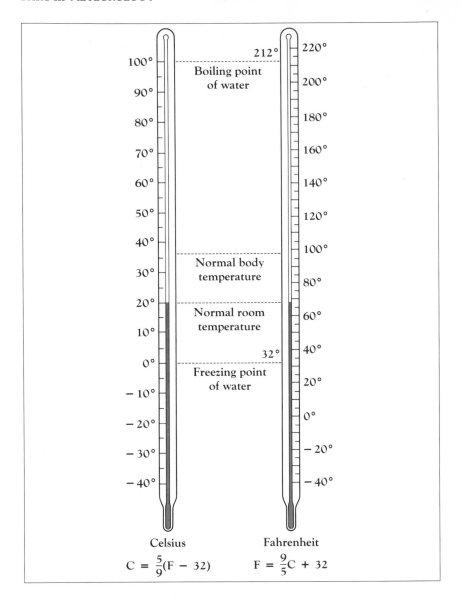

Celsius   Fahrenheit

$$C = \frac{5}{9}(F - 32) \qquad F = \frac{9}{5}C + 32$$

Conduction is probably the most familiar form of heat transfer. *In* **conduction,** *the heat travels through a material by molecular collisions.* In Figure 11-7, the molecules at the end of the rod that is in the flame absorb heat and so move more rapidly. These rapidly moving molecules collide with slower moving molecules, causing them to move faster, and these molecules, in turn, hit other molecules, and so on. In this way, the heat energy moves down the rod by means of molecular collisions. When a material is heated, the faster-moving molecules require more room in which to move, and this is why most materials expand when heated. A material that is *a poor conductor is called an* **insulator.**

F 11-7

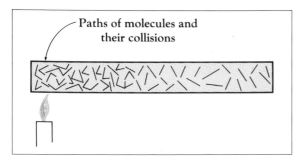

Figure 11-7
Heat transfer by conduction. Heating makes the molecules move faster, so they collide more often. The collisions cause the increased molecular motion to migrate down the rod, and so the heat energy in the form of increased molecular motion moves down the rod.

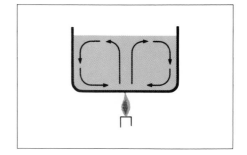

Figure 11-8
Heat transfer by convection. When a fluid is heated, it expands and thus is less dense. The warm, less-dense fluid rises, carrying the heat energy with it. Cool fluid from above sinks to complete the convection cell.

Convection can occur only in a liquid or a gas. *Heat is transferred in* **convection** *by actual movement of the fluid.* In Figure 11-8, the liquid is F 11-8 heated at the bottom. The liquid near the bottom absorbs heat, so its molecules begin to move faster, causing the heated part of the liquid to expand. The warmer liquid is lighter—that is, less dense—than the surrounding cool liquid, so it rises, carrying its heat with it. *As the warm liquid rises, cooler liquid takes its place, and a* **convective flow** *results,* as shown in Figure 11-8.

Radiation *is the transfer of energy by electromagnetic waves* (Figure 11-9). Light, X-ray, and radio and television waves are all types of F 11-9 electromagnetic waves. They can all travel through the near-vacuum of space, and different types of material are opaque to each. For example, television and radio waves pass through wooden buildings, but

Figure 11-9
Heat transfer by radiation. A hot body, such as the sun, radiates heat energy, a form of electromagnetic wave. This radiation can pass through the near-vacuum of space, and the heat is felt only when the waves interact with a material. A black object will not only absorb the most radiant heat; it is also the most efficient radiator.

wood is opaque to light waves. Most of the sun's energy is visible light. It passes through space, and it is converted into heat when it illuminates an object, especially if that object is dark colored. Dark objects not only absorb more radiation than light-colored objects, but dark objects also radiate more energy than light-colored objects.

HEATING OF THE ATMOSPHERE

Heat Balance

The earth as a whole is neither heating up nor cooling down but has had the same general temperature range for at least nearly a billion years. This conclusion is reached from the study of geologically dated rocks and the fossils they contain. This is not to say that temperatures have not varied somewhat overall or that some places that now have arctic conditions were not tropical in the past. The implication of these observations is that, in the long term, the energy received from the sun is balanced by the energy the earth radiates to space. If this were not so, the earth would have to be either heating or cooling.

The sun is a star, and, like all hot bodies, its temperature determines the wavelengths of the energy that it radiates. Figure 11-10 F 11-1● shows the wavelengths emitted by the sun, and it should not be a surprise that much of its energy is in the visible wavelengths and the somewhat shorter-wavelength **ultraviolet** as well as the somewhat longer **infrared** (or heat) wavelengths. All of these wavelengths do not reach the earth's surface. As we saw earlier, the ozone layer in the atmosphere absorbs much of the incoming ultraviolet energy and so heats that part of the atmosphere. Other gases in the atmosphere also absorb certain wavelengths, as shown in Figure 11-11. The ozone and F 11-1 oxygen absorb most of the ultraviolet, and the water vapor and carbon dioxide absorb much of the infrared. The other absorption bands are also caused by water vapor and carbon dioxide. About 20 per cent of the sun's radiation is absorbed by the atmosphere and the clouds. Most of the radiation that reaches the earth's surface is in the visible wavelengths.

Of the remaining 80 per cent of the sun's energy about 50 per cent reaches the earth's surface and is absorbed. The remaining 30 per cent is reflected back into space. *The amount of energy reflected by a surface is the* **albedo.** Earth's 30 per cent albedo is the result of 22 per cent reflected by clouds, three per cent reflected by the surface, and five per cent scattered by the atmosphere. **Scattering** *is the changing of the direction that a light ray travels by interaction with gas molecules and dust particles* (Figure 11-12). A light ray traveling toward earth may be F 11-1 scattered in any direction, and scattered light is what makes the sky bright during the day. The wavelength, or color, of the light determines the amount of scattering, and, because blue light is scattered

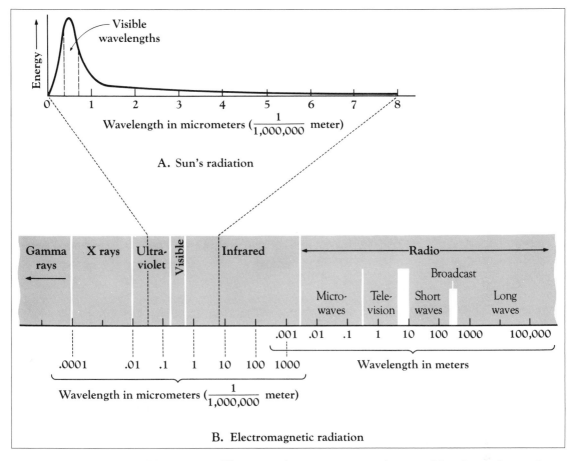

Figure 11-10 The sun radiates energy mainly in visible, ultraviolet, and infrared wavelengths. These very short wavelengths are a small part of the spectrum of electromagnetic radiation. Note that a simple scale could not be used in **B** because of the great differences in wavelengths.

Figure 11-11
Much of the sun's radiation outside of the visible wavelengths is absorbed by gases in the atmosphere. Only the radiation shown shaded reaches the earth's surface.

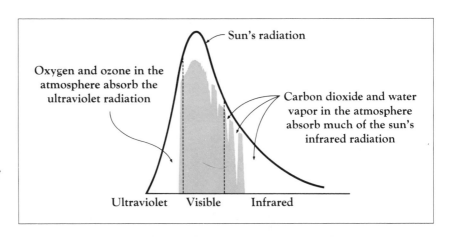

Figure 11-12
Scattering of light by molecules and dust in the atmosphere.

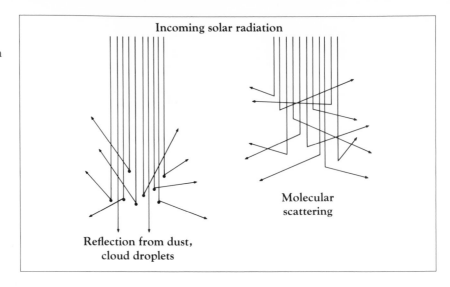

much more than red light, the sky appears blue. We see the scattered blue light coming from all over the sky, and the red light travels straight through the atmosphere.

All of these amounts of absorbed and reflected radiation are estimates averaged over the whole earth. At any point the amounts may be higher or lower, depending on whether the atmosphere is clear or cloudy. It is even possible for a place to receive more radiation than it would on a completely clear day. This can occur if the sky directly overhead is clear, so that the direct rays reach the point, and there are some clouds nearby that reflect and scatter more of the sun's energy to that place (Figure 11-13).

F 11-13

The earth, like the sun, is a warm body, and so it too radiates energy to space. The earth's temperature is much lower than the sun's, so it radiates much less energy and in longer wavelengths. The earth's radiation is in the infrared, or heat, wavelengths (Figure 11-14). The

F 11-14

Figure 11-13
Indirect radiation from clouds adds to the radiation received directly from the sun, so that a point on the surface could receive more radiation on a partly cloudy day than it can on a cloudless day.

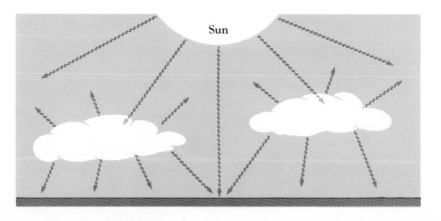

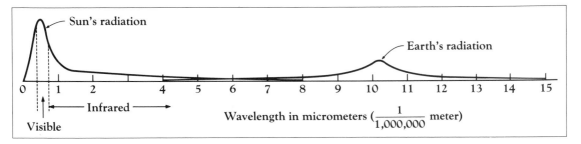

Figure 11-14 The earth is a cooler body than the sun, so its radiation is much less and at a longer wavelength than the sun's. The sun radiates a few hundred thousand times more energy than does the earth, so the amount of radiation by the earth is exaggerated here. The sun radiates mainly visible wavelengths, and the earth's radiation is at infrared wavelengths.

fact that the earth radiates long-wavelength radiation is very important because the atmosphere, particularly its water vapor and carbon dioxide, absorbs long-wavelength radiation. The atmosphere radiates this long-wavelength energy both back to earth and out into space. This process keeps a great deal of heat energy moving back and forth between the surface and the atmosphere. The process is called the **greenhouse effect** because it is similar to the way that the glass in a greenhouse was thought to let the short-wavelength radiation from the sun enter but to be opaque to the long-wavelength radiation from the interior of the warm greenhouse (Figure 11-15). The name is still F 11-15 used; but it is now known that the glass only keeps the warm air inside the greenhouse, and that glass is not opaque to long-wavelength radiation. The greenhouse effect keeps the average temperature of the earth's surface about 35°C (63°F) warmer than it would be without it. It is also the reason why the troposphere is heated from below and why the temperature falls with increase of elevation in the troposphere. If the troposphere were heated directly by the sun, the

Figure 11-15
The greenhouse effect. The short-wavelength radiation from the sun can penetrate the atmosphere, but the long wavelength radiation from the earth cannot and so is trapped, heating the atmosphere.

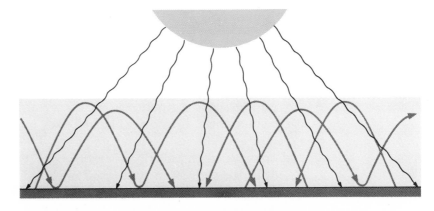

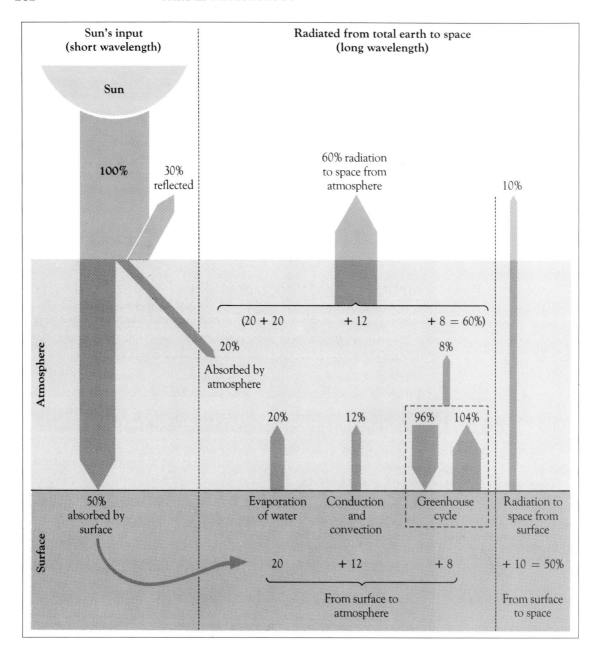

Figure 11-16
The earth's heat balance.

top would probably be the hottest part, just as the top of the thermosphere is the hottest part of the atmosphere.

Now we must account for the energy absorbed by the earth and the atmosphere. If it is not radiated back into space, the earth's temperature will increase (see Figure 11-16). Twenty per cent of the sun's short-wavelength radiation is absorbed by the atmosphere and 50 per cent by the surface, so 70 per cent must be radiated back to space.

The greenhouse effect produces another cycle. The earth radiates 104 per cent long-wavelength radiation that is absorbed by the atmosphere. Ninety-six per cent of this is, in turn, radiated back to the surface. This large amount of heat cycling back and forth between the surface and the atmosphere is what keeps them both warm. In this cycle, 8 per cent of the surface heat is transferred to the atmosphere.

Another 10 per cent of the surface's heat is radiated directly to space (Figure 11-16), leaving 32 per cent of the 50 per cent from the sun. Of this 32 per cent, 20 per cent is removed from the surface by evaporation, mainly of sea water. Evaporation of water is a process that absorbs heat, and the heat is released when the water vapor condenses in the atmosphere. (The process will be described in detail in Chapter 12.) In this way, 20 per cent of the heat is transferred from the surface to the atmosphere. The remaining 12 per cent of the surface heat is also transferred to the atmosphere, but by a different process. The air near the surface is warmed by conduction of heat from the surface, and the warm air rises, transferring the heat to the troposphere by convection. Thus conduction and convection transfer the last 12 per cent to the atmosphere.

The atmosphere now has the 20 per cent it absorbed directly from the sun, the 8 per cent from the greenhouse cycle, 20 per cent from evaporation and condensation, and 12 per cent from conduction and convection (Figure 11-16). The atmosphere's total, then, is 60 per cent, and this is radiated to space. This 60 per cent plus the 10 per cent radiated by the surface directly to space account for the 70 per cent of the sun's radiation originally absorbed by the atmosphere and the surface.

Sun's Illumination

The amount of energy received from the sun at any point on the earth varies on a yearly cycle, and this is the reason we have seasons. The cause of seasons is the tilt of earth's axis, as well as its orbit. Earth's **geographic axis** is the imaginary line connecting the north and south poles, and this is the axis about which it spins each day. Earth's spin axis is tilted about 23½ degrees from the vertical to the plane of its orbit around the sun (Figure 11-17). Earth's axis always points in the same direction as earth makes its yearly journey around the sun. The axis always points toward Polaris, the north star; of course, this is why Polaris is the north star.

Because of this tilt, the north pole is pointed at the sun when earth is at one point in its orbit; and a half-year later, the north pole is pointed away from the sun, as shown in Figure 11-17. This change in the amount of sunlight received at any point causes the seasons. Figure 11-17 shows four important dates in the year. *The day that either pole is most tilted (23½ degrees) toward the sun is called the* **solstice.** June 21 or 22, when the north pole is tilted toward the sun, is the **summer**

F 11-17

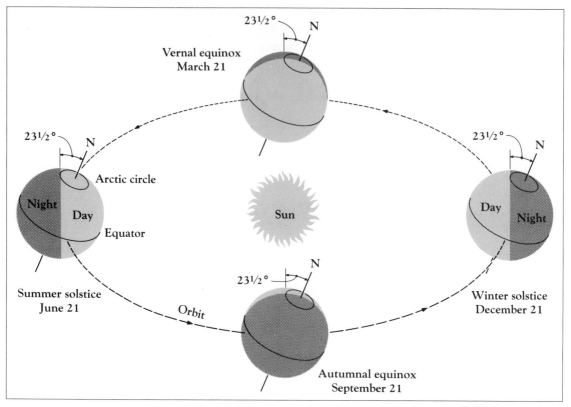

Figure 11-17
The earth's orbit and seasons.

solstice in the northern hemisphere and the **winter solstice** in the southern hemisphere. In the same way, December 21 or 22 marks the first day of winter and is the winter solstice in the northern hemisphere and the first day of summer and the summer solstice in the southern hemisphere. The other two important dates are the **vernal,** or **spring, equinox** and the **autumnal equinox.** *An equinox occurs when earth's axis is at a right angle to a line between the centers of earth and sun.* At equinox the sun is directly over the equator at noon. Equinoxes occur on March 21–22 and September 21–22; which is the first day of spring and which is the first day of fall depends on the hemisphere in which you live. If we look at how the sun illuminates earth on these four dates, their meaning and significance will become much clearer.

    Figure 11-18 shows the illumination of earth at both winter and summer solstice. For convenience in the rest of this discussion, only the northern hemisphere will be described. At the summer solstice, earth is tilted 23½ degrees toward the sun. In this position the area around the north pole stays in sunlight as earth makes its daily spin on its axis; and, in the same way, the area around the south pole stays entirely in the dark. The latitude circles on earth marking the boundaries of all-daylight and all-darkness are termed the **arctic circle** and the **antarctic circle,** and they are 66½ degrees north and 66½ degrees

F 11-18

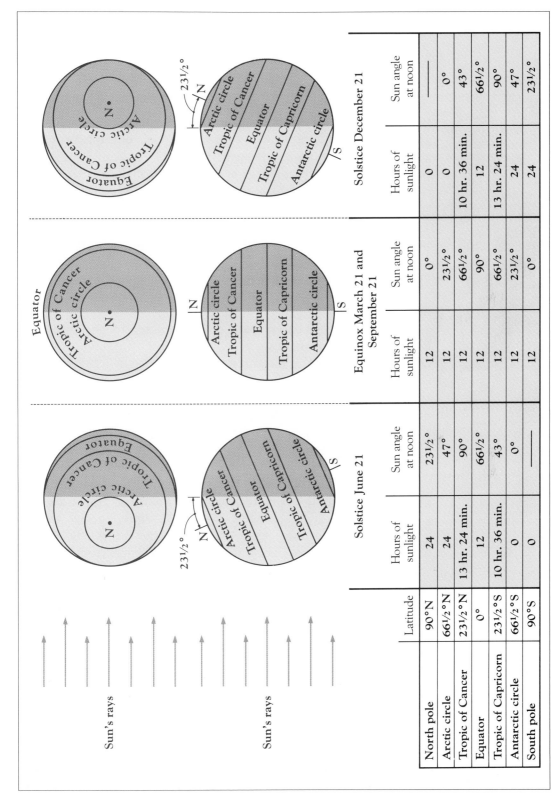

Figure 11-18 The illumination of the earth at solstice and equinox. (Refer also to Figures 11-17 and 11-19.)

Latitude	Solstice June 21		Equinox March 21 and September 21		Solstice December 21	
	Hours of sunlight	Sun angle at noon	Hours of sunlight	Sun angle at noon	Hours of sunlight	Sun angle at noon
North pole	24	23½°	12	0°	0	—
Arctic circle	24	47°	12	23½°	0	0°
Tropic of Cancer	13 hr. 24 min.	90°	12	66½°	10 hr. 36 min.	43°
Equator	12	66½°	12	90°	12	66½°
Tropic of Capricorn	10 hr. 36 min.	43°	12	66½°	13 hr. 24 min.	90°
Antarctic circle	0	0°	12	23½°	24	47°
South pole	0	—	12	0°	24	23½°

south latitude. The equator is 0 degrees latitude, and the poles are 90 degrees north and south latitude. The arctic and antarctic circles are at 90 degrees minus earth's tilt, 23½ degrees; that is, 90 − 23½ = 66½. Notice also that, at the summer solstice, the sun's rays fall perpendicularly to earth at latitude 23½ degrees north, and this is called the **Tropic of Cancer.** On that day at that latitude, there would be no shadows at noon. The same is true at the winter solstice at 23½ degrees south latitude on what is called the **Tropic of Capricorn.**

At either equinox, as shown in Figure 11-18, both poles receive sunlight. As earth spins on its axis, all parts of the planet have 12 hours of sunlight and 12 hours of darkness. The word *equinox* means equal night.

The variations in the number of hours of sunlight through the year are summarized in Figure 11-18. At each equinox every place gets 12 hours of sunlight. At each solstice the summer hemisphere gets more than 12 hours, and the winter hemisphere gets less than 12 hours. The greatest yearly variation is near the poles, and the least is near the equator.

The ecliptic, or apparent path of the sun seen by an observer on earth, changes with the seasons. The position of the ecliptic does not change relative to the distant stars, but its height as viewed from earth changes because of earth's tilt. At the winter solstice the sun rises in the southeast and sets in the southwest. At equinox it rises due east and sets due west, and its path through the sky is higher. At the summer solstice it rises in the northeast and sets in the northwest, and its path through the sky is at its highest point (Figure 11-19).   F 11-1

At equinox the sun's elevation angle at noon is 90 degrees minus the latitude (see Figure 11-18). At summer solstice its angle at noon is 23½ degrees more because, in effect, earth's axis has dipped 23½ degrees toward the sun (see Figure 11-18). At winter solstice the axis points 23½ degrees away from the sun, so the sun is 23½ degrees lower than at equinox. The total change in sun angle from solstice to solstice is 47 degrees. This change can be used to keep a house warm in winter and cool in summer by proper design of south-facing windows with overhanging roofs, as shown in Figure 11-20.   F 11-2

Heat Transport

As we have seen, all parts of the earth do not receive the same amount of radiation from the sun. The equatorial regions receive much more energy than the poles. Nevertheless, the geologic and historical records show that the poles are not getting colder and the tropics are not getting warmer.

Measurements show that the tropics receive more energy than they radiate to space and that the poles radiate more energy than they

Figure 11-19
The path of the sun through the sky at solstice and equinox. Note that the sun's height above the horizon at noon and the position of sunrise and sunset both change with season.

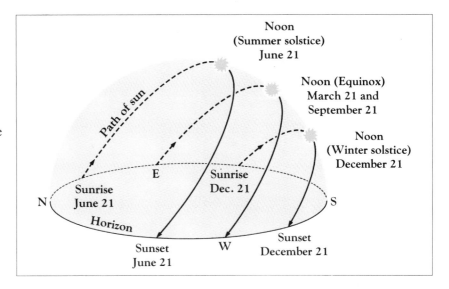

receive. The energy must be moved from the tropics toward the poles to keep the temperature distribution from changing. This circulation is the cause of weather and climate. The atmosphere and the oceans both transport heat from the tropics toward the poles. Most of the transporting is done by the atmosphere, but a significant amount is moved by the oceans. Some estimates of oceanic heat transport run as high as 25 per cent, but the total is probably less. However, oceanic currents move much slower than atmospheric currents. From the equator north and south to about 30 degrees latitude, large convection cells in the atmosphere carry the heat poleward. In the temperate regions, storm systems transport the energy. The details of these processes are the subjects of the next chapters.

Surface air temperatures reveal more about the sun's heating (Figure 11-21). The temperatures do not vary regularly from equator to pole, as diagrams of solar heating might suggest. Instead, the highest and lowest temperatures are found on the continents. This occurs

F 11-21

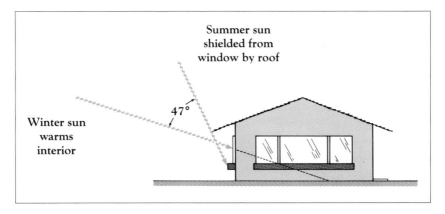

Figure 11-20
Proper design of roof overhang for a south-facing window.

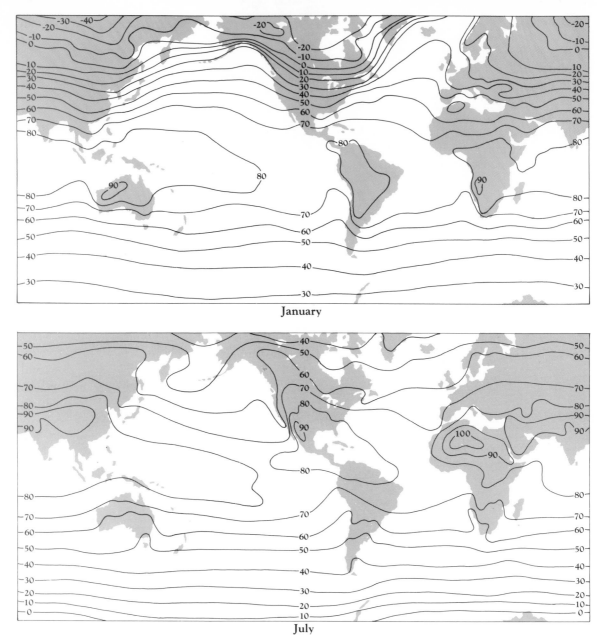

Figure 11·21 Average temperatures in January and July, in degrees Fahrenheit.

because the temperatures of rock and soil vary much more than the temperature of the oceans. Two variables are involved in determining the temperature of materials exposed to the sun. They are the **albedo,** a measure of how much of the radiation is reflected; and the **specific heat,** *a measure of how much heat is required to raise one gram of a material* *one degree Celsius* and thus a measure of how much heat a material can absorb. Sea water has a very low albedo (very little reflection) for vertical rays and a very high albedo (much reflection) for low-angle sun. Water has a very high specific heat, so its temperature increases a small amount when it absorbs much heat energy. Soil and rock have much lower specific heats, and so their temperature changes a great deal for a relatively small change in stored heat. The large heat capacity of the large bodies of water is the reason that temperatures change the least, both daily and seasonally, near oceans or other large water bodies.

Temperature Lag

The northern hemisphere receives its maximum amount of the sun's energy on June 21st and its minimum amount on December 21st, as shown in Figure 11-17. However, the hottest part of the summer in most places is in late July or in August, and the coldest part of winter is in January or February. A similar observation can be made on a daily basis; the hottest part of the day is in mid- to late afternoon, not at noon when the sun's input is highest. The reason for the lag is that the temperature of the air will continue to rise as long as the total heat input exceeds the outgo. As the day or the summer goes on, the ground is heated and its temperature rises. After the peak of the sun's input, at noon on June 21st, the ground radiates its heat to the air and so heats the air. The temperature continues to rise as long as the sun's input is more than the heat lost by radiation. After that point, the temperature begins to fall. Of course, other energy inputs are involved as well as ground radiation.

SUMMARY

- *Meteorology* is the study of the atmosphere and its changes and movements.

- *Weather* is the short-term changes in the atmosphere.

- *Climate* is the typical combination of weather elements that characterizes an area over a long time, taking into account seasonal changes as well as ranges and extremes of weather elements.

- *Air* is a mixture of 78 per cent nitrogen, 21 per cent oxygen, and nearly 1 per cent argon and other gases. The mixture is uniform up to about 100 kilometers (60 miles).

- Temperature differences at varying altitudes are used to define the layers of the atmosphere.

- The lowest layer is the *troposphere,* in which almost all weather systems and storms are found. The top of the troposphere is called the *tropopause.* In the troposphere the temperature falls as elevation increases, and at the tropopause the temperature is about −60°C (−75°F). The troposphere is about 10 kilometers (6 miles) thick in temperate regions and is higher in summer than in winter and higher in tropical areas than near the poles.

- The *stratosphere* extends from about 10 to 50 kilometers (6 to 30 miles). The temperature is constant in the lower stratosphere and increases in the upper part to about 0°C (32°F) at the *stratopause*, or top of the stratosphere.

- The *ozone layer* is in the stratosphere, and the maximum concentration of about 5 parts per million occurs at about 30 kilometers (18 miles). The ozone layer prevents ultraviolet radiation from the sun from reaching the earth's surface.

- The *mesosphere* is above the stratosphere and is a layer in which the temperature falls. The *mesopause* occurs at an elevation of about 85 kilometers (50 miles), where the temperature is about −85°C (−120°F).

- The *thermosphere* is the next layer, and it extends several hundred kilometers. The temperature rises in the thermosphere.

- The *kinetic theory* assumes that all matter is composed of molecules that are in motion.

- *Heat* is a form of energy and is proportional to the total motion or kinetic energy of the molecules of an object.

- *Temperature* is a measure of the intensity of heat and is proportional to the average energy of each molecule.

- In *conduction*, heat is transferred through a material by molecular collisions.

- In *convection*, heat is transferred by actual movement of a fluid. When warm fluid rises or cool fluid sinks, convective flow results.

- *Radiation* is the transfer of energy by electromagnetic waves.

- The earth's temperature has remained constant through much of geologic time, indicating that the earth is neither heating nor cooling. The earth maintains this *heat balance* because it radiates as much heat into space as it receives from the sun.

- The earth is a cooler body than the sun, so its radiation is at a longer wavelength than the sun's. Water vapor and carbon dioxide trap some of this long-wavelength radiation by the *greenhouse effect*.

- Seasons are caused by the earth's 23½-degree tilt and its orbit around the sun. At *solstice* the axis is inclined toward the sun so that the region near one pole is illuminated and the region near the other pole is in darkness. At *equinox* both poles are illuminated, so every place on earth has 12 hours of sunlight.

- Tropical regions receive much more energy from the sun than the polar regions. The atmosphere and, to some extent, the oceans transport heat from the tropics toward the poles. The resulting circulation is the cause of climate and weather.

- The hottest and coldest parts of the year or day do not occur at the times of maximum or minimum solar radiation, because heat absorbed and then radiated by the earth heats the air.

KEY TERMS

meteorology	mesosphere	radiation	equinox
weather	mesopause	ultraviolet radiation	vernal, or spring,
climate	thermosphere	infrared radiation	equinox
troposphere	heat	albedo	autumnal equinox
tropopause	temperature	scattering	arctic circle
stratosphere	kinetic theory	greenhouse effect	antarctic circle
stratopause	conduction	geographic axis	Tropic of Cancer
ozone layer, or	insulator	solstice	Tropic of Capricorn
ozonosphere	convection, or	summer solstice	specific heat
ozone	convective flow	winter solstice	

REVIEW QUESTIONS

1. What is the difference between weather and climate?
2. What gases compose most of the atmosphere?
3. In which layer of the atmosphere does almost all weather occur? Describe the temperature characteristics of this layer.
4. Describe how ozone forms.
5. Explain the effect ozone has on our planet. Why are we not poisoned by the gas?
6. Where is the stratosphere located? Contrast the temperature of the lower part of the stratosphere with that of the upper part.
7. Name the top two layers of the atmosphere. What temperatures occur in them?
8. Heat, the sun's radiation, and the motion of molecules are all forms of what?
9. What does temperature measure? Why does the thermosphere have high temperatures but low heat energy?
10. Heat energy always moves from what areas to what other areas? At what point does the movement stop?
11. Describe conduction. Do you think much heat is transferred within the atmosphere by conduction? Could some heat be transferred by conduction from the ground to the air along the surface?
12. Describe convection. Is convection likely to be an important method by which heat moves through the atmosphere?
13. Describe radiation.
14. Only 50 per cent of the sun's radiant energy is absorbed by the surface of the earth. What happens to the other 50 per cent?
15. Why is the sky blue?
16. (a) Draw a diagram of the greenhouse effect.
    (b) A small amount of the long-wave radiation escapes from the "greenhouse" and is radiated into space. If the amount of carbon dioxide in the atmosphere were to increase, what would be the effect on the earth's temperature?
17. Write a balance sheet to account for the energy absorbed by the atmosphere and the surface of the earth.

SUGGESTED READINGS

Barrett, E. C. *Viewing Weather from Space.* London: Longmans, 1967, 140 pp.

Gosnell, Mariana. "Ozone—The Trick Is Containing It Where We Need It," *Smithsonian* (1975), Vol. 6, pp. 49–55.

Hosler, C. L. "Of Wizardry, Witches, and Weather Modification," *Weatherwise* (1968), Vol. 21, pp. 110–113.

Inwards, R. *Weather Lore,* 4th ed. London: Reider, 1950, 251 pp.

Landsberg, H. E. *Weather and Health.* Garden City, N.Y.: Doubleday, 1969, 148 pp.

Lowry, W. P. *Weather and Life: An Introduction to Biometeorology.* Corvalis: Oregon State University Book Store, 1967, various paging.

Middleton, W. E. K. *A History of the Theories of Rain.* New York: Franklin Watts, 1965, 223 pp.

Oort, A. H. "The Energy Cycle of the Earth," *Scientific American* (September 1970), Vol. 223, No. 3, pp. 54–63.

Ross, F. X., Jr. *Weather: The Science of Meteorology from Ancient Times to the Space Age.* New York: Lothrop, Lee & Shepard, 1965, 200 pp.

Sloane, E. *Folklore of American Weather.* New York: Meredith Press, 1963, 63 pp.

Stewart, R. W. "The Atmosphere and the Ocean," *Scientific American* (September 1969), Vol. 221, No. 3, pp. 76–86.

Whitnah, D. R. *A History of the United States Weather Bureau.* Urbana: University of Illinois Press, 1961, 267 pp.

Woodwell, G. M. "The Carbon Dioxide Question," *Scientific American* (January 1978), Vol. 238, No. 1, pp. 34–43.

HUMIDITY, RELATIVE HUMIDITY, AND DEW POINT

Measurement of Relative Humidity

CHANGE OF STATE AND LATENT HEAT

LAPSE RATES AND VERTICAL STABILITY

CONDENSATION

CLOUDS

Fog

Vertical Motion and Cloud Formation

PRECIPITATION

Formation

Types of Precipitation

Measurement of Precipitation

CHAPTER 12

MOISTURE IN THE ATMOSPHERE

Typhoon (hurricane) in the Pacific Ocean. (Photo from NASA.)

In this chapter we will be dealing with topics familiar to us all: humidity, clouds, fog, and precipitation. But what is high humidity other than an uncomfortable and clammy feeling? How do ice, snow, dew, and rain form out of the atmosphere? How are clouds and fog formed? And what are the hidden phenomena that explain what we experience as our daily weather?

HUMIDITY, RELATIVE HUMIDITY, AND DEW POINT

Humidity is a measure of the amount of water vapor in the atmosphere. The amount of water vapor in the atmosphere varies between 0 to 4 per cent (4 grams per 100 grams air). There are several ways of expressing humidity. One is **absolute humidity,** the weight of water vapor in a given volume of air, expressed as grams of water vapor per cubic meter of air. A second, and more useful, way is **relative humidity,** *which is the ratio of the amount of water vapor actually in the air to the maximum amount of water vapor that the air can actually hold.* Relative humidity is expressed as a percentage. The confusing aspect of relative humidity is that the maximum amount of water vapor that the air can hold changes with the temperature. *Warm air can hold more water vapor than cold air.* Thus, suppose that a parcel of air is isolated from the rest of the atmosphere so that no water vapor can be added or taken away from it. If this parcel of air is heated, the relative humidity decreases because warm air can hold more water vapor than cooler air. In the same way, if the parcel is cooled, the relative humidity increases because cool air can hold less moisture than can warm air. In this example, the absolute humidity stayed the same, but the relative humidity changed because the temperature changed.

Another measurement related to humidity is the dew point temperature. The **dew point** *is the temperature at which the air becomes saturated, or, said another way, the relative humidity is 100 per cent.* This concept is easily understood from the last example. If the temperature of an isolated air parcel decreases, the relative humidity increases. If the cooling continues long enough, a temperature will be reached at which the relative humidity becomes 100 per cent. This temperature is the dew point temperature. At the dew point temperature the air is saturated; that is, it contains all the water vapor it can hold.

If the air temperature and the dew point temperature are known, the relative humidity can be determined; and if the relative humidity and the air temperature are known, the dew point can be determined. If the dew point and the air temperature are close to the same, the air is nearly saturated and the relative humidity is high. If the dew point and the air temperature are very different, the relative humidity is low.

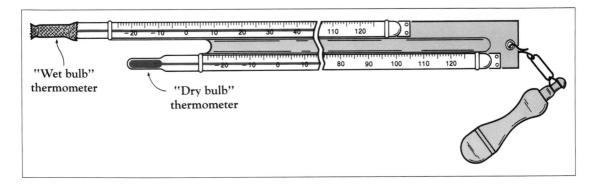

Figure 12-1 The sling psychrometer consists of wet bulb and dry bulb thermometers, and a means to spin them.

Measurement of Relative Humidity

Direct measurement of the amount of water vapor in the air is difficult, but relative humidity can be measured fairly easily. The simplest and least accurate instrument is the **hygrometer.** Most hygrometers use as their active element human hair or other fibers that change length. A system of levers moves a pointer over a scale calibrated from 0 to 100 per cent.

The instrument best suited to measure relative humidity is the **sling psychrometer** (Figure 12-1). This instrument consists of two F 12-1 identical thermometers, one of which is fitted with a piece of cloth wrapped around its bulb. The two thermometers are fastened together and have a handle arranged so that they can be spun. In use, the cloth is wet with water on what is called the *wet-bulb thermometer;* the other thermometer is called the *dry-bulb thermometer.* After wetting, the thermometers are spun. The dry-bulb thermometer simply reads the air temperature. The temperature of the wet-bulb thermometer is lowered by evaporation. The sling psychrometer is spun until the wet-bulb temperature stabilizes. The amount that the wet-bulb temperature is lowered or depressed depends on the amount of evaporation. The amount of evaporation is determined by the temperature and the amount of water vapor in the air. If the air has very little water vapor, more evaporation will occur and the wet-bulb temperature will be depressed more; but if the air is nearly saturated, the wet-bulb temperature will be depressed very little. Thus, if the relative humidity is high, the wet and dry bulbs will read nearly the same; but if the relative humidity is low, the temperatures will be very different. To determine the actual relative humidity, a table such as that in Appendix D is used. A similar table, also in Appendix D, can be used to determine the dew point from psychrometer readings.

CHANGE OF STATE AND LATENT HEAT

When the relative humidity reaches 100 per cent, the gaseous water vapor in the air condenses into tiny liquid-water droplets. As we will see in a later section, the process is somewhat more complicated, but for now, this will serve as an example of a **change of state.** The three states in which water exists on earth are gaseous, water vapor; liquid, water; and solid, ice. As water changes from one state to another, heat is either absorbed or released. Such heating and cooling are very important in producing weather.

A simple experiment will show that energy, in the form of heat, is involved in a change of state (Figure 12-2). If a container of ice and F 12-2 water is well mixed, the temperature of the mixture will be 0°C (32°F). If this container is heated, the temperature will remain at 0°C until all of the ice is melted. During this time heat is being added to the mixture. After the ice is melted, the temperature of the water will increase until boiling begins at 100°C (212°F). The water will remain at 100°C until it is entirely boiled away (changed to a gas). It will take much longer to boil the water than it did to melt the ice or to raise the water temperature to 100°C. The heat energy required to cause these changes of state is called **latent** (hidden) **heat** because the heat added does not cause the temperature to change.

It is not hard to understand latent heat. Ice is a solid, and the molecules are bonded together to form its crystal structure. To melt the ice, heat must be added to break the bonds. The heat causes the molecules to vibrate more and more until the bonds are broken (Figure 12-3). The added heat has gone into increased molecular move- F 12-3

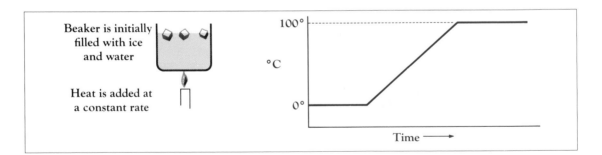

Figure 12-2 In this experiment heat is added at a constant rate. The temperature remains at 0°C until all of the ice is melted. Then the temperature of the water rises at a steady rate until boiling begins at 100°C. The water remains at 100°C until it boils away. The heat required for melting and boiling (change of state) is called *latent heat* because the temperature does not change.

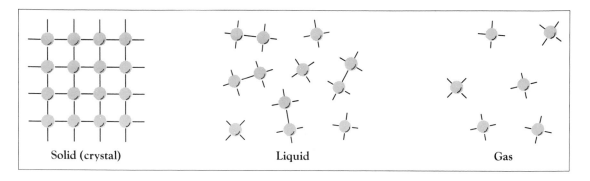

| Solid (crystal) | Liquid | Gas |

Figure 12-3
When a solid melts, the heat energy that is added causes the molecules to vibrate and so breaks many of the bonds. In a gas, the molecular motion has broken all of the bonds.

ment. To refreeze the water, this heat must be removed so that the molecules can bond together and form ice again. Thus, the latent heat is added when ice melts and is released when ice forms. A similar explanation can be given for boiling. Heat energy must be added to the water molecules until they have enough kinetic energy to escape from the liquid. This kinetic energy is released when the molecules are slowed in condensation back to liquid water.

The amount of energy involved in these processes is large. The unit of energy used in such studies is the **calorie.** The *calorie is the amount of energy necessary to increase the temperature of one gram of water one degree Celsius.* At 0°C, the freezing point, it requires 80 calories to melt 1 gram of ice or freeze 1 gram of water. At the boiling point, 540 calories must be added to vaporize 1 gram of water, and 540 calories are released when 1 gram of steam condenses.

Similar changes of state occur in the atmosphere, but at lower temperatures, especially in the case of **evaporation** and **condensation.** The amount of latent heat depends on the temperature, and, in general, about 600 calories are absorbed by each gram of water evaporated from the earth's surface, and 600 calories are released to heat the atmosphere for each gram of water vapor that condenses in the atmosphere. These processes and their terminology are summarized in Figure 12-4.   F 12–4

Evaporation and condensation are very important in atmospheric processes. Much of the earth's surface is covered with oceans, especially near the equator, where much of the sun's energy falls on the earth. The sun's energy that reaches these oceans is absorbed by the water near the surface. The temperature of the ocean surface would rise because of this energy input if evaporation did not occur. Evaporation removes heat from the ocean, just as your evaporating perspiration cools you on a warm day. The heat removed from the ocean warms the atmosphere when condensation occurs, and eventually the water is returned to the ocean by precipitation. The circulation of the atmosphere is much more rapid than the circulation of the oceans; so

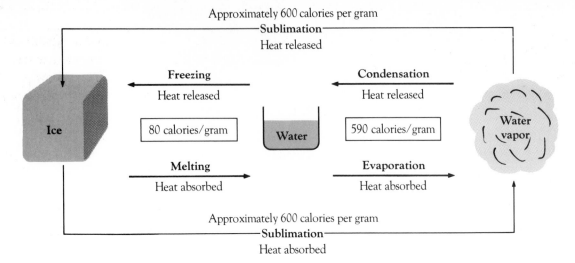

Figure 12-4 Energy absorbed and released with changes of state of water. The energy involved in changes directly between ice and water vapor depends on the temperature at which the change occurs.

once the sun's energy is transferred to the atmosphere by evaporation, it is moved away from the equator. Without this process, the equatorial regions would be warmer and the polar regions colder.

LAPSE RATES AND VERTICAL STABILITY

A temperature **lapse rate** *is the rate of decrease of temperature with increasing height in the atmosphere.* Several different lapse rates are of interest to us. The **environmental lapse rate** *is simply the average rate at which temperature drops with elevation.* The environmental lapse rate is 0.65°C per 100 meters (3.5°F per 1000 feet). This means that, if a thermometer is sent up with a balloon or by other means, this is the rate of change of temperature that would be recorded. The actual lapse rate at any place may depart widely from this average. At times the *temperature may increase with height in what is termed a* **temperature inversion.** In a temperature inversion, cold, heavy air near the surface may trap smoke and other pollutants.

The lapse rates of rising parcels of air are of importance in determining their stability. If an air parcel is warmer than the surrounding air, it will rise; if cooler, it will sink. As an air parcel rises, it expands because the atmospheric pressure decreases with elevation. When a gas expands, it also is cooled; if compressed, a gas is heated. *If no heat is gained or lost by the air parcel, the process is said to be* **adiabatic.** An **adiabatic lapse rate** *is the temperature change in a rising or falling air par-*

BOTTOM

When this lava flow cooled, it contracted, forming the joints or cracks that outline the columns. Devil's Postpile, California. Photo by James Behnke. (page 63)

TOP RIGHT

The top of the Devil's Postpile, showing the outlines of the columns. Glaciation caused the scratches and polish of this surface. (pages 63 and 167)

TOP LEFT

Petrified trees on Specimen Ridge in Yellowstone National Park, Wyoming. The trees were buried by pyroclastic rocks and were uncovered by erosion.
(pages 63 and 184)

TOP LEFT

Plunging wave. Photo from Hawaii Visitors Bureau. (page 250)

BOTTOM LEFT

Coastal scenery in Oregon. An arch is in the foreground. The cliffs and the flat probably wave-cut surfaces suggest uplift. (page 257)

TOP RIGHT

Stratocumulus clouds. Photo from The National Center for Atmospheric Research, Boulder, Colorado. (page 303)

TOP FAR RIGHT

Altocumulus clouds, forming in lenticular waves. Photo from The National Center for Atmospheric Research, Boulder, Colorado. (page 303)

BOTTOM RIGHT

A thunderhead and storm on the Great Plains in Colorado. Photo from The National Center for Atmospheric Research, Boulder, Colorado. (page 347)

TOP LEFT

The surface of Mars is strewn with rocks. This image was returned by Viking, a spacecraft that landed on Mars. Photo from NASA. (page 437)

BOTTOM LEFT

Jupiter, showing the bands and turbulence in its atmosphere. The Great Red Spot, probably a storm-like feature, is prominent. Photo from NASA. (page 443)

BOTTOM RIGHT

A composite photo showing Saturn and several of its satellites. Photo from NASA. (page 446)

TOP RIGHT

An erupting volcano on Io, a moon of Jupiter. Io is the only object, except for Earth, on which volcanic eruptions have been observed. The volcanoes erupt sulfur, giving Io its colors. Photo from NASA. (page 450)

ABOVE

Aurora borealis,
or northern lights.
Photo from the
National Center
for Atmospheric
Research, Boulder,
Colorado. (page
474)

cel; it is different from an environmental lapse rate, which is the change in temperature with height in the stationary atmosphere.

There are two adiabatic lapse rates. The **dry adiabatic lapse rate** is 1°C per 100 meters (5.4°F per 1000 feet). The dry adiabatic lapse rate is the rate of cooling of a rising air parcel as long as no condensation occurs in that air parcel. If condensation occurs, the latent heat of condensation, about 600 calories per gram, is released, and the lapse rate is much lower. The **wet,** or **saturated, adiabatic lapse rate** varies slightly and is about 0.6°C per 100 meters (3°F per 1000 feet).

We will consider the dry adiabatic lapse rate first. If the environmental lapse rate is at its average value of 0.65°C per 100 meters, and if a parcel of air rises and cools at the dry adiabatic lapse rate of 1°C per 100 meters, the rising air parcel will always be cooler than the

Figure 12-5
Stability of rising air depends on the lapse rate.

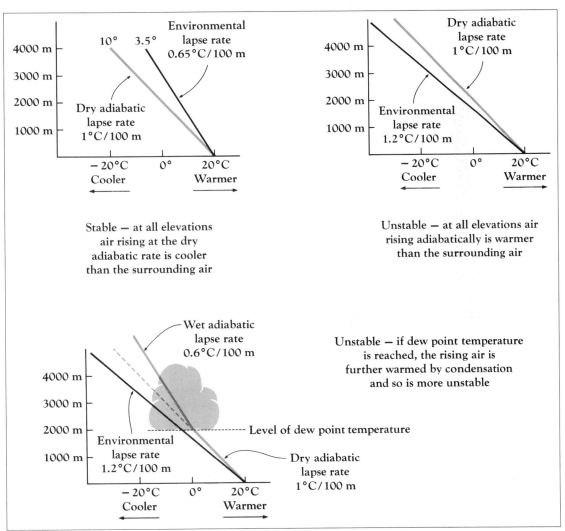

surrounding air. For every 100 meters the parcel rises, it drops 1°C, but the surrounding air drops only 0.65°C. The air parcel will be colder and heavier and, therefore, will tend to sink back to its original height. Such an air parcel is termed **stable,** because it tends to return to its original position. If the environmental lapse rate is greater than average—say, 2°C per 100 meters—a rising air parcel cooling at the dry adiabatic rate of 1°C per 100 meters will be **unstable.** The rising air parcel will be warmer than the surrounding air and so will continue to rise. Examples are shown in Figure 12-5. If the environ- F 12-5 mental lapse rate is greater than the adiabatic lapse rate, the air is usually unstable; if the environmental lapse rate is less than the adiabatic lapse rate, the air is usually stable.

As an air parcel rises, it is cooled. As it cools, the relative humidity rises because cool air can hold less water vapor than warm air. Eventually an elevation will be reached at which the relative humidity is 100 per cent. The temperature at this height is the dew point temperature, and, at or near this elevation, condensation occurs. This condensation level is also the elevation of the cloud base, or bottom. When condensation occurs, the latent heat of condensation is released, and this heat warms the air parcel. The warming, in turn, reduces the adiabatic lapse rate, and this is why the wet adiabatic lapse rate is less than the dry rate. For this reason, moist air is more likely to be unstable than dry air.

CONDENSATION

When the relative humidity reaches 100 per cent, condensation will occur if there is a surface available on which water vapor can condense. Near the ground many surfaces are available, and condensation is common in the form of **dew.** If the temperature of condensation is below freezing, **frost** forms (**frost point temperature**). A cold surface, such as a glass of ice water, will be the site of condensation if the air next to its surface is cooled below the dew point temperature.

In the atmosphere, surfaces are also necessary for condensation to occur. If the surfaces are few, the air will become **supersaturated;** that is, *the relative humidity will be more than 100 per cent and the temperature will be below the dew point.* The condensation surfaces in the atmosphere are in the form of tiny **condensation nuclei.** These nuclei come from salt spray from the ocean, dust carried aloft by winds, and smoke and other particulates, mainly from human activities. Typically the nuclei are very tiny, less than one one-thousandth (0.001) of a millimeter in diameter, and up to several thousand are present per cubic centimeter. Some of these *nuclei* are *composed of water-seeking materials such as salt* and *are called* **hygroscopic nuclei.** Some hygroscopic nuclei can initiate condensation when the relative humidity is less than 100 per cent, and this causes **haze.** Near industrial areas nuclei,

Figure 12-6
Relative sizes of rain drops and cloud droplets

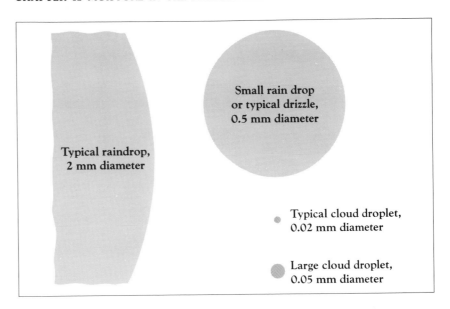

Typical raindrop,
2 mm diameter

Small rain drop
or typical drizzle,
0.5 mm diameter

Typical cloud droplet,
0.02 mm diameter

Large cloud droplet,
0.05 mm diameter

in some cases hygroscopic nuclei, may be abundant, and this may account for the thicker, more persistent fogs near some cities.

When condensation occurs in the atmosphere, droplets form around many nuclei at almost the same time. The water vapor around each condensation nucleus is attracted to the nucleus' droplet, and so no droplet can become much larger than the others. This is one reason why tiny droplets form and clouds are produced. If droplets heavy enough to fall were to form, rain or some other type of precipitation would be the result. Typically, the air in a cloud is rising, and this also prevents the droplets from falling. In ordinary clouds the droplets range in size from a few thousandths to 50 thousandths of a millimeter in diameter, and each cubic centimeter contains between 40 and 600 droplets (Figure 12-6).   F 12-6

CLOUDS

Clouds are classified by their shape or appearance and the elevation at which they occur (Figure 12-7). Only a few terms need to be F 12-7
learned for a simple classification of common clouds. As might be expected, this simple classification can be further subdivided, resulting in many terms for less common clouds or variants of common clouds. Clouds are divided into four families: the low, medium, and high clouds, and clouds of vertical development. There are three main shapes: stratus, cumulus, and cirrus. **Stratus clouds** *are layered.* **Cumulus clouds** *look like cotton, generally with flattish bottoms and rounded tops.* **Cirrus clouds** *are generally thin and wispy.*

Cloud classification is based on elevation above the earth's surface, not elevation above sea level. The elevations given here are for the

Stratus

Nimbostratus

Stratocumulus

Altostratus

Altocumulus

Altocumulus

Cirrus

Cirrocumulus

Cirrostratus

Cumulus

Cumulonimbus

Figure 12-7
Cloud types. (Cirrus, altostratus, and cumulonimbus photos from the National Center for Atmospheric Research, Boulder, Colorado; all others from NOAA.)

temperate regions. The low-cloud family is found from ground level up to 2 kilometers (6500 feet) above the surface. The low clouds form layers. *A cloud layer very close to the ground is called* **fog.** If it is higher, the term **stratus,** *meaning layered,* is used. *A stratus, or layered, cloud from which rain or other precipitation is falling is called a* **nimbostratus cloud.** (The prefix **nimbo-** *means rainstorm.*) Nimbostratus clouds are generally thicker and darker than stratus clouds. The last member of the low-cloud family is the stratocumulus cloud. **Stratocumulus clouds** *have flat bottoms and rounded, puffy tops.* They may appear as long ridges, and they commonly form in the clearing stages of a storm.

Medium clouds are found between 2 kilometers (6500 feet) and 7 kilometers (23,000 feet) above the ground. *The prefix* **alto-** *is used to designate members of the medium-cloud family.* **Altostratus clouds** *are generally gray layered clouds through which the sun can be seen.* In some cases they may be the source of precipitation. **Altocumulus clouds** *appear as a layer of puffy cumulus clouds.* They are generally associated with fair weather, although they may indicate an advancing cold front.

The high-cloud family is found above 7 kilometers (23,000 feet) from the ground and is made up of the various cirroform clouds. These thin clouds are composed of tiny ice crystals. **Cirrus clouds** *are thin, wispy streaks, commonly called mares' tails.* **Cirrocumulus clouds** *resemble a layer of cotton balls.* In some instances they look like fish scales and are termed mackerel sky. **Cirrostratus clouds** *are high, thin layers through which the sun can penetrate easily, although a halo may form.* The halo is created by refraction by the ice crystals that form the cloud. Cirrus clouds generally imply fair weather, but those that change to cirrostratus clouds may be the leading edge of a storm.

The vertical-cloud family is so called because its members may develop to great height and overlap one or more of the other cloud families. **Cumulus clouds** are the most common of the vertical clouds. Typical fair-weather cumulus clouds are the puffy cotton balls that may form on a warm summer afternoon. Cumulus clouds may grow to great heights, but all have flat bottoms. The biggest cumulus clouds are **cumulonimbus clouds** or thunderheads. These clouds may grow to 18 kilometers (11 miles) high and typically have anvil-shaped tops. Their presence implies strong vertical air currents aloft and thunder, lightning, and the possibility of sudden heavy rain and hail showers at ground level.

Fog

Fog is a cloud at or nearly at ground level. It can be a severe problem for any type of transportation if it is thick or dense and very close to the ground. Fog can form in a number of ways. **Radiation fog** (Figure 12-8A) has its start on clear, calm nights. The ground surface loses its   F 12-8 heat to the atmosphere by radiation, and the cool, heavy air near the ground drains downslope to the low areas. If the air is moist, it may be cooled to its dew point, and fog is the result. Radiation fog typically forms at night in valleys and burns off the next morning.

Advection fog (Figure 12-8B) forms when moist air moves across a cooler surface. The moist air is cooled to its dew point, and condensation results. The cool surface may be land or water, with water more common. Gentle winds are necessary for advection fog, because strong winds will dissipate the fog as fast as condensation occurs.

Evaporation fog (Figure 12-8C), also called **sea smoke** or **steam fog,** is similar to advection fog, except cold dry air moves slowly over a warm, moist surface. Evaporation brings water vapor into the cold air, where it condenses, forming fog.

Vertical Motion and Cloud Formation

Clouds generally form because air parcels rise and cool and so cause condensation. The air must be forced to rise to start this process.

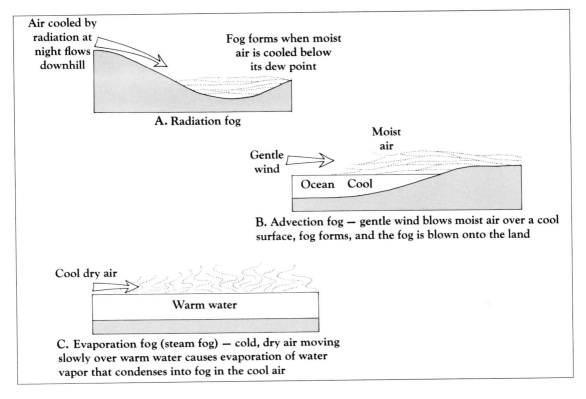

Air cooled by radiation at night flows downhill

Fog forms when moist air is cooled below its dew point

A. Radiation fog

Gentle wind

Moist air

Ocean   Cool

B. Advection fog — gentle wind blows moist air over a cool surface, fog forms, and the fog is blown onto the land

Cool dry air

Warm water

C. Evaporation fog (steam fog) — cold, dry air moving slowly over warm water causes evaporation of water vapor that condenses into fog in the cool air

Figure 12·8
Types of fog.

Convection created by differential heating of the ground is one cause and is especially important in the formation of cumulus and cumulonimbus clouds (Figure 12-9).

 The other ways in which air parcels are lifted depend on horizontal winds. **Orographic lifting** *occurs where wind forces air to rise over mountains* (Figure 12-10). The rising air is cooled, causing clouds or even

F 12–9

F 12–10

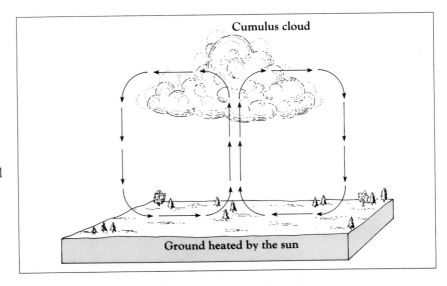

Cumulus cloud

Ground heated by the sun

Figure 12·9
Air in contact with a warm surface is heated and rises. Cool air from above descends to replace the rising air. The rising air is cooled, and condensation forms clouds.

Figure 12-10
Orographic lifting. Winds force warm air to rise over mountains. The rising air is cooled, and condensation causes clouds and, in some cases, precipitation.

Descending warm, dry air

rain to develop on the windward side of the mountain range. On the lee side of the mountains, the descending dry air is warmed by compression. Along the west coast of the United States, the prevailing winds are from the west off the Pacific Ocean. These moist winds rise over the mountains, producing clouds and rain on the west slopes; east of the mountains the climate is hot and dry.

Perhaps the most important way that clouds form is by warm air being driven by winds over cooler, heavier air (see Figure 14-4, p. 342). This process is involved in most storms in middle latitudes and will be discussed in Chapter 14.

PRECIPITA-TION

Formation

Once cloud droplets form, other processes must cause them to combine to form raindrops. Cloud droplets are too small to fall to earth, and water vapor was depleted in the air surrounding the droplet during its formation, so the droplet cannot grow by the same process that formed it. Cloud droplets are about 2 one-hundredths (0.02) of a millimeter in diameter, and only a few are larger than 5 one-hundredths (0.05) of a millimeter. Raindrops are usually about 2 millimeters in diameter, so about 1 million droplets must come together to form one raindrop.

The simplest process to understand in raindrop formation is **coalescence** (Figure 12-11). If some of the cloud droplets are bigger than others, they may fall and become larger by colliding and merging with smaller droplets. As a droplet enlarges, it falls faster, accelerating the process. Coalescence does not seem to be a common process in most clouds but is probably the way rain forms in warm, tropical areas. Away from the tropics, parts of most clouds contain ice crystals, and raindrops form in a very different way.

The tops of most clouds are high enough that their temperature is well below freezing. The droplets in clouds are nearly pure water be-

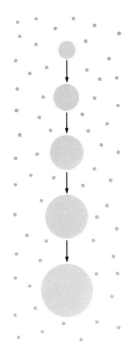

Figure 12-11
Formation of raindrops by coalescence. A somewhat larger cloud droplet falls through the cloud, colliding with other droplets. This is not a common way that raindrops form.

cause the volume of a droplet is very much larger than its condensation nucleus. Unless disturbed, pure water must be *cooled well below the freezing point* (**supercooled**) before freezing begins. An airplane flying through a supercooled cloud may experience icing on its wings because the plane agitates the cloud droplets, causing them to freeze. Unless disturbed, some cloud droplets may not begin to freeze until the temperature is as low as $-40°C$ ($-40°F$). Thus, if freezing begins near the top of a cloud, the ice crystals or snowflakes that form are surrounded by supercooled cloud droplets. When this situation exists, the water droplets evaporate and are deposited on the ice crystals. Thus, the ice crystals grow larger at the expense of the water droplets. This happens because air that is saturated with water vapor with respect to liquid water is supersaturated with respect to ice. The water vapor in the cloud deposits on the ice crystals, and this causes the droplets to evaporate because the air around them is no longer saturated with water vapor. This process is called the **ice-crystal process,** and it implies that most rain that falls to the ground started as snow and melted as it fell to earth (Figure 12-12). Early attempts at increasing rainfall by cloud seeding took advantage of this process by using dry ice and silver iodide, both of which promote ice-crystal formation. F 12-12

Types of Precipitation

Rain is drops over 0.5 millimeter in diameter and typically 1 or 2 millimeters in diameter.

    Drizzle is drops of 0.5 millimeter or less in diameter.

    Snowflakes are aggregates of ice crystals and can be up to several centimeters in diameter.

    Snowgrains are small ice crystals and are the solid equivalent of drizzle.

Figure 12-12
The ice-crystal process of raindrop formation. The water droplets deposit on the ice crystal because the air in the cloud is saturated with respect to liquid water but is supersaturated with respect to ice.

Ice crystal surrounded by supercooled water droplets

The supercooled water droplets evaporate and deposit on the ice crystal

    Hail is ice, generally with a layered structure and spherical shape, commonly 5 to 50 millimeters (0.1 to 1 inch) in diameter (Figure 12-13). Smaller grains may be termed *ice pellets*. F 12-13

Figure 12·13
Cross section of hailstone. (Photo from NOAA.)

Sleet in American usage is rain or drizzle that freezes before reaching the ground. Partially melted snow or a mixture of rain and snow are also called sleet by many.

Freezing rain and **freezing drizzle** are rain and drizzle that freeze on reaching the ground (Figure 12-14).   F 12-1

Dew is moisture deposited directly from the air onto surfaces near the ground.

Fog drip (mist) is moisture deposited on surfaces by fog or clouds moving past the surface.

Measurement of Precipitation

The **rain gauge** is used to measure precipitation, which is reported in centimeters or inches. Most rain gauges have a collecting area a few tens of centimeters in diameter. The water collected flows into a narrower tube, where the depth of rain is multiplied so that even minor amounts of rainfall are easily measured. Evaporation of the collected rain can be a problem if the gauge is not read often, so recording rain gauges such as a tipping bucket rain gauge have come into common use. Recording rain gauges give the rate of rainfall as well as its depth. Snow is measured as the depth in open areas and as the meltwater depth.

Figure 12·14
Freezing rain on power and telephone lines. (Photo from NOAA.)

SUMMARY

- *Relative humidity* is the ratio of the amount of water vapor actually in the air to the maximum amount of water vapor that the air can hold.

- Warm air can hold more water vapor than cold air.

- The *dew point* is the temperature at which the air becomes saturated, or, said another way, when the relative humidity is 100 per cent.

- The *hygrometer* and the *sling psychrometer* are used to measure relative humidity.

- *Latent heat* is the amount of energy absorbed or released when a substance *changes state*.

- The *calorie* is the amount of energy necessary to increase the temperature of one gram of water one degree Celsius.

- A temperature *lapse rate* is the rate of decrease of temperature with increase of height in the atmosphere.

- The *environmental lapse rate* is the average rate at which temperature drops as height increases, which is 0.65°C per 100 meters (3.5°F per 1000 feet).

- A *temperature inversion* occurs when the temperature increases as height increases.

- An *adiabatic process* is one in which no heat is gained or lost by the air parcel.

- An *adiabatic lapse rate* is the temperature in a rising or falling air parcel.

- The *dry adiabatic lapse rate* is 1°C per 100 meters (5.4°F per 1000 feet) and occurs if there is no condensation in the rising air parcel.

- The *wet*, or *saturated, adiabatic lapse rate* is less than the dry adiabatic lapse rate—about 0.6°C per 100 meters (3°F per 1000 feet)—and results if condensation occurs in the air mass.

- *Dew* and *frost* are types of condensation.

- Condensation occurs at 100 per cent relative humidity only if surfaces are available on which the moisture can condense. If the surfaces are sparse, the water vapor will become supersaturated; that is, the relative humidity will be more than 100 per cent and the temperature will be below the dew point.

- *Condensation nuclei* are necessary for clouds or fog to form. Typical *hygroscopic nuclei* form from salt spray from the ocean.

- *Stratus clouds* are layered clouds.

- *Cumulus clouds* are cotton-like clouds, generally with flattish bottoms and rounded tops.

- *Cirrus clouds* are generally thin, wispy clouds.

- The classification of clouds is shown in Figure 12-7.

- *Fog* is a cloud near the ground.

- *Radiation fog* develops when the surface loses its heat to the atmosphere by radiation and the air near the ground is cooled below its dew point.

- *Advection fog* develops when moist air moves across a cool surface.

- *Evaporation fog* (sea smoke, steam fog) develops when cold air moves over a warm, moist surface.

- *Orographic lifting* occurs where wind forces air to rise over mountains. The resulting cooling may cause condensation.

- Cloud droplets may become rain drops by *coalescence*. Most precipitation probably forms when water vapor is deposited on ice crystals in clouds. This is called the *ice-crystal process*.

KEY TERMS

humidity
absolute humidity
relative humidity
dew point
hygrometer
sling psychrometer
change of state
latent heat
calorie
evaporation
condensation
lapse rate
environmental lapse
 rate
temperature inversion
adiabatic

adiabatic lapse rate
dry adiabatic lapse
 rate
wet, or saturated,
 adiabatic lapse rate
stable air
unstable air
dew
frost
frost point tempera-
 ture
supersaturated
condensation nucleus
hygroscopic nucleus
haze
stratus cloud

cumulus cloud
cirrus cloud
fog
nimbostratus cloud
stratocumulus cloud
altostratus cloud
altocumulus cloud
cirrocumulus cloud
cirrostratus cloud
cumulonimbus cloud
radiation fog
advection fog
evaporation fog, or
 sea smoke, or steam
 fog
orographic lifting

coalescence
supercooled water
ice-crystal process
rain
drizzle
snowflakes
snowgrains
hail
sleet
freezing rain and
 freezing drizzle
fog drip, or mist
rain gauge

REVIEW QUESTIONS

1. Explain the difference between absolute humidity and relative humidity.

2. Does warm air hold more or less water vapor than cold air?

3. At 3:00 P.M. the temperature of the air was considerably warmer than it was at 9:00 A.M. During this time, the absolute humidity remained the same. Did the relative humidity decrease, increase, or remain the same? Explain. Did the dew point change? Explain.

4. Explain the role that evaporation and condensation play in distributing the earth's heat.

5. How does the environmental lapse rate differ from the adiabatic lapse rate?

6. What do meteorologists mean when they say that an air parcel is stable? Unstable?

7. Why is moist air likely to be more unstable than dry air?

8. Two conditions must be met before condensation can occur: First, the relative humidity must reach _____; second, a _____ must be available. What happens when the first condition is met but the second is not?

9. If a cloud has the term *cumulus* in its name, how is it shaped? Describe the shapes of stratocumulus and cirrocumulus clouds.

10. What is the scientific term for a thunderhead? To which cloud family do thunderheads belong? Why are they classified with that family?

11. Explain how each of the following is formed: radiation fog, advection fog, evaporation fog.

12. Describe the process by which rain probably forms in tropical areas. What is this process called?

13. In latitudes that lie above or below the tropics, rain originates as snow. What is the name of this process? Before the process can begin, the cloud, or part of it, must become supercooled. Why? Why does the water vapor condense on the ice instead of on the water droplets? What happens to the water droplets?

14. Fill in the blanks in the following chart with the specific names of clouds.

	Stratus (layered)	Cumulus (cottony)	Cirrus (thin)
Low elevation	(a) (very close to ground) _____	(d) _____	
	(b) (higher) _____		
	(c) (releasing precipitation) _____		
Medium elevation	(e) _____	(f) _____	
High elevation	(g) _____	(h) _____	(i) _____

SUGGESTED READINGS

Battan, L. J. *Cloud Physics and Cloud Seeding.* Garden City, N.Y.: Doubleday, 1962, 144 pp.

Battan, L. J. *Harvesting the Clouds.* Garden City, N.Y.: Doubleday, 1969.

Bentley, W. A., and W. J. Humphreys. *Snow Crystals.* New York: Dover, 1962, 266 pp.

Knight, Charles, and Nancy Knight. "Hailstones," *Scientific American* (April 1971), Vol. 224, No. 4, pp. 96–103.

LaChapelle, E. R. *Field Guide to Snow Crystals.* Seattle: University of Washington Press, 1969, 101 pp.

Ludlum, F. H., and R. S. Scorer. *Cloud Study: A Pictorial Guide.* London: Royal Meteorological Society, 1958, 80 pp.

Lynch, D. K. "Atmospheric Halos," *Scientific American* (April 1978), Vol. 238, No. 4, pp. 144–152.

Mason, B. J. *Clouds, Rain, and Rainmaking.* Cambridge, England: Cambridge University Press, 1962, 145 pp.

Myers, J. N. "Fog," *Scientific American* (December 1968), Vol. 219, No. 6, pp. 74–84.

Scorer, R. S., and H. Wexler. *Colour Guide to Clouds.* Oxford, England: Pergamon Press, 1963, 63 pp.

Small, Robert. "Terrain Effects on Precipitation in Washington State," *Weatherwise* (1966), Vol. 19, pp. 204–207.

World Meteorological Organization. *International Cloud Atlas.* Geneva, 1956, Vol. 1, 155 pp., Vol. 2, unnumbered.

ATMOSPHERIC PRESSURE

MEASURING ATMOSPHERIC PRESSURE

WIND

Pressure Gradient

Coriolis Effect

Geostrophic Wind

Friction

Measurement of Wind

WINDS IN HIGH-PRESSURE AND LOW-PRESSURE REGIONS

LOCAL WINDS

Sea-Land Breezes

Monsoons

Valley-Mountain Winds

Chinook (Foehn) Winds

PLANETARY CIRCULATION

Seasonal Changes and the Real Earth

CHAPTER 13

AIR IN MOTION: PRESSURE AND WIND

Trees deformed by prevailing onshore winter winds. Cannon Beach, northern Oregon coast. (Photo courtesy of Oregon Travel Information Section.)

ATMOSPHERIC
PRESSURE

We are familiar with two quite different types of pressure. The simpler is the pressure of a weight resting on a surface. The weight exerts a force on the surface; the pressure is that force divided by the area of the surface. We say the pressure is so many newtons per square meter (pounds per square inch). Such a pressure exerted by a solid body acts in only one direction (Figure 13-1).

F 13-1

   The pressure exerted by a fluid, such as the atmosphere or a body of water, is very different. On the bottom of a swimming pool, the pressure is again the force exerted by the weight of the water divided by the area of the bottom. So far, there is no difference between the two examples. The difference arises in the fact that the water exerts the same pressure on the sides of the pool at the bottom as it does on the bottom. This occurs because water is a fluid, and so it transmits force equally in all directions because all parts must be at the same pressure. A solid has rigidity and so can transmit a force in one or more directions; said another way, different parts of a solid can be under different pressures. Thus, at any level in a pool or in the atmosphere, the pressure is caused by the weight of the overlying fluid, and the pressure is equal in all directions (Figure 13-2).

F 13-2

   If you are not convinced of this, perhaps some examples will help. We live at the bottom of an ocean of air called the atmosphere. The pressure due to the weight of the atmosphere is a little over 100,000 newtons per square meter (about 15 pounds per square inch). If this **atmospheric pressure** acted only vertically, when you are standing, it would push only on your head and shoulders, which have an area of about 1/10 square meter (1 square foot). The atmospheric pressure acting on this area is, therefore, 1/10 times 100,000 newtons, or 10,000 newtons ($15 \times 12 \times 12 = 2160$ pounds). If you lie down, however, your area exposed to vertical pressure increases five or six times, and such pressure changes, if they existed, would surely be felt. If you still do not believe that atmospheric pressure acts in all directions, including vertically upward, ask yourself what keeps the roof of a house from collapsing under the force of atmospheric pressure. A typical home has a roof of about 130 square meters (1200 square feet), so the atmospheric pressure would exert a force of $130 \times 100,000$, or 13,000,000 newtons ($1200 \times 12 \times 12 \times 15 = 2,592,000$ pounds). Few materials could withstand such forces, but, because the same force acts on the top and bottom of a roof, there is, in effect, no net force acting on it except its own weight.

   Let us now explore the meaning of pressure a little further. According to the kinetic theory, gas pressure is caused by the constant impacts of the gas molecules on the faces of the container holding the gas. In a closed container, there are two ways to increase the pressure. If the gas is heated, the gas molecules will move faster and so will hit

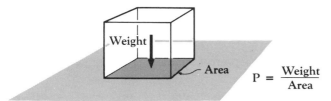

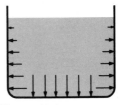

$$P = \frac{Weight}{Area}$$

Figure 13-1
The weight of a solid body exerts a pressure on the surface on which it rests. The pressure is the weight of the body divided by the area of the surface on which it rests.

Figure 13-2
The pressure acting on the bottom and sides of a container of fluid is caused by the weight of the fluid above any point.

the walls more often; therefore, the pressure will be increased. If the size of the container is reduced, again the gas molecules will hit the walls more often because they now have less room in which to move, and so the pressure will increase. Remember that these examples illustrate how a gas that is confined in a container behaves.

The atmosphere behaves somewhat differently because it is not confined. An increase in temperature may cause the air to expand as the molecules collide more often because their velocity increases. The expanded air will be less dense and so will weigh less, causing the atmospheric pressure to be reduced. Thus, in this case, heating has the opposite effect on the atmosphere.

Atmospheric pressure is caused by the weight of the air. This implies that, as one rises in the atmosphere, the pressure should diminish because there is less air above; measurements confirm this prediction. The cylinder in Figure 13-3 illustrates the changes in both pressure and density that occur in the atmosphere. As weight is added to the piston, the gas in the cylinder is compressed more and more, and the pressure increases because the gas molecules hit the cylinder walls more often. F 13-3

Because the atmospheric pressure is highest near the surface of the earth, more air molecules are found there; said another way, the density is greater. The pressure is due to the weight of these molecules, so, because there are more of them near the surface, the pressure falls

Figure 13-3
As the weight on the cylinder increases, so do the pressure and the density of the gas inside the cylinder.

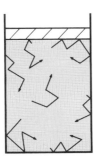
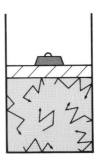
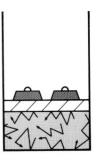

more rapidly with elevation near the surface than it does higher in the atmosphere.

Pressure changes in the vertical direction are thus quite marked, but much smaller changes in pressure from place to place near the surface are even more important in weather, because they are the cause of wind. The air tends to move from regions of high pressure to areas of low pressure, and this is the origin of wind. The high pressure near the surface does not cause a vertical wind toward the much lower pressures aloft, because the tendency of air to move upward from high pressure to low pressure is balanced by the gravitational force. Gravitation is the attraction between the air molecules and the earth. Gravity is, therefore, the reason that the earth retains an atmosphere and is the cause of atmospheric pressure.

MEASURING ATMOSPHERIC PRESSURE

The classical way to measure atmospheric pressure is with the **mercury barometer.** This is a simple device that balances the atmospheric pressure against the weight of a column of mercury. It is easy to make a mercury barometer. All that is required, besides the mercury, is a glass tube at least 90 centimeters (3 feet) long, sealed at one end, and a small dish (Figure 13-4). Carefully pour the mercury into the tube F 13-4 so that it is completely filled, and fill the dish to a depth of a few centimeters. Then, hold the end of the tube so that no mercury can escape, and place it in the dish in the inverted position. The mercury will fall a short distance in the tube, leaving a near-vacuum at the top of the tube. This is now a mercury barometer, because the atmospheric pressure on the surface of the mercury in the dish is exactly balanced by the pressure due to the weight of the mercury in the tube (Figure 13-4). The earliest barometers were made in this way, and modern mercury barometers are similar but with refinements to allow accurate reading of the height of the mercury column.

Mercury barometers are not very convenient and are not particularly portable, so a need existed for a smaller and less fragile barometer. This need was filled by the **aneroid** (without liquid) **barometer** (Figure 13-5). It consists of a small, sealed, partially evacuated cham- F 13-5 ber. When the pressure increases, the pressure causes the chamber to become smaller, and when the pressure decreases, the chamber enlarges. These minor changes in size are transmitted to an indicating needle by means of a series of levers that magnify the movements of the chamber. Aneroid barometers are also used as altimeters. Because the changes in pressure with altitude are well known, the scale on a

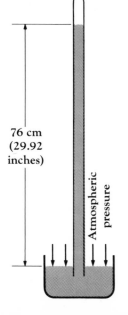

76 cm (29.92 inches)

Atmospheric pressure

Figure 13-4 The mercury barometer. The pressure caused by the weight of the mercury in the tube is balanced by the atmospheric pressure acting on the mercury surface in the dish.

Figure 13-5
Aneroid barometer.

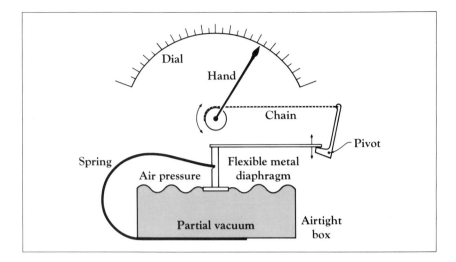

barometer can be calibrated in altitude. Such altimeters read elevation above sea level and not above ground, so they must be adjusted for atmospheric pressure and used with caution by pilots over mountainous terrain.

Several different units are used in the measurement of atmospheric pressure. The height of mercury in the mercury barometer has been in use for many years. The average sea-level pressure in such units is 760 millimeters (29.92 inches) of mercury. In the earlier discussion, pressure was described in newtons per square meter. In these units, the average sea-level pressure is 101,325 newtons per square meter. This large number is awkward to use, so a larger unit is employed. The **bar** is a unit of pressure equal to 100,000 newtons per square meter. The pressure unit in common use is the **millibar,** or one one-thousandth (0.001) of a bar. Thus, one millibar is 100 newtons per square meter. In millibars, the average sea-level pressure of 101,325 newtons per square meter is 1013.25 millibars (101,325 divided by 100). From here on, we will use millibars (mb) as the unit of pressure.

All weather stations report the sea-level pressure. Because most stations are at elevations above sea level, the readings must be corrected. Using all sea-level pressures simplifies interpretation of surface weather maps. Only in very unusual weather, especially during severe storms, does the sea-level pressure deviate more than about 30 millibars above or below the standard sea-level pressure.

WIND

Pressure Gradient

Uneven heating of the earth by the sun is the cause of horizontal pressure differences in the atmosphere. **Wind** *is the movement of air parallel to the surface.* The force that causes wind is the tendency to

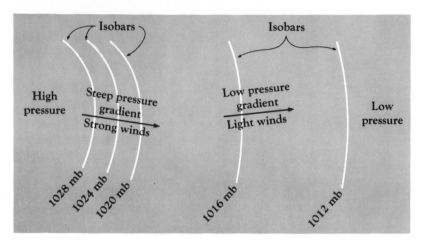

Figure 13-6 Isobars are lines on a weather map all points on which are at the same pressure. Where the isobars are close together, the pressure gradient, or rate of change, is high and winds are strong.

equalize horizontal pressure differences. This implies that winds blow from areas of higher pressure to areas of lower pressure, but the process is not quite that simple. The effects of the earth's rotation and, near the ground, friction affect both the speed and direction of wind.

The **pressure gradient** *is the measure of how much the air pressure changes with distance.* It is best seen on weather maps where the pressure is shown by isobars (Figure 13-6). **Isobars** *are lines drawn through* F 13-6 *points of equal pressure.* Where the isobars are close together, the pres-

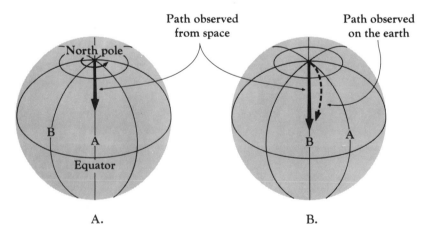

Figure 13-7 The Coriolis effect. A rocket fired from the north pole travels in a straight line, and the earth turns beneath it. An observer in space would see the rocket's path as a straight line. An observer on the earth would see a curved path because of the earth's rotation.

sure gradient is high because large pressure changes occur in a short distance; where the isobars are widely spread, the pressure gradient is low. The higher the pressure gradient, the stronger or faster the wind will blow. The force that results from the pressure gradient is what starts the movement of the air. This force acts at a right angle to the isobars, and the initial wind is at a right angle to the isobars and blows from high pressure toward low pressure. As soon as the air begins to move, other forces act on it to change its direction and speed.

Coriolis Effect

As we noted in Chapter 9, the effect of the earth's rotation on objects moving above it is called the **Coriolis effect.** It was first described and analyzed by Gaspard Gustave de Coriolis (1792–1843), a French engineer who dreamed of a military career but was too frail. Instead, his contribution was to improve the accuracy of artillery shells by calculating the effect of the earth's rotation on a shell's trajectory. *The concept of the Coriolis effect is quite simple: while an object is moving above the earth, the earth turns underneath that object.* Viewed from above the north pole, the earth turns counterclockwise. If a rocket, an artillery shell, or an air mass at the north pole starts moving southward along a longitude line toward the equator (Figure 13-7), it will travel directly south in a straight line if viewed by an observer in space above the north pole. While that object is in motion, the earth will be turning on its axis, and the object will reach the equator at a point to the west of the longitude line on which it started. The object's path through space is a straight line, and this is what an observer above the north pole will see. An observer on the earth, however, would see the object travel a curved path. Thus, the Coriolis effect is an apparent change of direction, but to us living on the earth, it appears real.   F 13-7

Some other examples may help to make the Coriolis effect clear. Imagine a merry-go-round with two people riding on opposite sides. If one of them attempts to throw a ball to the other, he must aim ahead of the other. If he throws directly toward the other, that person will have been rotated away from the point at which the ball was thrown. So far, all of our examples have involved north-south movement, but the same deflection occurs with motion in any direction, even east-west. In Figure 13-8A, a wind is moving to the west (an east wind) along a latitude line. A few hours later, the position of the latitude line has changed because of the earth's rotation (Figure 13-8B). The wind is still moving in the same direction, as viewed from space, but the latitude line is now at an angle to that direction, so that the original westward-moving wind is now moving toward the northwest (a southeast wind).   F 13-8

Figure 13-8
Deflection of an east wind by the Coriolis effect.

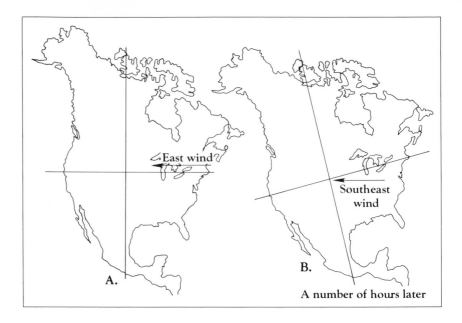

Let us summarize how the Coriolis effect acts on wind. The Coriolis effect begins to affect wind only after motion begins. It alters only the direction of motion and not the speed. The magnitude of the Coriolis effect is proportional to the speed of the wind. It acts at right angles to the direction of motion. The effect is greatest at the equator and least at the poles. *In the northern hemisphere, the deflection is to the right of the direction in which the wind is moving, and it is the opposite in the southern hemisphere.* (Figure 13-9).

F 13-9

Figure 13-9
Summary of the Coriolis effect. It is maximum at the poles and zero at the equator. Deflection is to the right in the northern hemisphere and to the left in the southern hemisphere.

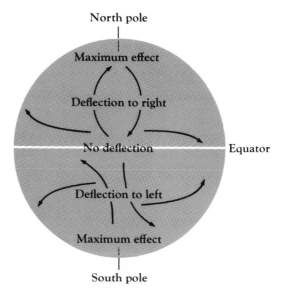

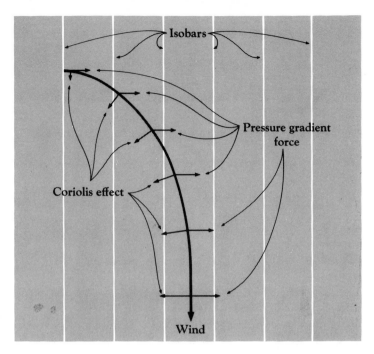

Figure 13-10 Geostrophic wind. The wind starts as a pressure-gradient wind; but as soon as the wind moves, the Coriolis effect deflects it. The deflection continues until the Coriolis effect equals the pressure gradient force and the wind travels parallel to the isobars.

Geostrophic Wind

The Coriolis effect profoundly changes the direction of winds. When air begins to move, it initially moves at right angles to the isobars and from high pressure toward lower pressure. As soon as the air is in motion, the Coriolis effect begins to act on it, and in the northern hemisphere deflects the wind to its right (Figure 13-10). The Coriolis effect again deflects the wind to the right of this new direction. This process continues to turn the wind to its right until it is moving parallel to the isobars. When the wind is parallel to the isobars, the Coriolis effect acts in the opposite direction to the pressure gradient force. The Coriolis effect balances the pressure gradient force, and the wind continues to move parallel to the isobars. A **geostrophic wind** is one that *moves parallel to the isobars*. This can only occur above the level at which friction along the ground affects winds—that is, above about a kilometer.

Note that we started our discussion of winds by saying that air tends to move from high pressure toward lower pressure, but now we

F 13-10

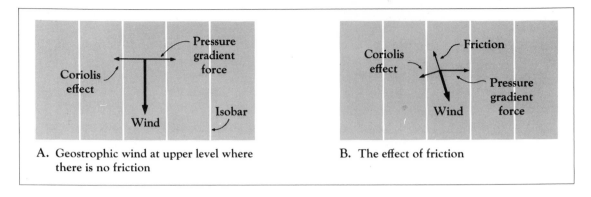

A. Geostrophic wind at upper level where there is no friction

B. The effect of friction

Figure 13-11
Friction near the surface weakens winds and deflects them.

see that, at least aloft, this does not occur. In the next section we will see that it does not occur near the ground either, although there the deflection is less. Geostrophic winds occur because the pressure gradient force and the Coriolis effect are nearly equal in most cases.

In understanding and predicting weather, knowledge of the upper atmosphere is important. At ground level it is relatively easy to measure wind and pressure and to construct a map with isobars on it. Similar measurements of the upper atmosphere are much more difficult and expensive. However, it is fairly easy to measure wind direction and speed, using weather balloons. From this information, and knowing what causes geostrophic winds, it is possible to construct good weather maps of the upper atmosphere.

Friction

The other force that acts on wind as soon as it begins to move is friction. Friction affects winds up to an elevation of about a kilometer. Over smooth surfaces, such as oceans, the effect may not reach that high. Friction always acts in the opposite direction from that in which the wind is moving (Figure 13-11), slowing the wind down. F 13-1▶ Because the wind is slower, the Coriolis effect is reduced. The pressure gradient is not affected by friction; so in the friction layer, the pressure gradient force is always greater than the Coriolis effect. Thus, the wind crosses the isobars at an angle. As we move upward through the friction layer, the effect of the friction is reduced, and the winds cross the isobars at smaller and smaller angles until we reach the geostrophic winds. At the surface, wind crosses the isobars at about 15 degrees over smooth oceans, where the wind is slowed about a third. Over rough terrain, wind may cross the isobars at a 45-degree angle and be slowed to half the speed of the corresponding geostrophic wind. Other effects of friction that lead to vertical movements will be discussed in the next sections, where specific winds are described.

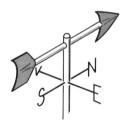

Figure 13-12
Wind direction is measured with a wind vane.

Measurement of Wind

Both the direction and the speed of wind must be measured. *Wind direction is measured with a* **wind vane** (Figure 13-12). *Winds are named for the direction from which they come;* that is, a northwest wind comes from the northwest and is blowing toward the southeast. Wind vanes are designed to point into the wind. *Wind speed is measured with an* **anemometer.** A typical anemometer is shown in Figure 13-13.

F 13-12

F 13-13

<table>
<tr><td>

WINDS IN HIGH-PRESSURE AND LOW-PRESSURE REGIONS

</td><td>

The flow of air in high-pressure and low-pressure regions is important in understanding weather, in part because such flow can generate vertical motion. A *low-pressure system is called a* **cyclone.** The term should not imply a violent storm, although in parts of the world hurricanes or typhoons are called cyclones. To avoid confusion, middle-latitude storms are sometimes called either *extratropical cyclones* or *wave cyclones,* and the more violent tropical storms are called *tropical cyclones.* In the northern hemisphere, the pressure gradient is radially inward in a low-pressure system, and the Coriolis effect is to the right of the wind (Figure 13-14). Above the friction zone, the resulting geostrophic wind is counterclockwise around the low. In the southern hemisphere, the winds are clockwise around a cyclone.

</td></tr>
</table>

F 13-14

A *high-pressure system is called an* **anticyclone.** In the northern hemisphere, the winds are clockwise around a high, or anticyclone; in the southern hemisphere, the winds are counterclockwise (see Figure 13-14). In the earlier discussion of geostrophic wind, the isobars were straight and the pressure gradient was balanced by the Coriolis effect. The flow around high- and low-pressure systems is curved, so an acceleration is involved. Around a low-pressure area, or cyclone, the Coriolis effect is reduced; and, at an anticyclone, the Coriolis effect is greater than it would be for winds in a straight line. This point is not critical to understanding winds, but it would be a source of confusion to anyone who has studied physics.

Near the surface, friction changes this simple flow. In the northern hemisphere, winds spiral inward counterclockwise around a cyclone, and they spiral outward clockwise around an anticyclone (Figure 13-15). An implication of such flow is that air is piled up or accumulated at the center of a cyclone. Such *an inflow of air is called* **convergence.** Convergence causes air to accumulate at the center of a low-pressure area. If nothing else intervenes, convergence would cause the pressure to increase and thus destroy the low-pressure system. Commonly, then, if a low-pressure system continues to exist, vertical upward flow is occurring, and, aloft, air is moving away from the area. *The movement of air outward from an area is called* **divergence** (Figure 13-16). The vertical flow is much slower than the horizontal

F 13-15

F 13-16

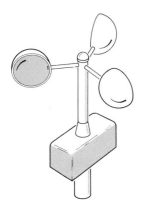

Figure 13-13
Three-cup anemometer records wind speed.

Figure 13-14
Winds around high-
pressure and low-
pressure systems in
the northern and
southern hemispheres.

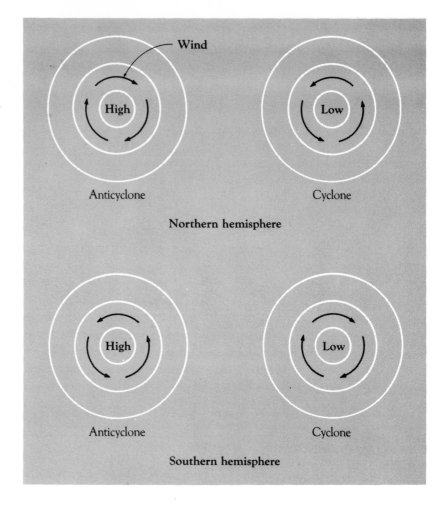

winds and probably averages about 1 kilometer per day or less. This rising air generally produces cloudy skies over low-pressure areas or cyclones.

At an anticyclone the opposite conditions develop. At the surface the air is diverging, and aloft it is converging. Thus, if the high-pressure area is to persist, vertical flow downward must occur (see Figure 13-16). Downward-flowing air is heated by compression, and so clear skies prevail.

Most storms in the middle latitudes are associated with low-pressure systems. A cyclone may form, but it will not persist without divergence aloft. If the divergence aloft is greater than the convergence at the surface, the cyclone will intensify, and a storm may develop. This is one of the reasons why upper-atmosphere data are important in forecasting surface weather.

When you experience a wind, it is interesting to be able to know where the pressure systems are that cause the wind. **Buys Ballot's law,**

Figure 13-15
Friction near the ground causes winds to spiral inward into an area of low pressure. The resulting pileup of air is called *convergence.* For the same reason, divergence forms at an anticyclone.

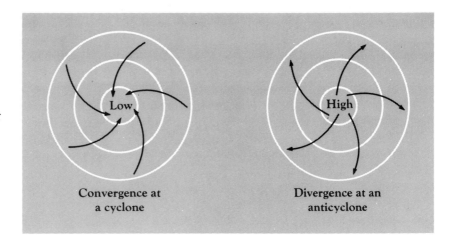

Convergence at
a cyclone

Divergence at an
anticyclone

Figure 13-16
Convergence or divergence at the ground causes vertical flow of air.

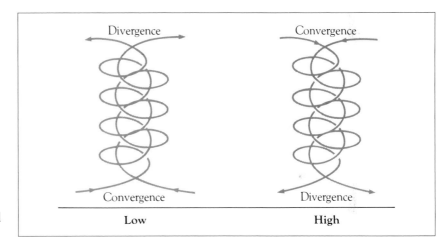

named for a 19th-century Dutch meteorologist, is an easy way to determine where the low pressure is. *If you stand with your back to the wind, low pressure is to your left in the northern hemisphere.* In the southern hemisphere, the low is to your right. This rule works well for geostrophic winds aloft but is less accurate for surface winds.

LOCAL WINDS

Sea-Land Breezes

At a coastline during the day, the temperature of the land rises more than the temperature of the ocean. The air in contact with the land is heated by conduction, and, because warm air is lighter than cold air, it rises. Cool air from the ocean takes its place, and thus a convective flow is established (Figure 13-17). This is called a **sea breeze** (remember that winds are named for the direction from which they come). On a clear night, the opposite conditions occur. The land loses its heat by radiation, and the air in contact with it becomes cooler than the air over the ocean. The warmer air over the ocean rises, and the breeze flows from land toward ocean. This is called a **land breeze.** F 13-17

Friction near the surface affects sea-land breezes. Wind flowing over smooth ocean experiences little friction, but when it encounters the coast, it is slowed by friction. This pileup of air at the coast can cause the moist air from the ocean to rise, and clouds may develop. This is one reason that some coasts are cloudy.

Monsoons

The **monsoons** of Asia are similar to sea-land breezes, but the size of the area involved is much greater and the winds are seasonal, not daily. In winter the Asian continent is cool, and the heavy, cool air flows across the Indian subcontinent toward the warmer ocean. These winter winds are cool and dry. In summer, the land becomes much warmer than the ocean, and moist winds come from the Indian Ocean. These winds rise as they approach the Himalaya Mountains, and the cooling produces heavy rainfall.

Valley-Mountain Winds

In areas where large differences in elevation occur, the valley slopes are heated by the sun during the day, and the warm air rises, producing the **valley breeze.** At night, the air in the highlands loses its heat by radiation, and the resulting cool air flows down the slopes. This is called the **mountain wind** (Figure 13-18). F 13-18

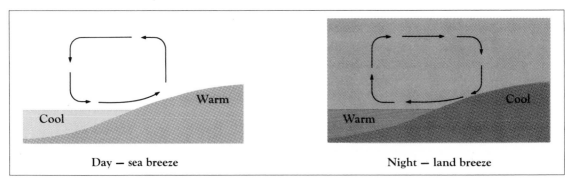

Day — sea breeze   Night — land breeze

Figure 13·17 Sea and land breezes. During the day, the sun warms the land; the resulting convective flow causes the sea breeze. At night, the land loses its heat by radiation and becomes cooler than the water, reversing the breeze.

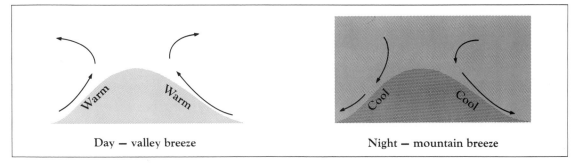

Day — valley breeze　　　Night — mountain breeze

Figure 13-18 Valley and mountain breezes. During the day, the sun warms the valley sides, and the rising warm air creates the valley breeze. At night, the valley sides lose their heat by radiation, and the falling cool air forms the mountain breeze.

Chinook (Foehn) Winds

Warm winds that flow downslope are called **chinooks** in western North America and are called **foehns** in the Alps. They result when a low-pressure system develops on the lee side of an upland or mountain area. The air rushes downslope toward the low pressure and so is heated by compression. The resulting warm, dry winds can cause dramatic temperature increases in a very short time. Increases of 10° to 20°C (18° to 36°F) in 15 minutes are common. The record may be 17°C (31°F) in three minutes at Havre, Montana. These warm, dry winds can remove much snow in a few hours.

In southern California, the **Santa Ana winds** are caused by the downslope, westward movement of generally already warm, dry air from the high deserts. The dry air is heated further by compression as it moves westward toward low pressure. These strong, drying winds often increase the danger of forest fires.

PLANETARY CIRCULATION

The **planetary circulation** of the earth is the largest possible scale of air movement. We do not yet understand all aspects of this important circulation, and this is one reason why long-range weather forecasts are not more accurate. Figure 13-19 shows the generally accepted model of planetary circulation. Vertical and upper-air movements are shown along the edge, and surface winds are indicated on the globe. The forces that cause the planetary circulation are the sun's heating and the earth's rotation. Much of the sun's radiation falls near the equator, and the planetary circulation distributes this heat to the rest of the earth.

Near the equator, the sun's heat warms the air, causing it to rise. Much of this part of the earth is covered by oceans, so the overlying

F 13-19

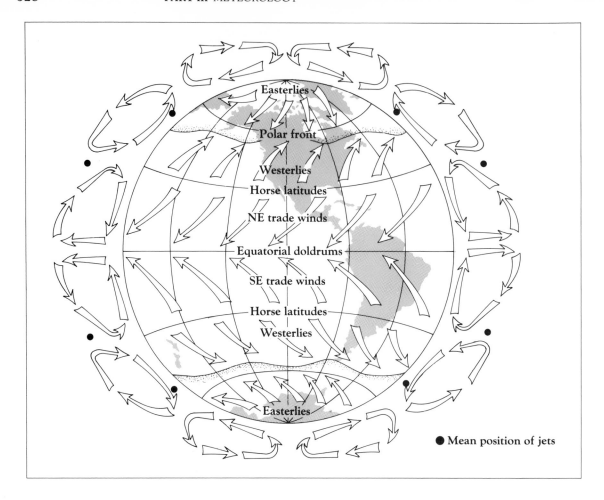

Figure 13-19
The earth's planetary circulation. The jet streams and the polar front are also shown.

air is moist. As the warm, moist air rises, it cools, producing clouds and rain. Most of the tropical rain forests are in this belt. The light surface winds near the equator give this region the name **doldrums.** The rising air moves north and south from the equatorial regions. By the time it approaches latitudes 30 degrees north and south, the air has lost much of its heat by radiation to space, and the cold, heavy air descends. Another factor that contributes to the descent is that, as the air travels poleward, the Coriolis effect acts on it, and when it has reached 30 degrees latitude, it is moving eastward. Because the air is no longer traveling poleward, it piles up, and this contributes to its descent. The descending dry air is warmed by compression, making this region near 30 degrees warm, clear, and dry. Many of the warm desert areas are near 30 degrees north and south of the equator.

The descending air divides; some continues toward the poles, and the rest returns to the equatorial regions. The surface winds are light, and the region near 30 degrees is called the **horse latitudes** because, in the days of sailing ships, horses on ships becalmed in this area had to be destroyed when their food and water ran out.

The surface air returning toward the equator is acted on by the Coriolis effect that turns it westward, producing the easterly **trade winds.** The trade winds were used by sailing ships on their way toward the new world.

In summary, in the region from the equator north and south to 30 degrees latitude, the surface winds are easterly, and the winds aloft are westerly. At 30 degrees north and south, the winds aloft move from west to east, and their speeds commonly exceed 160 kilometers per hour (100 miles per hour) and at times reach 400 kilometers per hour (250 miles per hour). These winds are called the **jet streams,** because they are narrow rivers of high-speed winds. Generally they do not form a continuous belt circling the earth (Figure 13-19). Large temperature contrasts at the surface are believed to be an important factor in their formation.

In the region from 30 to 60 degrees, the surface winds are westerlies. In our model, some of the air descending near 30 degrees moves poleward along the surface, and the Coriolis effect deflects it to the east, producing westerly winds. In this same model, the poleward-moving surface air rises at 40 to 60 degrees latitude, where it meets heavy, cold air moving down from the poles. The rising air divides, with some continuing poleward and some returning to 30 degrees. The air returning aloft to 30 degrees latitude is also a westerly wind because it has enough westerly momentum to overwhelm the Coriolis effect. Thus, in the region between 30 and 60 degrees latitude, the winds are westerlies at all elevations.

The place where the westerly winds of the temperate areas meet the cold polar air is called the **polar front.** Its location changes, especially with seasons, but it generally varies between 40 and 60 degrees latitude. This area of rising air is cloudy and the site of much precipitation. It is also a region of strong horizontal temperature changes and contains the **polar front jet stream.** This jet stream flows from west to east, and its location, like that of the polar front, is quite variable (Figure 13-19). The polar front is very important in the development of the weather of the temperate zones, as we will see in the next chapter.

Some of the air rising at the polar front continues aloft toward the poles. As this air moves poleward, it loses its heat by radiation to space and so becomes cold and heavy. The cold, heavy air descends near the poles. The descending air is warmed somewhat by compression and, like the similar descending air near 30 degrees latitude, produces little precipitation. The descending air moves along the surface toward the equator. The Coriolis effect deflects this air toward the west, thus forming the polar easterly winds.

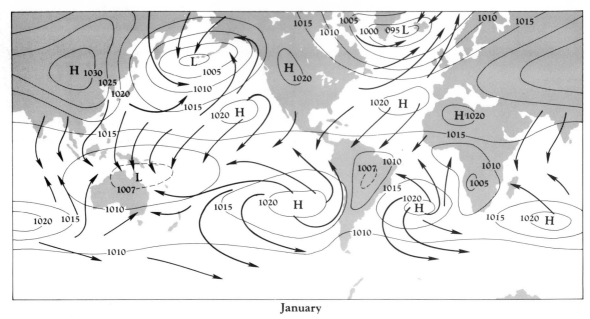

January

July

Figure 13-20 Average sea-level pressures in January and July, and the resulting winds.

Seasonal Changes and the Real Earth

So far the discussion of the planetary winds has not taken into account the real earth that is composed of both continent and ocean nor the effects of the seasonal changes in the sun's heating. The planetary circulation predicts the following distribution of pressure:

• an equatorial low, near the equator
• subtropical highs at 30 degrees north and south
• subpolar lows at the polar fronts
• polar highs

Figure 13-20 shows the average sea-level pressure for both January **F 13-20**
and July and the resulting surface winds. The distribution of pressure
and winds is similar to our prediction, differing mainly in that the
highs and lows do not form continuous belts around the globe. The
subpolar low in the southern hemisphere is the only one that circles
the earth, and it is the only one that is entirely over ocean. Thus, the
distribution of land and water affects the surface temperatures (see
Figure 11-21, p. 288), and these temperatures affect the pressure and
the winds.

The most prominent features on the maps are the subtropical
highs, indicated by the highs over the oceans near 30 degrees north
and south. The highs tend to form on the east side of the oceans, and
they are most intense in the hemisphere experiencing summer. They
also migrate about 10 degrees poleward in the summer hemisphere.

The most pronounced seasonal changes are in the northern hemisphere because of the concentration of continents there. In January,
the cold continents of the northern hemisphere are high-pressure
areas covered by cold, heavy air; in July, rising warm air creates low
pressure. On the January map, intense low-pressure areas develop at
the northern polar front in the Atlantic and Pacific oceans. These
two lows, the Aleutian and Icelandic lows, are areas where cyclonic
storms form on the polar front, as will be described in the next chapter.

SUMMARY

- Atmospheric pressure is caused by the weight of the overlying air, and the pressure is equal in all directions. Because atmospheric pressure is caused by the weight of the overlying air, the pressure decreases markedly as elevation increases.

- Pressure is expressed in force divided by area.

- A *barometer* is used to measure pressure.

- The average sea-level pressure is 760 millimeters of mercury, 29.92 inches of mercury, 101,325 newtons per square meter, or 1013.25 millibars.

- *Wind* is the movement of air parallel to the surface. Winds blow from areas of high pressure to areas of low pressure.

- The *pressure gradient* is a measure of how much the air pressure changes with distance.

- *Isobars* are lines drawn through points of equal pressure on a weather map. Where the isobars are close together, the pressure gradient is high and the winds are strong.

- The *Coriolis effect* is caused by the earth's rotation; as a result of this effect, a moving object or air mass in the northern hemisphere is deflected to the right of the direction in which it is moving, and in the southern hemisphere it is deflected to the left.

- A *geostrophic wind* is one that moves parallel to the isobars. This occurs when the Coriolis effect is balanced by the pressure-gradient force.

- Near the earth's surface, friction tends to slow the winds and change their direction so that they cross the isobars.

- Wind direction is measured with a *wind vane*.

- Winds are named for the direction from which they come.

- Wind speed is measured with an *anemometer*.

- A low-pressure system is called a *cyclone*.

- In the northern hemisphere, winds move counterclockwise around a cyclone. In the southern hemisphere, they move clockwise.

- A high-pressure system is called an *anticyclone*.

- In the northern hemisphere, the winds move clockwise around an anticyclone. In the southern hemisphere, they move counterclockwise.

- An area where air is flowing inward is called a *convergence*.

- An area where air is flowing outward is called a *divergence*.

- Buys Ballot's law states that, if you stand with your back to the wind, low pressure is to your left if you are in the northern hemisphere and to the right if you are in the southern hemisphere.

- During the day at a coast, air warmed by contact with the warm land rises and cool air from the ocean replaces it. This is called a *sea breeze*. At night, the land is cooler than the ocean, so the wind is in the opposite direction and is called a *land breeze*.

- *Monsoons* are similar to sea-land breezes, but they are seasonal, not daily.

- Air warmed by valley sides rises during the day, creating a *valley breeze*. At night, cool air from the mountains flows down the valleys, creating a *mountain breeze*.

- *Chinook (foehn) winds* are warm winds flowing downslope. Because they are flowing downslope, they are compressed and so are heated.

- The earth's planetary circulation is caused by its rotation and the sun's heating. This circulation is shown in Figure 13-19.

- The planetary circulation predicts the following distribution of pressure systems:

  An equatorial low
  Subtropical highs at 30 degrees north and south
  Subpolar lows at the polar fronts
  Polar highs

KEY TERMS

atmospheric pressure	geostrophic wind	sea breeze	Santa Ana wind
mercury barometer	wind vane	land breeze	planetary circulation
aneroid barometer	anemometer	monsoon	doldrums
bar	cyclone	valley breeze	horse latitudes
millibar (mb)	anticyclone	mountain wind	trade winds
wind	convergence	chinook (western	jet stream
pressure gradient	divergence	North America), or	polar front
isobar	Buys Ballot's law	foehn (Alps)	polar front jet stream
Coriolis effect			

REVIEW QUESTIONS

1. What prevents our atmosphere from floating away to outer space? What causes atmospheric pressure?

2. Why is the densest air found closest to the surface of the earth?

3. State the principle that accounts for the origin of wind. Atmospheric pressure decreases with altitude, yet we do not experience strong, constant vertical winds. Why not?

4. Think about how a mercury barometer works. When the atmospheric pressure increases, will the mercury column rise or fall? Why?

5. Explain how the aneroid barometer demonstrates that the force exerted by the weight of the atmosphere is directed in all directions.

6. Define the millibar. What is the average air pressure at sea level in millibars?

7. What is the cause of horizontal pressure differences in the atmosphere? Is a low-pressure area likely to be warmer or colder than a high-pressure area along the same latitude? Explain your answer.

8. In area A the isobars are quite close together; in area B they are widely spaced.
   (a) Which area has the higher pressure gradient? Explain your answer by defining the terms *pressure gradient* and *isobar*.
   (b) Will the wind be stronger in area A or area B?

   (c) Initially the wind blows perpendicularly to the isobars. Why?
   (d) What alters the wind's original direction?
   (e) What alters the wind's original speed?

9. Explain why the Coriolis effect appears real to us on earth but is only an apparent change in direction.

10. Because of the Coriolis effect:
    (a) Is a wind blowing in the northern hemisphere deflected to its right or its left?
    (b) Is a wind blowing in the southern hemisphere deflected to its right or its left?
    (c) Are winds at the Tropic of Cancer deflected to a greater or lesser degree than winds of the same velocity blowing at the arctic circle?

11. Explain how a geostrophic wind occurs.

12. Explain why geostrophic winds do not occur below an elevation of about 1 kilometer?

13. A low-pressure system lies over the middle of the United States.
    (a) Draw several concentric circles to indicate the isobars and label the areas of highest and lowest pressure.
    (b) Add dashed arrows to your drawing to show the direction in which the wind starts to move. Explain in one sentence why your dashed arrows are pointing in this direction.

(c) Next, add solid arrows to show the direction in which the wind is blowing at high altitude. What has caused the direction of the wind to shift from its initial course?

(d) Finally, choose another color to draw arrows showing the approximate direction of the wind along the surface of the earth. Is the wind converging or diverging along the surface?

14. Draw a horizontal line to indicate the ground; in the middle of the line draw an X to indicate the center of the low-pressure system described in Question 13. Add arrows to show the vertical movement of the air. Label the areas of convergence and divergence. Using what you learned in Chapter 12, explain why cloudy skies usually lie over low-pressure areas.

15. An anticyclone lies over the United States.
(a) Are the high-altitude winds blowing in a clockwise or counterclockwise direction? Explain your answer either in words or by drawing and labeling a sketch.
(b) Make a drawing to show the vertical flow of the air and label the areas of convergence and divergence.
(c) Why do high-pressure systems usually bring clear skies?

16. Are winds called by the direction from which they come or by the direction in which they blow?

17. Explain why people living along the shore frequently observe a sea breeze during the day and a land breeze during the night. *Note:* As you describe each step of the energy transfer, state whether radiation, conduction, or convection is involved.

18. How are the monsoon seasons of India similar in principle to sea–land breezes? What causes the heavy rains of the monsoon?

19. What conditions are required for the development of a chinook? Describe what is happening during a chinook. Are the winds moving down the mountain cool or warm?

20. What two phenomena are responsible for the planetary circulation?

21. The following questions will help you review atmospheric circulation between the equator and 30 degrees latitude:
(a) The equator is an area of little surface wind and lush tropical rain forests. Why?
(b) Is the air moving north and south from the equator at a high or low altitude? Because of the Coriolis effect, what direction is it moving in as it approaches a latitude of 30 degrees north?
(c) Like the equator, 30 degrees latitude has little surface wind; however, the land is apt to be desert-like. Why?
(d) Where are the trade winds found? Is this air flowing from, or returning to, the equator? From what direction do the trade winds blow?

22. The winds between 30 and 60 degrees north are westerlies at all elevations. Why?

23. Explain why the polar front is an area of heavy precipitation.

24. The poles receive no more precipitation than do many of the desert areas of the world. Explain why the north pole's climate is so dry.

25. For each of the following four areas, state whether air is ascending or descending and whether you would expect generally low or generally high air pressures:
• equator
• 30 degrees north or south
• polar fronts
• poles

26. In January the northern polar front tends to have high pressures over continental areas and low pressures over oceanic areas. Explain this variation.

SUGGESTED READINGS

Heidorn, K. C. "Land and Sea Breezes," *Sea Frontiers* (1975), Vol. 21, No. 6, pp. 340–343.

Newell, R. E. "The Circulation of the Upper Atmosphere," *Scientific American* (March 1964), Vol. 210, No. 3, pp. 62–74.

Oort, A. H. "The Energy Cycle of the Earth," *Scientific American* (September 1970), Vol. 223, No. 3, pp. 54–63.

Reiter, E. R. *Jet Streams.* Garden City, N.Y.: Doubleday, 1967, 189 pp.

Stewart, R. W. "The Atmosphere and the Ocean," *Scientific American* (September 1969), Vol. 221, No. 3, pp. 76–86.

AIR MASSES

FRONTS

Cold Fronts

Warm Fronts

Occluded Fronts

Stationary Fronts

CYCLONIC STORMS OF THE MIDDLE LATITUDES

SEVERE STORMS

Thunderstorms

LOCAL THUNDERSTORMS

ORGANIZED THUNDERSTORMS

LIGHTNING AND THUNDER

Tornadoes

Hurricanes

FORECASTING

CHAPTER 14

STORMS

Lightning.
(Photo from NOAA.)

AIR MASSES

Air masses *are large volumes of air that have uniform characteristics at any given altitude.* Generally, an air mass is at least 1600 kilometers (1000 miles) in diameter, and at any altitude the temperature and humidity are more or less similar throughout the mass. The characteristics of an air mass are determined by the place where it originates. Air masses are most likely to form in areas where winds are gentle, so that very little mixing with outside air occurs. Source areas of air masses are more likely to be high pressure than low pressure because low-pressure systems are areas of convergence and, therefore, of mixing. An air mass that forms over a cold land surface is cold and dry, and one that forms over warm ocean water is warm and humid. When the air mass moves away from its area of origin, it carries such characteristics with it.

Air masses are named for the areas where they originate, as follows:

A Arctic (Antarctic) form near the poles
P Polar   form near the polar front
T Tropical   form under the subtropical high-pressure
                       areas near 30 degrees north and south
E Equatorial   form near the equator

Notice that air masses do not in general form in the temperate regions because these regions do not have uniform characteristics over large areas. The arctic and polar regions are the source of cold air masses, and the tropical and equatorial regions form warm air masses. Note, too, that polar air masses form near the polar front and not near the geographic poles, where the air masses are called arctic or antarctic.

A second letter is used to further describe an air mass, as follows:

c   for air masses that form over continents
m   for maritime air masses that form over oceans

Continental air masses tend to be dry, and maritime air masses generally are humid.

A third letter may be used to indicate whether the air mass is warmer or colder than the area over which it is moving. If the air mass is colder than the surface it is above, the letter k is added; if it is warmer, w is used. This notation tells us something about the stability of the air mass. If it is colder than the underlying surface, its lower levels will be warmed and so may rise, probably forming clouds and perhaps precipitation. If the air mass is warmer than the surface, its lower levels will be cooled, preventing vertical movements unless forced uplift occurs.

These letter symbols are used on weather maps to identify air masses. The designation cPk would indicate a polar, continental air mass that is colder than the area over which it is traveling.

Figure 14-1
Source areas and
movements of typical
air masses on North
America.

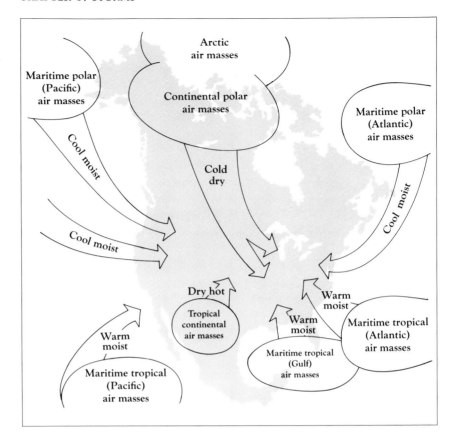

The sources of the air masses of most interest in North American
weather and their general routes of movement are shown in Figure
14-1. The Pacific maritime polar (mP) air masses originate in the area   F 14-1
of the Aleutian low and move to the east or southeast. The air is cool
and moist, and even in winter is above the freezing point. As it moves
over the mountain ranges of western North America, the resulting
uplift may cause heavy rain or snow. Once across the mountains, the
air is cool and relatively dry and so may bring moderate weather to
eastern North America in both summer and winter.

The continental arctic and polar (cA, cP) air masses that form over
northern Canada generally move to the south into the midcontinent
area between the Rocky Mountains and the Appalachians. They
bring cool, dry weather into the area, and in winter they may bring
bitter cold weather. During the winter, some of these air masses cross
the Great Lakes, where they obtain water vapor, and thus the lower
part of these air masses are warmed. On reaching the southern or
southeastern shore of the lakes, the now-moist, unstable air may deposit substantial amounts of snow (Figure 14-2).   F 14-2

The maritime polar (mP) air masses that form in the western Atlantic Ocean generally move to the south and east. A few, however,
move southwesterly into New England and the Maritime Provinces

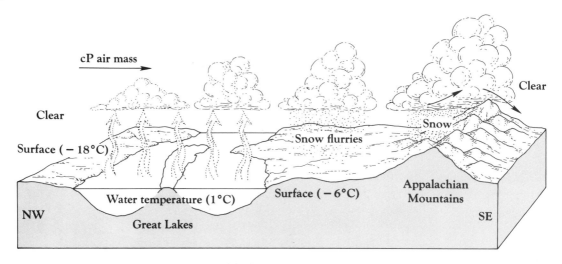

cP air mass →

Clear

Surface (−18°C)

Snow flurries

Water temperature (1°C)

Surface (−6°C)

Snow

Clear

Appalachian Mountains

NW

Great Lakes

SE

Figure 14-2 A cold, dry air mass (cP) passing over the Great Lakes obtains water vapor, and its lower portion is warmed. The resulting unstable air may deposit much snow.

of Canada, causing the "northeasters" of that region. The cool, moist air brings drizzle in summer and heavy snow in winter.

The maritime tropical (mT) air mass in the Pacific Ocean is active mainly in the winter months and may cause extreme rain in southern California and the rest of the southwest.

The Sonoran Desert of northern Mexico and the adjacent southwestern United States is the source of continental tropical (cT) air masses during summer months. This hot, dry air dominates the summer weather of this region.

The tropical maritime (mT) air masses that originate in the Pacific Ocean, the Gulf of Mexico, and the Caribbean Sea move northward, northwestward, and northeastward into the United States. This warm, humid air may come in contact with arctic and polar air masses in the midcontinent region, causing storms and precipitation.

Air masses eventually mix with the surrounding air, or their characteristics are modified so much by their new surroundings that they eventually lose their identity. The mixing processes cause new weather systems in some cases. In the next section, we examine what happens when air masses of different characteristics come together.

FRONTS

The boundary separating two air masses of different temperatures is called a **front.** Air masses of different temperatures have different densities; because the cold air stays near the ground and the warmer air aloft, they do not tend to mix. The width of a front is between 15 and 200 kilometers (9 and 125 miles), and this is narrow compared with the minimum 1600-kilometer (1000-mile) diameter of a typical air mass. On sea-level weather maps, fronts are shown as lines where they meet the ground surface, although fronts are really surfaces, not lines.

Cold Fronts

In both cold fronts and warm fronts, a warm air mass and a cold air mass are involved. Whether a front is warm or cold is determined by which air mass is moving into an area previously occupied by the other air mass. A **cold front** *develops when a cold air mass moves into a region previously occupied by warm air.* The cold air mass is heavier than the warm air, so it moves along the surface. The lower part of the cold air mass is slowed by friction as it moves along the surface. For this reason, a cold front has a steep slope, generally between 1 in 40 and 1 in 80 (Figure 14-3). (A slope of 1 in 40 means that, 40 kilome- F 14-3 ters behind where the front touches the ground, the cold air mass is 1 kilometer thick.) As the cold air moves into the warm air, the warm air is forced sharply upward along the steep front. This produces rapid cooling of the warm air, and cumulus and cumulonimbus clouds typically develop in a narrow zone along the front. If the warm air is moist, a line of thunderstorms from 300 to as much as 800 kilometers (180 to 500 miles) long may develop. Cold fronts typically move at speeds between 30 and 50 kilometers per hour (20 and 30 miles per hour), with the higher speeds more common in winter. Thus, a typical cold front passes a point on the ground in a few hours, and the weather at that point is a number of short but intense rain showers, perhaps with thunder and lightning. After the front has passed, the temperature is generally cooler, and the weather is clear with a few cumulus clouds. A cold front is shown on weather maps as a heavy line with triangles pointing in its direction of movement (see Figure 14-6, p. 345).

Figure 14-3
Cold front. Cumulus clouds are associated with cold fronts. A narrow band of heavy showers from cumulonimbus clouds generally forms at or just behind the cold front.

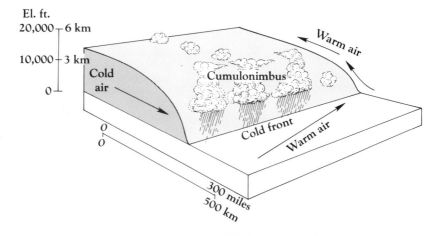

Warm Fronts

A **warm front** *develops when a warm air mass moves into an area previously occupied by cold air.* The warm air, being lighter than the cold air, moves over the cold air and so experiences little friction. For this reason, warm fronts have much shallower slopes than cold fronts (Figure 14-4). Typical slopes are 1 in 100 to 1 in 200; that is, at 100 to 200 kilometers ahead of where the front touches the ground, the bottom of the warm air is 1 kilometer above the surface. Warm fronts tend to move more slowly than cold fronts, and they typically travel 15 to 30 kilometers per hour (10 to 20 miles per hour).   F 14-4

Because warm fronts have much shallower slopes than do cold fronts, they affect the weather over a very much larger area. The first indication that a warm front is approaching is the appearance of cirrus clouds that form at elevations above 7 kilometers. These cirrus clouds may be seen first 1600 or more kilometers (1000 miles) in advance of where the warm front is passing on the surface (Figure 14-4). As the warm front approaches, the clouds become thicker and lower as their types progress from cirrus to cirrostratus to altostratus to stratus and generally to nimbostratus (Figure 14-4). Thus a fairly continuous layer of stratus-type clouds is associated with a warm front. The approach of the front may take several days from the time when the first cirrus clouds appear. An area of steady, gentle rain, up to 650 kilometers (400 miles) in diameter, precedes the front. Thus the weather in a typical warm front is a day or two of thickening, lowering stratus clouds, followed by a day or two of steady, gentle rain.

If, however, the warm air is unstable and so rises rapidly, cumulus and cumulonimbus clouds may develop on a warm front. The resulting weather may resemble that of a cold front.

A warm front is shown on weather maps as a heavy line with half circles pointing in the direction it is moving (see Figure 14-6).

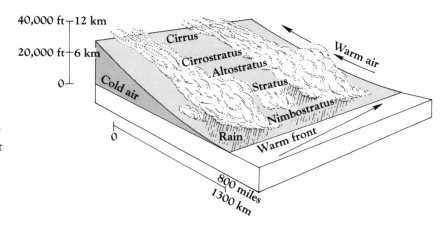

Figure 14-4
Warm front. The slope of a warm front is much less than that of a cold front. Stratus clouds are generally associated with warm fronts.

Figure 14-5
Occluded fronts.
A. Cold-type
occlusion. **B.** Warm-
type occlusion.

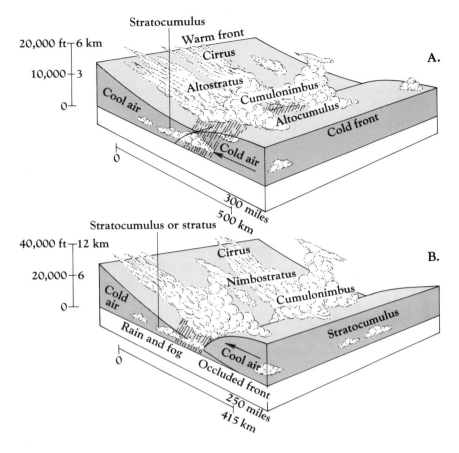

Occluded Fronts

Occluded fronts *form when a more rapidly moving cold front overtakes a warm front.* A warm front is occluded when it is no longer in contact with the ground. Two types of occluded fronts are possible. If the invading cold air is colder than the cool air ahead of the original warm front, a **cold-type occluded front** forms (Figure 14-5A). If the invading cold air is not so cold as the cold air ahead of the original warm front, a **warm-type occlusion** is the result (Figure 14-5B). The weather and clouds that are found in occluded fronts are a mixture of those associated with both warm and cold fronts. An occluded front is indicated by a heavy line with alternate triangles and half circles on the same side (see Figure 14-6).

F 14-5

Stationary Fronts

A boundary between warm and cold air masses that does not move apprecia-bly is a **stationary front.** This situation may develop if the flow within the cold air mass is more or less parallel to the front. A stationary front is indicated on weather maps by a heavy line with triangles on the warm-air side, alternating with half circles on the cold-air side (see Figure 14-6).

CYCLONIC STORMS OF THE MIDDLE LATITUDES

Cyclones are low-pressure areas in which the winds move counterclockwise in the northern hemisphere and clockwise in the southern hemisphere. Most of the storms in the middle latitudes occur in low-pressure systems. The typical cyclonic storms of the middle latitudes are low-pressure systems that contain both a warm front and a cold front. They are commonly called *cyclones,* although the term *cyclone* only means a low-pressure system. To reduce confusion, the cyclonic storms of the middle latitudes are also called **wave cyclones,** or **extratropical cyclones.** The latter term also serves to differentiate these storms from the destructive tropical cyclones (hurricanes).

These storms are called *wave cyclones* because they form as waves or perturbations along a stationary front such as the polar front (Figure 14-6). Along the polar front in the northern hemisphere, cold air is F 14-6 to the north, and warm air is to the south. The wind in the cold air is easterly, and in the warm air it is westerly. When these winds are parallel, they create a mutual shearing or dragging that prevents smooth flow. At some point a shallow, wave-like undulation, up to several hundred kilometers long, develops in the position of the front. In a map view, these waves resemble cross-sections of ocean waves. The cold air from the north moves southwesterly under the warm air, and the warm air from the south moves northeasterly over the cold air. Once such a wave is started, it quickly develops as tongues of warm air move to the northeast and cold air to the south and west. These tongues form active warm and cold fronts.

Minor waves may develop at many places along the polar front, but only a few of these grow into cyclonic storms. Studies of the air aloft suggest that certain conditions favor the formation of cyclonic storms. A low-pressure system, or cyclone, at the surface is an area of convergence, and divergence aloft is necessary to maintain that cyclone. When conditions aloft create divergence, a cyclonic storm can form at the surface. At times such conditions aloft can create a cyclone on the surface, even where there was no stationary front. Divergence aloft can intensify or deepen a storm once cyclonic flow has been established. This is why upper-atmosphere data are important in forecasting surface weather.

Once a cyclonic storm has developed, it goes through a definite sequence of stages. At first it is *open,* with warm and cold fronts meeting at the center of the low-pressure area (Figure 14-6C). Because cold fronts move somewhat faster than warm fronts, the cold air pushes under the warm air at the center of the storm, producing the *occluded*

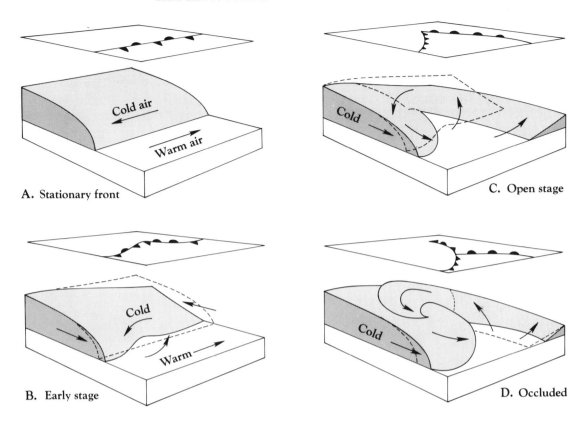

A. Stationary front

C. Open stage

B. Early stage

D. Occluded

Figure 14-6
Life cycle of a cyclonic storm of the middle latitudes.

stage (Figure 14-6D), with its characteristic curved occluded front. The final stage is the dying of the storm by the mixing of the two air masses. The occluding cold air reunites with the rest of the cold air mass, reforming the polar front and isolating the remaining parcel of warm air above the cold air. A cyclonic storm may last from a few days to over a week. The storm center travels generally eastward at between 30 and 60 kilometers per hour (20 and 40 miles per hour). Some such storms travel halfway around the world, although most move much less far (Figure 14-7).

F 14-7

When the conditions are right, a series of storms generally forms in the same area. This family of cyclonic storms travels more or less along the same route, so that a place on its path experiences several storms, one after the other. Cyclonic storms form mainly in only a few places. Those that form along the western Pacific Ocean move toward North America. Those that form in the western Atlantic move toward Europe. Those that form in the plains of Canada and the United States move eastward and northeastward. Along the polar front in the southern hemisphere, the storms form and move in the zone between 40 and 50 degrees south.

A better understanding of the typical mid-latitude storm will be gained from study of the sequence of the events that occur as such a

Figure 14-7
Common paths of
low-pressure systems
(cyclones) across the
United States.

storm passes a point (Figure 14-8). First we will consider what hap- F 14-8
pens in the area south of the center of the storm (in the northern
hemisphere) and then in the storm center where the occlusion develops.

In the area south of the path of the storm center, the first indication that the warm front is approaching is the appearance of cirrus
clouds 1000 kilometers (600 miles) or more ahead of the front. Falling
barometer readings and thickening and lowering clouds herald the
approaching front. Rain, snow, or sleet begins a half-day to a full day
after the first sighting of cirrus clouds. As the warm front nears, the
temperature increases, as does the amount of precipitation. When the
warm front passes, the winds shift from easterly to southerly, and the
precipitation ceases. The weather in the warm air mass between the
fronts is generally clear, with warm southerly winds and perhaps a
few cumulus clouds. This weather may last up to a day, depending on
how far apart the fronts are. As the cold front approaches, towering
cumulonimbus clouds are seen, and the winds become gusty. A line
of thunderstorms, called a **squall line,** may precede the actual cold
front if it is an active, fast-moving front. Showers, hail, and even tornadoes may be associated with a squall line. Heavy showers and gusty
winds mark the cold front. As it passes, the winds shift from southerly to westerly or northwesterly, the temperature falls, and the pressure rises. The weather in the cold air mass is generally clear, with
bright blue skies and cool temperatures.

An observer in the path of the storm center, or the occluded part
of the storm, will not see so many changes as the storm passes. The
ground surface in this part of the storm remains in the cold air mass,
so the temperatures remain cool throughout the storm. The approaching storm will be preceded by easterly winds and thickening

and lowering clouds, as well as by falling barometer readings. The first clouds will be cirrus, and stratus types will quickly follow. Precipitation will occur sooner because here the sequence of clouds associated with the warm front aloft will be compressed. As the storm center approaches, occlusion also will begin, and this slows the passage of the storm and gives the occluded front its backward curve. The storm is more severe in this area, both because it is slowed and so spends more time there and because the center is the most intense part of the storm. The transition from the steady rain of the warm front into the intense showers of the occluded front occurs without passing through a clear area. When the storm center passes to the north, the pressure begins to rise, the winds change from easterly to southerly, and the precipitation continues. As the precipitation dies out, the winds become more northwesterly or westerly, and the clear, crisp weather of the cold air mass begins.

Most middle-latitude storms proceed more or less in this way. If you understand the changes in temperature, pressure, clouds, and wind that occur as a storm moves past, you can better predict what will probably happen next. Just seeing the changes occur and knowing what will happen next gives one a good feeling. Unfortunately, most cyclonic storms are followed by other cyclones in sequence. Thus you may see the whole sequence, or at least parts of the sequence, repeated several times as a number of cyclones march in line past your location.

SEVERE STORMS

The severe storms described in this section are thunderstorms, tornadoes, and hurricanes (tropical cyclones). They are termed severe because they are the most destructive storms. That is not to say that ordinary mid-latitude cyclonic storms do not cause any damage. Heavy snow storms, blizzards, hail, and ice storms surely cause damage and discomfort, as do prolonged heat waves or cold waves.

Thunderstorms

Thunderstorms are the most widespread severe storms. They are commonly associated with cold fronts but also occur in warm fronts and as local storms. Hail and lightning that accompany thunderstorms cause much damage, and lightning kills more people in the United States each year on the average than either tornadoes or hurricanes. Local thunderstorms have been studied in some detail, and their development illustrates the processes involved in all thunderstorms.

LOCAL THUNDERSTORMS Thunderstorms are most likely to form in warm, moist air masses. They are most common in the afternoon or

February 1, 1979

February 2, 1979

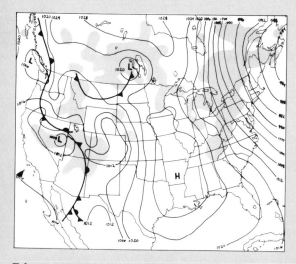

February 1, 1979

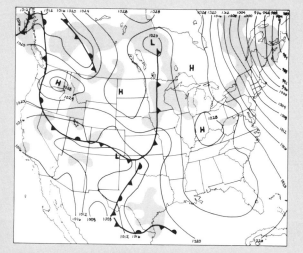

February 2, 1979

Figure 14-8 Typical weather patterns. Surface weather maps and satellite images for consecutive days. Shaded areas indicate precipitation. (Maps and photos from NOAA.)

February 3, 1979

February 4, 1979

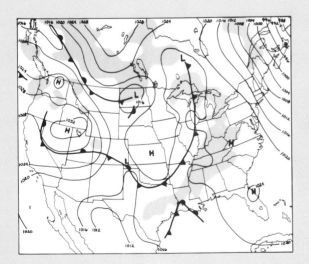

February 3, 1979

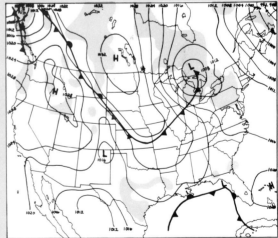

February 4, 1979

June 28, 1979

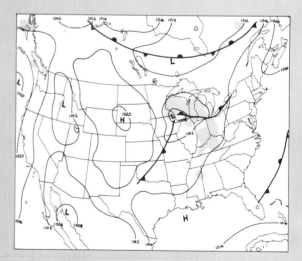

June 29, 1979

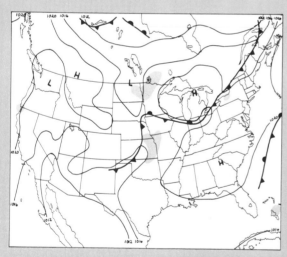

June 28, 1979

June 29, 1979

Figure 14-8 (continued) Typical weather patterns. Surface weather maps and satellite images for consecutive days. Shaded areas indicate precipitation. (Maps and photos from NOAA.)

June 30, 1979

July 1, 1979

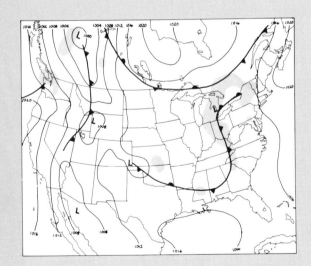

June 30, 1979

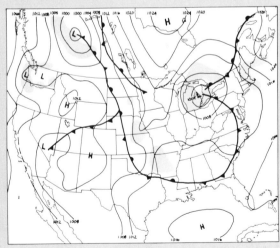

July 1, 1979

Figure 14-9
Development of a
thunderstorm.

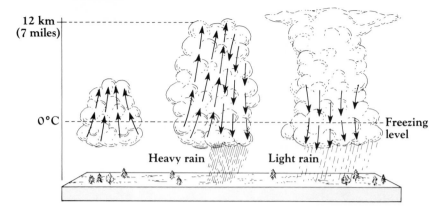

early evening in spring and summer, because this is when differences
in surface heating are greatest. To start a thunderstorm, the moist air
must rise, and the rising air must be warmer than the surrounding air.
The latent heat of condensation is generally important in maintain-
ing this unstable situation as the cumulus cloud develops. In most
cases surface heating alone is not enough to start a thunderstorm; the
air must be forced to rise by hills or mountains or by a cold front.
The rising moist air first forms a cumulus cloud. As more moist air is
added to the cloud, its height grows, and the dominant motion in the
cloud is upward. This is called the *cumulus stage* (Figure 14-9). If the   F 14-9
cloud grows to above the freezing-temperature elevation, precipitation develops in it by the processes discussed in Chapter 12. The precipitation falls and creates downdrafts. The downdrafts are inten-
sified by adjacent cold air, which is also brought down by the falling
precipitation. This stage, with both rising and falling air in what is
now a cumulonimbus cloud, is called the *mature stage*. At this stage
heavy rain or perhaps hail is falling below the cloud, and the winds
are gusty. The downdrafts from the cloud are the cause of the cool,
gusty winds that often precede a thundershower. As the downdrafts
become more dominant, the *dissipating stage* begins. The incoming
cold air and the lack of a continuing supply of warm, moist air combine to cause the cloud to evaporate. Generally, as this stage is entered, the cloud top has grown to over 10 kilometers (6 miles) in
height, and the winds there move the ice crystals in the cloud downwind, creating the familiar anvil-shaped thunderhead, or cumulonimbus cloud.

ORGANIZED THUNDERSTORMS Organized thunderstorms are those
associated with fronts, and they form in lines called *squall lines*.
Organized thunderstorms tend to be more persistent and more intense than local thunderstorms. Along active cold fronts, tornadoes
also may form. The thunderstorms that form at cold fronts were described earlier. One feature of active cold fronts that was only men-

tioned previously is the squall line that may develop in front of the cold front. One way this squall line is created is by the downdraft from the thunderheads on the cold front. Some of the cold air in the downdraft surges ahead of the front and so is, in effect, another cold front. Thunderstorms form along this new cold front, creating a squall line. Once formed, this squall line can be reformed by the downdrafts in the dissipating stage of its thunderstorms, and in this manner the squall line is self-perpetuating. Such a squall line can travel faster than the cold front that first formed it, and so it may be as much as 300 kilometers (180 miles) ahead of the cold front.

Thunderstorms are most common in the area to the east of the Rocky Mountains, where moist tropical air from the Gulf of Mexico meets cool air from the north. The area of the United States that has the most thunderstorms is Florida, where they are formed by converging winds from the Gulf of Mexico and the Atlantic Ocean.

Hail is common with thunderstorms. It apparently forms at the mature stage, when updrafts and downdrafts cause precipitation to be cycled through the cloud many times. A new layer is deposited on the hailstone during each cycle. Hail tends to fall in streaks a few hundred meters wide by 8 kilometers or so long.

LIGHTNING AND THUNDER Perhaps the most frightening and fascinating aspect of thunderstorms is lightning and thunder. **Lightning** is produced because the processes that occur in a thunderhead separate the static electrical positive and negative charges. When the resulting electrical field between the earth and the cloud becomes great enough, the electric charges flow between them in the form of a spark (the lightning), and this flow reduces the electrical field.

The process of charge separation that occurs in a cloud is believed to be related to the freezing of the water drops in the cloud. Other processes must also be at work, however, because lightning can occur in tropical clouds, no part of which is below the freezing point. Most lightning does come from clouds whose tops are below freezing. When a cloud drop freezes, it freezes from the outside inward. The freezing process concentrates positive charge on the surface and negative charge in the still-liquid core. When the core does freeze, it expands and shatters the drop. The core, with the negative charges, is heavier than the shattered pieces of the surface, and so it falls toward the base of the cloud. This gives the cloud base a negative charge. The lighter pieces of the surface of the drop are carried upward by the rising currents in the cloud, giving the top a positive charge.

The negative charge at the base of the cloud causes a positive charge to develop on the ground surface below the cloud. A **lightning flash** begins when the negative charge moves down from the cloud. It moves downward in a series of advances and so is called a **stepped**

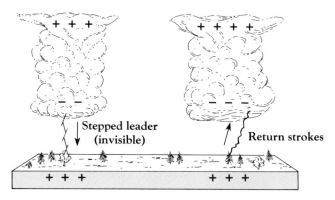

Figure 14-10 Lightning. The stepped leader moves downward in a series of steps. The visible lightning is the return strokes, which move upward from the ground in air that was ionized by the stepped leader.

leader. The stepped leader is almost invisible. The visible parts of a lightning flash are the **return strokes.** These occur after the leader has ionized a path between the cloud and the ground. The return strokes move from the ground to the cloud, even though it appears just the opposite to an observer (Figure 14-10). A lightning flash is generally F 14-1• made up of three or four strokes, although many more are possible. Lightning appears to flicker when we observe it because of the successive return strokes that carry electrical charge from higher and higher parts of the cloud. Most fires started by lightning are caused by exceptionally long strokes.

Safety in a lightning storm is a matter of seeking shelter inside a building and away from electrical conductors, especially if they are exposed to the lightning. Lightning naturally tends to strike tall objects, so trees and towers should be avoided. Most people struck by a nearby lightning stroke, although unconscious and not breathing, can be saved by artificial respiration.

The **thunder** that accompanies lightning is caused by the very high temperatures created by the lightning stroke. The high temperatures cause the air to expand explosively. Thunder can be heard up to about 20 kilometers (12 miles) away. The light from a lightning flash travels with the speed of light, so we see it almost as it occurs. The sound of the thunder travels much slower—about 330 meters per second (about 1000 feet, or 0.2 mile, per second) in air. Thus, if you count the number of seconds between the time when you see a lightning flash and the time when you hear the thunder, you can easily calculate the distance to the lightning.

Lightning may also go from one cloud to another. All lightning is similar, and the phenomenon called **heat lightning,** commonly seen on warm evenings, is merely lightning too far away for the observer to be able to hear the thunder.

Tornadoes

Tornadoes, also called **twisters, cyclones,** or **funnel clouds,** are the most destructive storms (Figure 14-11A). They are associated with F 14-11 thunderstorms on active cold fronts and so are common in the Great Plains of North America where cold polar air comes in contact with warm, moist air from the Gulf of Mexico. This area has more tornadoes than any other place on earth. They tend to be most common in spring or early summer, when temperature contrasts are greatest, but they occur in all months.

A tornado is a small, rapidly spinning body of rising air. Its diameter is usually a few hundred meters. Wind velocities are estimated to reach as high as 800 kilometers per hour (500 miles per hour), but the average is probably less than 500 kilometers per hour (300 miles per hour). Needless to say, few weather stations could survive such conditions and provide accurate data; even if they could, tornadoes are such small storms that it would be difficult to place a weather station in a tornado's path. The atmospheric pressure in the center of a tornado is estimated to be up to 100 millibars lower than the surrounding pressure. As a result, a building in the path of a tornado explodes because the air trapped in the building expands explosively when the outside pressure is reduced by the passing tornado. Opening windows may reduce tornado damage.

The high winds and explosions cause the severe damage of tornadoes. Generally, the area of destruction is small, but the damage in that area is nearly complete. The damage is typically confined to a strip a few hundred meters wide and a few tens of kilometers long. Tornadoes move at speeds of 40 to 65 kilometers per hour (25 to 40 miles per hour) and so do their damage in a few minutes.

A tornado is a funnel-shaped cloud, extending downward from a cumulonimbus cloud. The air in the cloud is not only spinning rapidly but also rising rapidly. A tornado is believed to start with the violent updraft associated with some thunderstorms. This updraft is several kilometers in diameter, and the rising air has appreciable horizontal speed. If the updraft is strong enough, the air converges on the center of the updraft. As the air converges, its angular momentum is conserved, and its horizontal spin speed increases greatly. (**Angular momentum** is mass times spinning speed times radius.) The process is similar to what happens to the skater who is spinning with arms outstretched. As the arms are brought in to the body, the spin becomes much faster. The air in a funnel cloud expands because the pressure is reduced, and, as a result, it is cooled below its dew point, causing condensation. This releases the latent heat of condensation and so adds to the energy in the tornado by heating the air, causing it to rise faster.

A.

Figure 14-11
Tornadoes. **A.** Funnel-
shaped cloud of a
tornado. **B.** Tornado
damage. (Photos from
The National Center
for Atmospheric
Research, Boulder,
Colorado.)

B.

Tornadoes cause great damage where the funnel-shaped clouds touch the surface (Figure 14-11B). They form and cause destruction in a few minutes. Fortunately, the severe thunderstorms that spawn tornadoes can be recognized on radar screens. When such thunderstorms are noted, a **tornado watch** is announced. If a tornado is actually reported, a **tornado warning** is instituted. Tornadoes also produce lightning and emit other forms of electrical waves. You can see this electrical energy on a television screen. Tune the set to one of the high-frequency channels, such as channel 13, and reduce the brightness until the screen is dark. Then switch to a low-frequency channel, such as channel 2, and if severe thunderstorms or tornadoes are in the vicinity, the screen may glow.

Waterspouts are similar to tornadoes but occur over water. They are smaller, averaging about 50 meters (150 feet) in diameter, and they have winds averaging 80 kilometers per hour (50 miles per hour), although these figures are doubled in some. They tend to form over warm, shallow water during fair weather and are associated with cumulus clouds whose tops are below the freezing-point elevation. They last only about 15 minutes and so may more closely resemble dust devils on land than tornadoes. Their dark color is caused by condensation droplets, not by water sucked up by the cloud. Most waterspouts cause damage only in a very small area and have nowhere near the destructive power of tornadoes. Occasionally a tornado will move from land into water, and the **tornadic waterspouts** created in this way are quite destructive.

Hurricanes

Hurricanes are tropical storms without fronts, and they depend mostly on latent heat of condensation for their energy. The other types of storm that we have considered have fronts and get much of their energy from air masses of different temperatures. Hurricanes are big destructive storms. They are called **typhoons** in the Pacific, **cyclones** or **tropical cyclones** in the Indian Ocean, and **willy-willys** in Australia. A tropical storm is called a *hurricane* if the wind speed exceeds 120 kilometers per hour (75 miles per hour). Hurricanes form over warm oceans in the regions of the tropical easterly winds, but not in equatorial waters. Most form in the Pacific Ocean, but none forms in the southeastern Atlantic Ocean, mainly because the surface temperatures are too low there.

Most hurricanes form in late summer, when the surface waters are warmest and the air is very humid. Such conditions favor the rise of the warm, moist air, and cumulonimbus clouds develop. These clouds must enlarge or join to form a hurricane. As in the formation of mid-latitude storms, the conditions in the upper atmosphere must favor surface convergence. Once surface convergence is established, the

Figure 14-12
Satellite image of Hurricane Allen approaching the Texas coast on August 8, 1980. This storm was responsible for at least 272 deaths and much property damage. (Photo from NOAA.)

newly formed hurricane supplies its own energy by condensation. The warm, moist air spirals inward toward the center of the developing low-pressure area, where it rises. The rising air cools, condensation occurs, and the latent heat of condensation heats the rising air, causing it to rise faster. The rising air lowers the pressure and causes more air to spiral in and rise. In this way, the hurricane builds, getting its energy from condensing water vapor.

The high wind speeds come from this spiraling inward. The angular momentum of the wind is conserved, so, as its radius (distance from the center) is reduced, its speed must increase. The process is the same as that described for tornadoes. Winds as high as 325 kilometers per hour (200 miles per hour) have been observed in hurricanes. This is near the upper limit imposed by the friction of the wind passing over the ocean. When the inward-spiraling wind gets near this speed, it can no longer accelerate because of friction and therefore begins its spiraling rise. This limit is reached about 10 or 15 kilometers (6 or 9 miles) from the hurricane center, so there is a calm area about 20 to 30 kilometers (12 to 18 miles) in diameter in the center of a hurricane.

A typical hurricane has a diameter of 600 kilometers (370 miles), as shown in Figure 14-12. The pressure drop at its center can be as much as 100 millibars but is usually less, and between 15 and 30 centimeters (6 and 12 inches) of rain falls as it passes a point. Hurricanes form in the zone of tropical easterly winds and so move generally westward or northwestward at speeds of about 25 kilometers per hour (15 miles

F 14-1

Figure 14-13
Typical hurricane paths.

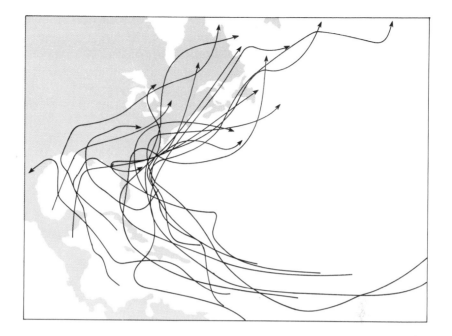

per hour). Eventually, most move northward into the westerly winds, and there they travel at speeds up to 100 kilometers per hour (60 miles per hour). Their tracks are very irregular (Figure 14-13), making F 14-13 prediction of their future location very difficult. Satellite observation has aided in this phase of hazard reduction.

Hurricanes die out by mixing with cold air or by moving over land. Above land, their source of energy in the form of condensing water vapor is reduced or removed, and the increased surface friction reduces the wind speed. The worst damage generally occurs where they hit the coastline. The high winds cause much damage, but the associated **storm surge** usually causes even more destruction. The storm surge is a very high tide caused by the combination of strong onshore winds and the reduced pressure at the hurricane center. Even after moving inland, a dying hurricane may cause very heavy rainfall.

Some attempts have been made to reduce hurricane damage by cloud seeding. The idea is to seed the clouds and cause condensation and freezing before they have moved into the high-wind part of the hurricane. In this way the hurricane's energy, which comes from condensation, would not reach the center of the hurricane. This could prevent a storm from growing into a hurricane or perhaps cause a hurricane to die out.

FORECASTING

Weather forecasting began as an art and then progressed to a science. The keys to accurate forecasts are the timely gathering of accurate data and the analysis of them in time to make a useful prediction. The telegraph was the first step, and as communications improved, so did forecasting. Today, most countries are members of the World Meteorological Organization. Every day about 10,000 surface stations report current weather information four times a day, and many fewer stations report upper-atmosphere conditions using radiosondes. These data are reported every six hours to three centers—Washington, D.C.; Melbourne, Australia; and Moscow—where synoptic charts are prepared. *A* **synoptic chart** *is a "snapshot" of the weather at a particular moment* (see Figure 14-8). These centers prepare and distribute surface weather maps, as well as charts of the upper atmosphere. In addition, weather satellites report cloud cover and other data.

Until the 1960s synoptic charts were the main source of weather forecasts. Fronts and other features were carefully analyzed, and a series of empirical rules was developed for how rapidly they would move or dissipate. This method is generally accurate only for a few hours to a day or so, but it is still used for local forecasts. With the advent of high-speed computers, it has become possible to use **numerical weather prediction.** In this method the atmospheric data are fed into computers programmed to apply physical laws to the behavior of the atmosphere. The resulting prognostic charts are more accurate for up to a few days. Other approaches to weather forecasting include **persistence forecasts,** which are somewhat similar to synoptic forecasts and are used locally for short-range prediction; **analog forecasts,** which use similar situations in the past on which to base prediction; and **statistical forecasts,** in which predictions are made according to the most probable weather. At best, accurate, detailed forecasts are possible for only a few hours to a few days, and long-range forecasts are still quite general.

SUMMARY

- *Air masses* are large volumes of air that have uniform characteristics at any given altitude.

- Air masses are named for the place in which they originate:

  A Arctic (Antarctic)   form near the poles
  P Polar   form near the polar front
  T Tropical   form near the subtropical high
  E Equatorial   form near the equator

- The boundary separating two air masses of different temperatures is called a *front.*

- A *cold front* develops when a cold air mass moves into a region previously occupied by warm air. Cold fronts have steep slopes, move at speeds between 30 and 50 kilometers per hour (20 to 30 miles per hour), and have cumulus clouds associated with them.

- A *warm front* develops when a warm air mass moves into an area previously occupied by cold air. Warm fronts have gentle slopes, move somewhat slower than cold fronts, and have stratus clouds associated with them.

- Precipitation associated with warm fronts is generally gentle rain (or snow) over a large area, and that with cold fronts is a narrow band of heavy showers.

- *Occluded fronts* form when a more rapidly moving cold front overtakes a warm front.

- A *stationary front* is a boundary between warm and cold air masses that is not moving appreciably.

- *Cyclonic storms of the middle latitudes* are also called *wave cyclones* or *extratropical cyclones*. They form in areas of low pressure and have counterclockwise circulation in the northern hemisphere.

- Cyclonic storms of the middle latitudes have both a warm and a cold front. As the storm passes a point, first the warm front and then the cold front influences the weather. They die as the result of the mixing of the two air masses, and the occluded stage is part of the mixing.

- *Thunderstorms* develop from tall cumulus clouds. *Organized thunderstorms* are associated mainly with cold fronts and tend to form in *squall lines*.

- *Lightning* is produced when static charges in clouds are separated and the charge flows between the cloud and the earth. A lightning *flash* begins when a negative charge moves down from the cloud in a series of advances called a *stepped leader*. The visible part of the flash consists of the *return strokes* that follow the same path and move from the earth to the cloud.

- *Tornadoes* (twisters, cyclones, funnel clouds) are the most destructive storms. They are only a few hundred meters in diameter but may have winds as high as 800 kilometers per hour (500 miles per hour). Where the funnel-shaped cloud touches the ground, the path of destruction is only a few hundred meters wide. Low pressure in the cloud causes buildings to explode.

- *Waterspouts* are similar to tornadoes but occur over water. Normally they are not destructive, but a tornado that moves onto water may be destructive.

- *Hurricanes* (typhoons, cyclones, willy-willys) are tropical storms with winds over 120 kilometers per hour (75 miles per hour). Their energy comes from condensing water vapor. Typically they are 600 kilometers (360 miles) in diameter. At coastlines they may cause flooding by their *storm surge*.

- Weather forecasting depends on accurate, widespread data gathering.

- A *synoptic chart* is a snapshot of the weather at a particular moment.

KEY TERMS

air mass	wave cyclone, or extratropical cyclone	heat lightning	cyclone (Indian Ocean); **willy-willy** (Australia)
front		tornado, twister, or funnel cloud	
cold front	squall line	angular momentum	storm surge
warm front	thunderstorm	tornado watch	synoptic chart
occluded front	organized thunderstorms	tornado warning	numerical weather prediction
cold-type occluded front		waterspout	
warm-type occluded front	lightning, lightning flash	tornadic waterspout	persistence forecast
		hurricane; typhoon (Pacific Ocean);	analog forecast
stationary front	stepped leader	cyclone, tropical	statistical forecast
cyclone	return strokes		
	thunder		

REVIEW QUESTIONS

1. Define an air mass. About how large are air masses?

2. At what latitudes are air masses most likely to form? Do they generally form in high-pressure or low-pressure areas? Explain your answers.

3. The following are the names of the air masses that most affect North American weather. For each mass, state the place of origin, characteristics (hot, cold, wet, dry), direction(s) in which the mass tends to move, and the kind of weather it brings to the continental United States:
   - Pacific mP
   - cA and cP
   - Atlantic mP
   - Pacific mT
   - continental cT
   - Gulf and Caribbean mT

4. Air masses have very narrow fronts and tend not to mix. Why is this so?

5. At both cold fronts and warm fronts, the warm air mass lies over the cold air mass. True or false?

6. Explain what the following means:
   (a) A cold front has a slope of 1 in 60.
   (b) A warm front has a slope of 1 in 140.

7. The weather forecaster says that we can expect heavy showers this morning, followed by clear skies and cooler temperatures. What kind of front is probably moving in? Explain how this front can cause the weather predicted for today.

8. Describe the cloud patterns that are likely to appear during the approach of a warm front. Explain why this series of clouds forms.

9. When is a warm front occluded? What causes the occlusion? What is the difference between a warm-type occluded front and a cold-type occluded front?

10. When is a stationary front most likely to develop?

11. Write the abbreviation or symbol for each of the following meteorological terms:
    - Cold continental arctic
    - Maritime polar
    - Warm continental tropical
    - Maritime equatorial
    - Cold front
    - Warm front
    - Occluded front
    - Stationary front

12. Do most storms of the middle latitudes occur in low-pressure or high-pressure systems?

13. Differentiate among these three terms: cyclone, wave (or extratropical) cyclone, and tropical cyclone.

14. Stages in the life of a middle-latitude storm in the northern hemisphere are listed below. Describe what is happening during each step. In your descriptions, state the positions of cold and warm air masses along the fronts and the location of any low-pressure system.
    (a) Parallel winds along polar front create a wave.
    (b) Divergence aloft.
    (c) Open cyclonic storm forms.
    (d) Occlusion occurs.
    (e) The storm dies out.

15. Name three major areas in which cyclonic storms of the northern hemisphere form. Toward what areas do these storms move?

16. While a cyclonic storm is in the occluded stage, its southern portion passes over your area. During the days that the storm is passing through, you make the following observations:
    - Day 1: Cirrus clouds in the early morning. By midday, clouds thickening and lowering and barometer falling. Winds easterly.
    - Day 2: Rain all day. Temperature rose late in afternoon.
    - Day 3: Clear; warm; southerly winds.
    - Day 4: Heavy showers. Umbrella turned inside-out.
    - Day 5: Clear; cool. Barometer risen; winds from northwest.
    (a) Draw a sketch of the air circulation within a cyclonic storm during the occlusion stage.
    (b) On your sketch mark the part of the storm that is passing over your area on each of the days described above.

(c) Explain what is causing each of your observations.

17. Explain why the formation of a thunderstorm requires both warm, moist air and either mountains or an approaching cold front. Where in the continental United States are thunderstorms most common? Why?

18. (a) Describe the mature stage of a thunderstorm.
    (b) What causes the gusts of wind just before and during the downpour?
    (c) Why is hail commonly associated with thunderstorms?

19. Describe the dissipating stage of a thunderstorm. Explain how the dissipation process can maintain the squall line of an incoming cold front.

20. (a) Write an outline of the steps involved in the formation of lightning. Or, if you prefer, make a series of sketches to show the process.
    (b) What causes thunder?

21. Tornado damage is caused by very high winds and a sudden, extreme drop in air pressure.
    (a) Explain how these damaging conditions develop.
    (b) What causes condensation within the tornado? What effect does the condensation have on the storm?
    (c) Where and at what time of year are tornadoes most likely to occur? Why?

22. (a) Unlike extratropical cyclones, thunderstorms, and tornadoes, hurricanes require no front (or mountain) to get them started. How, then, does a hurricane originate?
    (b) How does condensation cause the cyclone to build to hurricane force?
    (c) Why does a calm exist in the eye of a hurricane?
    (d) Where do most of the world's hurricanes form? Why?
    (e) Why do hurricanes die out soon after they move over land?

SUGGESTED READINGS

Battan, L. J. *Radar Meteorology.* Chicago: University of Chicago Press, 1959, 1961 pp.

Battan, L. J. *Nature of Violent Storms.* Garden City, N.Y.: Doubleday, 1961, 160 pp.

Battan, L. J. *The Thunderstorm.* New York: Signet Science Library, 1964, 128 pp.

Changnon, S. A., Jr., and R. G. Semonin. "A Great Tornado Disaster in Retrospect," *Weatherwise* (1966), Vol. 19, pp. 56–65.

Dunn, G. E., and B. I. Miller. *Atlantic Hurricanes.* Baton Rouge: Louisiana State University, 1964, 377 pp.

Flora, S. D. *Tornadoes of the United States.* Norman: University of Oklahoma Press, 1954, 221 pp.

Flora, S. D. *Hailstorms of the United States,* 2nd ed. Norman: University of Oklahoma Press, 1956, 201 pp.

Fujita, T. T. "Tornadoes Around the World," *Weatherwise* (1973), Vol. 26, pp. 56–62, 79–83.

Idso, S. B. "Tornado or Dust Devil: The Enigma of Desert Whirlwinds," *American Scientist* (September–October 1974), Vol. 62, No. 5, pp. 530–541.

Lane, F. W. *The Elements Rage.* Philadelphia: Chilton, 1965, 346 pp.

Ludlum, D. M. *Early American Hurricanes 1492–1870.* Boston: American Meteorological Society, 1963, 200 pp.

Ludlum, D. M. *Early American Winters 1604–1870.* Boston: American Meteorological Society, 1966, 285 pp.

McClement, F. *The Anvil of the Gods.* Philadelphia: Lippincott, 1964, 272 pp.

Schonland, B. *The Flight of Thunderbolts.* Fair Lawn, N.J.: Oxford University Press, 1964, 182 pp.

Stewart, G. R. *Storm.* New York: Modern Library, 1947, 348 pp.

Viemeister, P. E. *The Lightning Book.* Garden City, N.Y.: Doubleday, 1961, 316 pp.

CLIMATE

CLIMATIC CHANGE

External Causes of Climatic Change

GEOLOGIC CAUSES

ASTRONOMICAL CAUSES

ATMOSPHERIC TRANSPARENCY CHANGES

Human Causes of Climatic Change

URBAN WEATHER

AIR POLLUTION

Gaseous Pollutants

Oxides of Carbon

Oxides of Sulphur

Oxides of Nitrogen

Hydrocarbons

Concentration of Pollutants—Inversions

Types of Pollution

Industrial Smog

Photochemical Smog

OTHER INADVERTENT CAUSES OF CLIMATIC CHANGE

Deliberate Climate Modification

Changes in the Grinnell Glacier in Glacier National Park, Montana, reflect climatic changes. (Photos from U.S. Geological Survey.)

CLIMATES AND CHANGING CLIMATES

1900

1911

1935

1956

CLIMATE

Climate *is the typical combination of weather elements that characterizes an area over a long term; it takes into account seasonal changes as well as ranges and extremes of weather elements.* The necessary weather data for climatic classification are not available for much of the earth, so climates are classified by other means, mainly by vegetation types.

The most widely used climatic classification was devised in 1918 by Wladimir Köppen. He used the monthly and yearly precipitation and temperature of an area and determined the relationship between the precipitation and the potential evaporation in that area. He devised five major headings: A, tropical; B, dry; C, humid with mild winter; D, humid with cold winter; and E, polar. These headings are subdivided into 14 secondary headings, and these, in turn, are further subdivided into a total of 31 climatic types (Table 15-1). Köppen believed that the vegetation of an area was the best overall indicator of climate, so he used plant associations to draw climatic boundaries on maps. The boundaries shown on such maps are not exact but rather indicate average locations.   T 15-1

The Köppen climate classification has been widely used because it is relatively easy to use, but many other classifications have also been proposed. The climate is the product of many factors. Temperature and precipitation are the main factors used by Köppen. These and the other elements of weather are generally controlled by many factors, such as the latitude, whether the area is near the ocean or inland, the prevailing wind pattern, and whether the area is in a zone of rising (low-pressure) or descending (high-pressure) air.

Figure 15-1 shows average precipitation and temperatures for a number of climates, and Figure 15-2 shows the worldwide distribution of climate. Fewer climatic types are used in these figures than are used in the Köppen classification, and the approximate correspondence is indicated in Table 15-2.   F 15-1 F 15-2

T 15-2

CLIMATIC CHANGE

The first question about climate is: Does it change? The answer is clearly yes; climate does change over long lengths of time. The geologic record provides many examples, perhaps the best of which is the most recent ice ages, or glacial times. Other examples of climatic changes in the geologic past are described in Chapter 7. The next question is whether climate changes in shorter time intervals, especially intervals short in terms of human lifetimes. Here the evidence is nowhere nearly so clear. Although changes can be seen and documented, it is not at all clear whether these are long-term changes or merely fluctuations of more or less constant climates. We will look at a few examples of recent climatic change and then search for the possible causes of changing climate.

Table 15-1   KÖPPEN CLIMATE CLASSIFICATION

A. Tropical

Af	wet	
Am	short dry season	
Aw	winter dry season (wet and dry)	
As	summer dry season (rare)	

B. Arid (desert) and semiarid (steppe)

BSh	semiarid hot	
BSk	semiarid cold	
BWh	arid hot	
BWk	arid cold	

C. Humid mild winter

Cwa	winter dry	subtropical
Cwb	winter dry	marine
Cwc	winter dry	marine cool
Csa	summer dry	subtropical
Csb	summer dry	marine
Csc	summer dry	marine cool
Cfa	no dry season	subtropical
Cfb	no dry season	marine
Cfc	no dry season	marine cool

D. Humid severe winter

Dwa	winter dry	warm summer
Dwb	winter dry	cool summer
Dwc	winter dry	subarctic
Dwd	winter dry	subarctic cool
Dsa	summer dry	warm summer
Dsb	summer dry	cool summer
Dsc	summer dry	subarctic
Dsd	summer dry	subarctic cool
Dfa	no dry season	warm summer
Dfb	no dry season	cool summer
Dfc	no dry season	subarctic
Dfd	no dry season	subarctic cool

E. Polar

ET	tundra	
EF	ice cap	

Temperature records have been kept for the last hundred years, and they show a slight gradual increase in temperatures up to about 1940 and then a similar gradual decline. The shorter-term fluctuations of a few years' duration are of almost the same magnitude, and this makes interpretation difficult. These short-term fluctuations probably correspond to local floods or droughts. The advance or re-

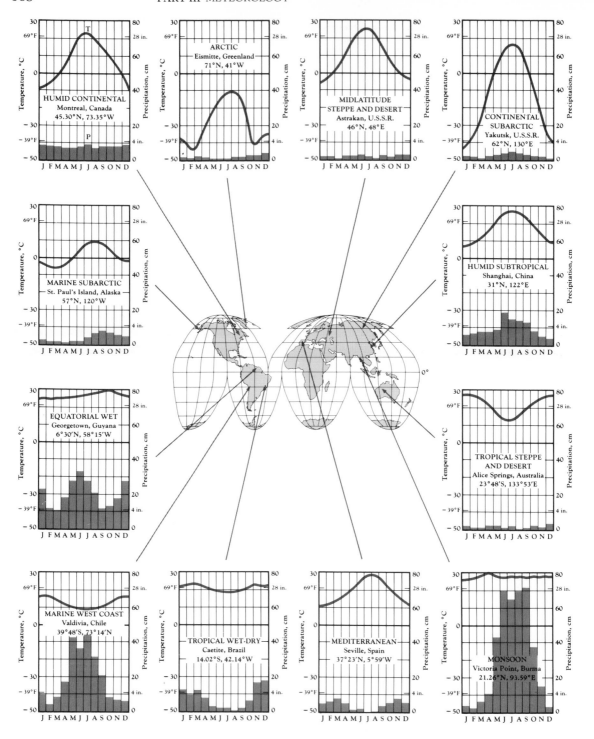

Figure 15-1 Average precipitation and temperature plots for a number of points. Temperature is shown by the curve and precipitation by the bar graph. (After Marsh and Dozier, 1981.)

Table 15-2 CLASSIFICATION OF CLIMATES USED IN FIGURES 15-1 AND 15-2, AND THE APPROXIMATE CORRESPONDENCE TO THE KÖPPEN CLASSIFICATION OF CLIMATES

Tropical

Equatorial wet	Af Am
Tropical wet-dry	Aw
Monsoon	Aw
Tropical steppe and desert	BWh BSh
West-coast desert	BWh

Mid-latitude

Mediterranean	Csa Csb
Marine west coast	Cfb Cfc
Humid subtropical	Cfa
Humid continental	Dfa Dfb Dwa Dwb
Mid-latitude steppe and desert	BWk BSk

Subarctic and arctic

Continental subarctic	Dfc Dfd Dwc Dwd
Marine subarctic	ET
Arctic	E
Ice cap	EF

Figure 15-2
World climates. (After Marsh and Dozier, 1981.)

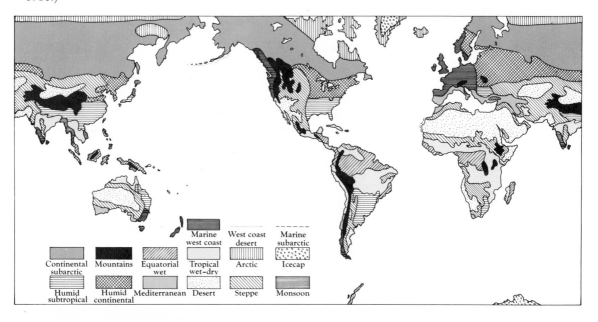

Figure 15-3
Between 1890 and 1973, the Muir Glacier at Glacier Bay, Alaska, retreated 34 kilometers (21 miles). (Data from U.S. Geological Survey.)

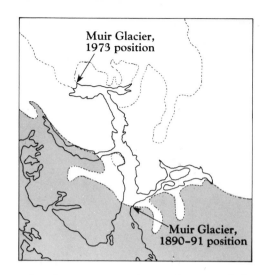

Muir Glacier, 1973 position

Muir Glacier, 1890–91 position

treat of mountain glaciers is another indicator of climatic change. If summers are warm and/or winters are mild with less snowfall, the glaciers melt back, or retreat (Figure 15-3). Under the opposite condi-   F 15-3 tions, of course, they advance down their valleys. These changes are averaged over many years by the glacier, because it takes that long for snow that fell at the glacier's source area to move all the way down the valley. In Europe, mountain glaciers advanced three times between 1540 and 1890, and there are historic accounts of alpine villages damaged by these glacial advances. During the first half of this century, the glaciers in the northern Rocky Mountains were in gradual retreat. About 1945 they began to advance, and they have since fluctuated. Climatic changes are also recorded in the growth rings of trees and in deep-ocean sediments.

External Causes of Climatic Change

GEOLOGIC CAUSES The uplift of a mountain range can change the climate locally. Air rising on the windward slopes is cooled, causing precipitation; and the dry air sinking on the lee side is warmed by compression. The moving of continents described in Chapter 7 can change the climate of a whole continent. These geologic causes of climatic change are very slow, requiring millions—probably tens of millions—of years.

ASTRONOMICAL CAUSES Changes in either the energy output of the sun or in earth's orbit could affect climate. As described in Chapter 18, there are many variable stars, whose energy ouput changes from time to time. Our measurements of the sun's radiation show no such change in output, but it is difficult to measure accurately the amount of energy actually received at the top of our atmosphere. We can, however, easily see and count sunspots. **Sunspots** *are cooler and there-*

fore darker areas on the sun that are associated with the sun's magnetic activity. Since 1700 the number of sunspots has varied in an approximately 11-year cycle; however, between 1645 and 1715 sunspots were virtually absent. This suggests that the sun, at least in part, is a variable star. The period of low-sunspot activity corresponded roughly with the glacial advances in Europe discussed above. Thus, changes in the sun's radiation could be a cause of climatic change.

Changes in earth's orbit would not materially change the amount of energy received from the sun, but they would change where the radiation is received on earth. The changes that occur are in the shape of the orbit, the tilt of earth's axis, and the direction in which earth's axis points. These changes all occur periodically. The orbital shape changes in a cycle that requires about 100,000 years. The cycle for the axial tilt is about 41,000 years, and that of the axial direction is about 26,000 years. Recently, calculations have shown that all of these orbital factors would have had their maximum effect during the last ice or glacial age. This does not prove that they were the cause, but it is suggestive. If these orbital changes did cause the ice age, why have not ice ages occurred periodically throughout geologic time? One possible answer is that glaciers can form only on continents, and so a glacial age can only occur if there are continents at high latitudes.

ATMOSPHERIC TRANSPARENCY CHANGES Changes in the amount of solar radiation that reaches the earth's surface can be caused by dust or other materials in the atmosphere. A volcanic eruption can inject much fine-grained material into the atmosphere. Careful studies of a number of recent volcanic eruptions have shown that they caused the average temperature to fall a few tenths of a degree Celsius. One striking example of a much more dramatic change occurred in 1816. That year is known in New England as "the year without a summer." In 1815, in what is now Indonesia, Mount Tambora erupted, throwing much fine ash into the atmosphere; this resulted in an unusually cool summer in both North America and Europe. Some geologists have attempted to show that volcanic activity was high during the last glacial age and so have suggested that the eruptions caused the glaciation.

Human Causes of Climatic Change

It has also been suggested that human activities, especially farming, have produced so much dust that it is affecting the climate. So far, attempts to prove such an effect have been unsuccessful. Dust and other particulates in the atmosphere may cool the earth's surface by shielding it from the sun, but the dust also absorbs and scatters some of the sun's energy and so warms the atmosphere.

Table 15-3 EFFECTS OF URBANIZATION	
Temperature	Slightly higher
Relative Humidity	About 6% lower
Cloud Cover	5% to 10% more
Fog	Much more common
Dust	Up to 10 times more
Wind	About 25% lower

URBAN WEATHER Large cities change the weather locally (Table 15-3). The effect that was first noted is that *temperatures are higher in cities than in the surrounding countryside.* This effect is called the **urban heat island** (Figure 15-4). Although the maximum temperatures are slightly higher, the minimum temperatures show this effect best. On clear, calm nights, rural areas lose—through radiation to the sky—the heat accumulated during the day. In the city, concrete buildings and asphalt streets and parking lots hold more heat than the countryside and release it more slowly. Many other causes probably contribute to the urban heat islands. Winds are reduced by buildings, perhaps as much as 25 per cent. Cities with strong, persistent winds do not have heat islands. Most of the precipitation in a city runs off through street drains, so that the evaporative cooling of the ground is much less than in rural areas. Cities may also be covered by a blanket of pollutants, especially particulates such as smoke, and this may reduce radiative cooling; of course, such a cover would also limit the amount of the sun's radiation reaching the ground. Much heat is generated in a city by industry, domestic heating and cooling, and other processes, and much of this heat escapes to the atmosphere.

Precipitation is also greater in cities. About 9 per cent more occurs over cities, but up to 25 per cent more falls downwind from them. Studies have also revealed that much of this excess precipitation occurs on weekdays, not on weekends, thus showing its relationship to industry and transportation in the city. There are several possible reasons for this precipitation. The heat island over the city may cause

T 15-3

F 15-4

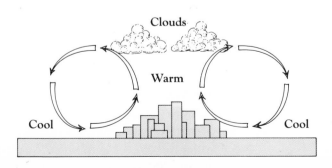

Figure 15-4
Urban heat island.

air to rise and so promote convective showers. In any event, the rising air over a city does cause a breeze to blow from the country into the city on many nights. The city air contains many condensation nuclei, and this too may cause increased precipitation. These nuclei do cause more fog to form in cities than in the country. Other studies have also shown that more thunderstorms and hail occur over cities and downwind from cities.

AIR POLLUTION The first problem in understanding air pollution is to decide what is, and what is not, an air pollutant. Many of the things generally considered pollutants are present in natural air— that is to say, in air with nothing added to it by human activities. The amount of a substance locally present in the air is clearly important in defining a pollutant. Equally important are the amount of harm or inconvenience caused by the substance, and how long it remains in the atmosphere. These three factors are the three Ts: tonnage, toxicity, and time in the atmosphere. The eventual fate of air pollutants is to be washed out of the air by rain and returned to the ground.

Natural dry air is composed mainly of nitrogen and oxygen, but it contains minor amounts of many gases. Carbon dioxide, methane, the oxides of nitrogen, and ozone are gases present in natural air that, when produced in larger concentrations by human activities, cause air pollution.

Particulates of both natural and human origin also cause pollution. Smoke, both natural from forest fires and human-caused from industry and other sources, is common. Dust can come from windstorms and volcanoes as well as from industry and agriculture. Pollen in the air is considered a pollutant by most hay-fever sufferers, but it and bacteria, fungi, and spores are natural constituents of the atmosphere.

Gaseous Pollutants

Oxides of Carbon The pollutant introduced into the atmosphere in the largest quantities by human activity is **carbon monoxide** (CO). It is the product of incomplete combustion, and the largest contribution comes from automobile exhaust. Carbon monoxide is a colorless, odorless, tasteless gas. In high enough concentrations it is toxic. Exposure to small or moderate concentrations can cause dizziness and headache, and large concentrations cause unconsciousness and death.

Carbon dioxide (CO_2) is also a product of combustion of fossil fuels. It is a minor constituent of natural air, but the increasing use of fossil fuels may be causing an increase in the amount of carbon dioxide in the atmosphere. Because carbon dioxide prevents the escape of radiation from the earth, the long-term effect of increased carbon dioxide in the atmosphere could be an overall warming of the earth. If the climate becomes warmer, glaciers in Antarctica and Greenland may melt, raising sea level and flooding our coastal cities, where a large fraction of the population lives and works. An increase in carbon dioxide has been observed, but in the long term it may eventually be dissolved in the oceans.

Oxides of Sulfur   **Sulfur oxides** are the most toxic and dangerous of air pollutants. They have been associated with all the fatal air pollution disasters. Their origin is the burning of coal and oil that contain minor amounts of sulfur. In combustion, the sulfur combines with oxygen to form *sulfur dioxide* (SO_2). This sulfur dioxide in the atmosphere combines with more oxygen to form *sulfur trioxide* (SO_3). The sun's energy causes this reaction to occur, so the process is called a **photochemical reaction.** The sulfur trioxide, in turn, reacts with water in the atmosphere to form sulfuric acid (H_2SO_4). The oxides of sulfur and sulfuric acid all cause damage to humans, animals, plants, and many other materials as diverse as marble statues and nylon hose. Areas downwind from industrial areas that produce these gases experience acid rainfalls. The major producers of sulfur oxides are industrial plants, electricity generating plants, and domestic heating devices that use fuels that contain minor amounts of sulfur.

Oxides of Nitrogen   The main source of **nitrogen oxides** is automobile exhaust, but these gases are also produced by industry and electrical power generation. In any high-temperature combustion, *nitrogen oxide* (NO) is produced. The nitrogen oxide reacts with oxygen in the atmosphere to form *nitrogen dioxide* (NO_2), a red-brown gas that gives the air a brown color. The oxides of nitrogen are also produced naturally by certain bacteria, but the lesser amounts produced by human activity are much more concentrated. Nitrogen dioxide has an unpleasant odor and causes some lung irritation, but its most serious effect is its role in causing photochemical smog, which will be examined in a later section.

Hydrocarbons   **Methane** (CH_4) is a **hydrocarbon** produced naturally by decaying plant material. Human-produced hydrocarbons come mainly from incomplete combustion in automobile engines and from the evaporation of solvents, fuels, and paints. Hydrocarbons, too, play a key role in photochemical smog.

These are not the only gases involved in air pollution, but, in general, the others are only important locally. This may be of little consolation to people who must live with obnoxious-smelling gases such as *hydrogen sulfide* (H_2S) (rotten egg smell) or who live near smelters whose gases have destroyed vegetation over wide areas.

Figure 15-5
A temperature inversion traps smoke and other pollutants in the lower atmosphere.

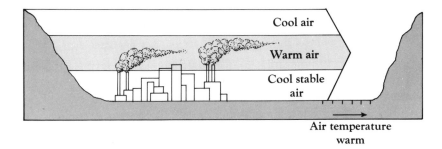

Figure 15-6
Convection caused by solar heating of the ground can move pollutants from the top of the inversion layer to the surface.

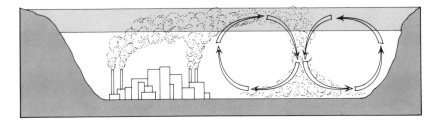

Concentration of Pollutants—Inversions If the gases and particulates produced by humans were swiftly dispersed by winds, serious levels of pollution would rarely develop. Short of removing pollutants either before they are formed (for example, removing sulfur from coal) or before they are released to the atmosphere, the most effective way to reduce local air pollution is by using high smokestacks to aid in dispersal. Geographic and climatic conditions at many places make even this ineffective.

Cities in valleys or surrounded by mountains are most susceptible to air pollution because temperature inversions may develop. Normally, the temperature becomes lower with elevation above the surface, but, in a **temperature inversion,** *cold air is near the surface and the temperature increases in the inversion layer; above it the temperatures are normal and fall with increasing elevation* (Figure 15-5). In an inversion, the cold, heavy air near the ground does not circulate well and traps pollutants. Inversions tend to develop in valleys on clear nights because the ground cools by radiation; the air in contact with the cool ground is also cooled, becoming heavier and flowing downhill to fill the valley. On the west coast gentle nighttime breezes bring cool air from the ocean to cities, such as Los Angeles, that are surrounded by mountains that trap the cool air over the cities.

Inversions trap the nighttime and early-morning pollutants in the inversion layer. As the morning sun heats the ground, the air near the ground is also warmed and so rises. This starts a convective flow within the inversion layer (Figure 15-6) and may bring the pollutants closer to the surface, making them more annoying. Eventually the sun's warming may increase the height of the convective circulation and destroy the inversion. At other times, high-pressure systems may

F 15-5

F 15-6

move into an area and cause the inversion to persist for many days. The cold air at ground level in an inversion may also cause fog to develop if the temperature and humidity are in the appropriate range. Fog prevents the sun from heating the ground, and under such conditions the inversion may persist, trapping air pollutants for days.

Types of Pollution   Air pollution comes in many forms, and many particular instances can be described. We will examine two contrasting types, because they are the types that affect major cities and so affect many people and cause much damage. The term **smog,** a contraction of smoke and fog, is commonly used to describe air pollution. Although descriptive of what severe air pollution may look like, it is a poor term: a mixture of smoke and fog will obscure the sky, but air pollution generally involves photochemical reactions. The chemicals that are most troublesome in air pollution are formed in the atmosphere from waste gases.

Industrial Smog   The most serious type of air pollution is what has been termed **industrial,** or gray, **smog.** It is caused mainly by smoke and the oxides of sulfur that are released by burning coal and oil containing minor amounts of sulfur. In the atmosphere, the oxides of sulfur form sulfuric acid, which is both toxic to life and damaging to many materials; and the smoke gives the air a gray color. This type of smog is associated with industrial cities, generally in cool, somewhat humid areas where domestic heating using the same fuels adds to the smog. It has long been associated with such cities as London, Pittsburgh, Chicago, and New York.

   Industrial smog has caused air pollution disasters in a number of places. One of the worst occurred in London in December of 1952, when five days of stagnant air brought about by a high-pressure system caused between 3500 and 4000 deaths. In 1948 a similar instance occurred at Donora, Pennsylvania, where 20 died and 6000 became ill.

Photochemical Smog   Brown smog, or **photochemical smog,** is caused largely by automobile exhaust gases. It is more common in cities in warm, dry areas with much sunshine, such as Los Angeles, Denver, and Salt Lake City (Figure 15-7). Los Angeles was the first major city F 15-7 to try to reduce this type of air pollution, and it was as a result of studies there that photochemical smog was first understood. By 1947, the smog in Los Angeles was obscuring vision, causing plant damage, irritating eyes, and damaging materials. In 1948, a law was passed prohibiting outdoor burning in dumps and home incinerators and otherwise limiting smoke and other particulates from many sources. The effect was to improve visibility somewhat, but the eye irritation and

Figure 15-7
Photochemical smog over Salt Lake City, Utah. (Photo from U.S. Department of the Interior, Bureau of Reclamation.)

crop damage continued. It was then observed that the compounds in the smog that were causing these problems were not in the waste gases discharged into the atmosphere. By the early 1950s, the photochemical reactions forming these compounds were recognized, and it was realized that the automobile was the main cause.

The gases in automobile exhaust that are involved are the oxides of nitrogen and the unburned hydrocarbons. The energy that causes these gases to react and form new compounds comes from the sun. Such reactions are called *photochemical reactions*, and it is the blue to ultraviolet wavelengths of the sun's energy that are absorbed by the smog reactions. Nitrogen dioxide and the hydrocarbons react photochemically to form ozone (O_3), peroxyacetylnitrate (PAN), and formaldehyde. PAN and ozone cause plant damage, and PAN and formaldehyde are eye irritants.

OTHER INADVERTENT CAUSES OF CLIMATIC CHANGE A change from forest to row crops may affect the local climate. These two types of vegetation absorb different amounts of solar energy, and they also give off different amounts of water vapor into the atmosphere.

Major engineering works also cause climatic change. Large dams change the regime of rivers, both above and below the dam, a topic described in more detail in Chapter 6. The lakes created by such dams also increase evaporation and so affect climate. Irrigation is also commonly associated with dams, and this, too, changes climate over a wider area than the newly cultivated region. Land reclamation projects of all kinds have the same effect.

Deliberate Climate Modification

Weather modification is a goal toward which people are moving. The potential benefits of such a program are almost limitless, and so are the possible dangers. Large-scale climatic changes could upset the whole planet, so weather modification must be approached carefully with detailed long-range planning.

Cloud seeding to increase rainfall is the most direct approach to weather modification. In this process, clouds are treated with silver iodide or dry ice to make snow that melts, forming raindrops. Raindrop formation is a complex physical process in which the tiny droplets that compose the cloud must coalesce to form drops large enough to fall to earth. The results so far suggest that the amount of rainfall can be increased somewhat, and perhaps localized somewhat, from clouds that would have produced some rain anyhow. Thus redistribution of rainfall is one result. An area where no rain falls also results downwind from the seeding. Such redistribution of rain could result in many lawsuits within a country and even wars between countries. Another problem is to determine how toxic silver iodide would be if used over a long period of time in an area.

Cloud seeding has been somewhat successful in reducing hail damage by causing precipitation before dangerous hail can form.

Cloud seeding has also apparently succeeded in weakening hurricanes. Hurricane experiments are difficult, because storms must be chosen that will remain over open ocean. If the storm goes inland and causes damage, the experimental seeding might be blamed for at least part of the damage. Storms get much of their energy from the latent heat that is released by the condensation of water vapor. The energy needed to evaporate water is released when the resulting water vapor condenses. The energy of hurricanes comes largely from the condensation of water vapor, and they are continually resupplied with water vapor from the underlying ocean. When hurricanes move over land, they weaken because they no longer have a large source of water vapor for energy. The idea behind cloud seeding of a hurricane is to cause condensation and freezing to occur in such a way that the energy is redistributed so that the wind velocity is reduced. In 1969, Hurricane Debbie's winds at 3660 meters (12,000 feet) were reduced from 182 kilometers per hour (113 miles per hour) to 125 kilometers per hour (78 miles per hour) by seeding. After seeding was stopped, the storm built up to winds of 184 kilometers per hour (114 miles per hour), and seeding then reduced the winds to 156 kilometers (97 miles per hour). Hurricanes cause great damage in terms of lives and property, but one adverse effect of hurricane suppression might be a climatic change in the eastern United States. About one-sixth of the moisture received there between June and October comes from hurricanes.

SUMMARY

- *Climate* is the typical combination of weather elements that characterizes an area over a long term and that takes into account seasonal changes as well as ranges and extremes of weather elements.

- Vegetation has been used to classify and recognize climates.

- The Köppen classification is widely used and depends on precipitation and temperature, which are used to compare precipitation and potential evaporation.

- Climatic changes are difficult to detect in weather records but are clearly shown by the advance and retreat of glaciers.

- Geologic processes such as the uplift of mountain ranges or the movement of continents can cause climatic change.

- Some climatic changes may correspond with sunspot cycles or perhaps other changes in output by the sun. The periodic changes in the earth's orbit may also affect climate.

- Volcanic eruptions and major dust storms may also be the cause of climatic changes.

- The *urban heat island* is the term used to describe the higher temperatures that occur above cities as compared to the surrounding countryside.

- Precipitation and cloudiness are greater in cities and in areas downwind from them, especially on weekdays, probably because of increased condensation nuclei.

- The main gases involved in air pollution are:

  oxides of carbon—carbon dioxide and carbon monoxide
  oxides of sulfur—sulfur dioxide and sulfur trioxide

  oxides of nitrogen—nitrogen oxide and nitrogen dioxide
  hydrocarbons—methane

- *Temperature inversions* may trap pollutants near the earth's surface. A temperature inversion occurs when cold air is near the surface, the air above it is warmer, and above that the temperatures are normal.

- *Industrial smog* is gray smog that is caused mainly by smoke and oxides of sulfur released by burning coal and oil. It forms near industrial cities, generally in cool, somewhat humid areas. It is associated with cities such as London, Pittsburgh, Chicago, and New York. Industrial smog has been the cause of a number of air-pollution disasters.

- *Photochemical smog* is brown smog that is caused mainly by automobile exhaust gases. Ultraviolet energy from the sun creates the brown smog from the oxides of nitrogen and unburned hydrocarbons. It is common in warm, dry areas such as Los Angeles, Denver, and Salt Lake City.

- Climatic changes may result from changes in vegetation such as those that result from clearing forests or damming a river to form a lake.

- *Cloud seeding* to increase rainfall in an area may affect the climate both locally and downwind.

- Cloud seeding may also reduce hail damage and may weaken hurricanes.

KEY TERMS

climate	carbon dioxide	oxides of nitrogen	smog
sunspot	oxides of sulfur	methane	industrial smog
urban heat island	photochemical reac-	hydrocarbon	photochemical smog
carbon monoxide	tion	temperature inversion	cloud seeding

REVIEW QUESTIONS

1. On what two measurements is Köppen's climate classification based? What determined where Köppen chose to draw the boundaries between different groups?

2. Describe some ways in which the following can cause climatic change:
   (a) geologic processes
   (b) astronomical changes
   (c) fluctuations in the atmosphere's transparency
   Which of the methods that you have described take the most time; which take the least time?

3. Cities can change climate locally.
   (a) What contributes most heavily to the urban heat island?
   (b) Cities have greater precipitation than have their surrounding countrysides. What are the suspected causes of this difference?

4. What are the three Ts used in defining an air pollutant?

5. List five gases that have become serious pollutants because they are produced in large or concentrated quantities by human activities. Also state the chief sources of these gaseous pollutants and explain why each of the gases is potentially dangerous.

6. Explain why cities that are surrounded by mountains are most susceptible to air pollution. How do the following affect a temperature inversion: Sun? High-pressure system? Fog?

7. What is industrial smog? In what areas is it most likely to occur? Why?

8. Describe how photochemical smog forms.

9. Review Chapter 12; then explain why cloud seeding with silver iodide can speed up rain formation and cause more rain to fall. What are the potential risks of cloud seeding?

10. Briefly state why cloud seeding can weaken a hurricane. What are some advantages and disadvantages of experimenting with hurricane seeding?

SUGGESTED READINGS

Ackerman, A. E. "The Köppen Classification of Climates in North America," *Geographical Review* (1941), Vol. 31, pp. 105–111.

Hare, F. K. "Future Climates and Future Environments," *Bulletin of the American Meteorological Society* (1971), Vol. 52, pp. 451–456.

Landsberg, H. E. "Man-Made Climatic Changes," *Science* (December 18, 1970), Vol. 170, No. 3964, pp. 1265–1274.

Madden, R. A., and V. Ramanathan. "Detecting Climate Change Due to Increasing Carbon Dioxide," *Science* (August 15, 1980), Vol. 209, No. 4458, pp. 763–768.

Marsh, W. M., and Jeff Dozier. *Landscape: An Introduction to Physical Geography.* Menlo Park, Ca.: Addison-Wesley, 1981, 637 pp.

Massachusetts Institute of Technology. *Inadvertent Climate Modification: A Report of the Study of Man's Impact on Climate.* Cambridge: Massachusetts Institute of Technology Press, 1971, 308 pp.

National Academy of Sciences. *Understanding Climatic Change.* Washington, D.C.: U.S. Government Printing Office, 1975.

Newell, R. E. "Climate and the Ocean," *American Scientist* (July–August 1979), Vol. 67, No. 4, pp. 405–416.

Oliver, J. E. *Climate and Man's Environment: An Introduction to Applied Climatology.* New York: John Wiley, 1973.

Spence, C. C. *The Rainmakers. American "Pluviculture" to World War II.* Lincoln: University of Nebraska Press, 1980, 182 pp.

Stommel, Henry, and Elizabeth Stommel. "The Year without a Summer," *Scientific American* (June 1979), Vol. 240, No. 6, pp. 176–186.

U.S. Department of Commerce, Environmental Data Service. *Climatic Atlas of the United States.* Washington, D.C.: 1968.

U.S. Department of Commerce, Environmental Data Service. *Climates of the World.* Washington, D.C.: 1969.

Woodwell, G. M. "The Carbon Dioxide Question," *Scientific American* (January 1978), Vol. 238, No. 1, pp. 34–43.

PART IV

ASTRONOMY

To understand the development of modern astronomy we will embark on a trip through time. Our time capsule will start in ancient Greece where we will study under its great scholars. We will then stop at the Renaissance to observe the Copernican revolution. When we arrive back in the 20th century, we will accompany the space probes to investigate our neighboring planets and to look for life in our solar system. Finally, our ship will move out into the galaxy to explore the almost science fictional world of red giants, white dwarfs, and black holes.

Perhaps the most amazing thing about astronomy is that we know so much about places we have never visited and can barely see. Chapter 16 explains how we know what we know. It discusses the observations and deductions made first by the ancient Greeks and later by such Europeans as Copernicus, Galileo, Kepler, and Newton. It describes the motions of earth and moon, and finally tells us how electromagnetic radiation—which includes visible light—is used to explore the planets and the stars.

Chapter 17 tours the solar system. It describes the geology, meteorology, and history of the planets and some of their satellites. It looks at the asteroid belt, at those blazes of ice that we call comets, and at meteors and meteorites. The chapter also describes how the solar system is believed to have formed nearly 5000 million years ago.

Chapter 18 shows us how astronomers, with their feet firmly planted on earth, are able to measure a star's distance, diameter, mass, and magnetic field, find out how fast a star rotates, take its temperature, and analyze its composition. Stars have life cycles, and the chapter discusses the evolution of stars into red giants or supergiants, white dwarfs, black dwarfs, and, possibly, black holes. The chapter concludes with descriptions of galaxies, nebulae, and quasars—those puzzling signals from very far away.

ANCIENT ASTRONOMY

THE BEGINNINGS OF MODERN ASTRONOMY

EARTH'S MOTIONS

Revolution

Precession

Calendars

Rotation and Time

THE MOON'S MOTIONS

LIGHT, SPECTRA, AND TELESCOPES

Light and Other Electromagnetic Radiation

Spectra

Doppler Effect

Lenses, Mirrors, and Optical Telescopes

Telescopes for Invisible Radiation

CHAPTER 16

THE EARTH AS A PLANET

Eleven and one-half years were required to shape and polish the mirror of the 200-inch (5-meter) Hale telescope. (Photo from Palomar Observatory, California Institute of Technology.)

As we begin the astronomy unit, be sure to watch for several points, such as the sizes of and distances to the various objects we will study. For instance, if you know the sizes of the earth, the moon, and the sun, as well as their distances apart, you will realize that most of the diagrams in this or any other astronomy book cannot be drawn to scale. The distance to the nearest star beyond the sun is almost incomprehensibly great, but the distance across our own Milky Way galaxy is many, many times farther. The diameter of the Milky Way and the number of stars in it are equally difficult to imagine.

The real wonder of astronomy is how astronomers have been able to measure these vast distances and how they can determine the composition, size, temperature, and mass of stars to which we can never reasonably hope to send spacecraft or to visit. In the study of celestial objects, astronomers work with very little information. They can observe the position of an object; its apparent motion, if any; its brightness; and the nature of its radiation. From these limited observations (although made in many cases with very sophisticated instruments) and an understanding of physics, almost all of our knowledge of astronomy is obtained.

We will begin our study of astronomy with a brief historical survey of how the early observers of the sky deciphered the motions of the planets. Today we all accept the fact that earth and the other planets orbit the sun. The ancients did not have satellites to photograph earth and to send to other planets, so it took many years to discover that earth is spherical and that it orbits the sun. It took a long time to discover the proofs of these now commonly accepted statements.

ANCIENT ASTRONOMY

The true beginning of astronomy may never be known. Early peoples spent much of their lives in the outdoors and so became keen observers. Probably the moon's monthly phases, or perhaps the seasons, were the first to be recognized. Seasons controlled people's lives, whether they were hunters and gatherers or farmers, and so seasonal changes, heralded by changing patterns of stars in the night sky, were of great importance to them. Some cultures worshipped the sun, moon, or stars, and many had legends to explain changing seasons or the patterns of stars. As civilizations developed, the priests were the ones who observed and kept records, and they developed first calendars and then the ability to predict eclipses of the sun and moon. In the minds of these early people, the cause and effect of celestial events were not clear, and the idea that the changes in the night sky controlled their lives probably became current. The idea persists to this day as *astrology,* and many people still believe that events and their lives are determined by the stars.

Figure 16-1
As viewed from the earth, the sun and the planets appear to travel in a narrow path through the sky. The path of the sun is called the *ecliptic*, and the zodiac is a band about 9 degrees on each side of the ecliptic.

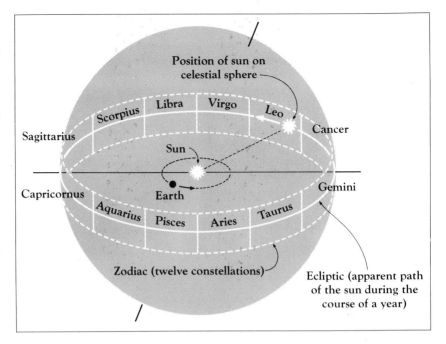

We will begin our story with the ancient Greeks. It is amazing how accurate they were in their measurements and reasoning, especially if we remember that they could observe only changes in position and gross changes in brightness. They gave us the name *planetes*, meaning "wanderers," because the planets move among the stars. They named the planets for their gods. They noted that the planets and the moon travel across the same band of the sky. The ancients named this path through the sky the *zodiac* (circle of animals), and the constellations (groups of stars) in the zodiac are mostly named for animals—Taurus, the bull; Aries, the ram; and so on. *The apparent path of the sun in front of the stars is called the* **ecliptic,** *and the* **zodiac** *is a zone about 9 degrees on each side of the ecliptic* (Figure 16-1). The reason that the sun, moon, and planets move across a narrow path in the sky is that their orbital planes are nearly parallel.

F 16-1

Aristotle, in the fourth century B.C., reasoned that earth was at the center of the universe and did not move. He thought that, if earth did move around the sun, we should have constant winds because when we move we feel a wind. He devised a further test. If earth orbits the sun, then the position of a nearby star should shift its apparent position relative to the more distant stars as earth moves from one side of its orbit to the other (Figure 16-2). No such movement can be detected with the naked eye; indeed, it was not until 1838 that instruments accurate enough to measure this shift were made. Aristotle's test failed, not because it was a poorly reasoned idea, but because no one in his day had any inkling how vast stellar distances are.

F 16-2

Figure 16-2
Aristotle's test. If the earth orbits the sun, the near stars as seen from earth will shift in position relative to the more distant stars. The apparent shift is so slight that it can be detected only with telescopes.

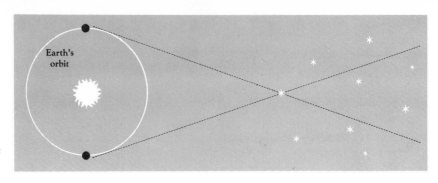

Examples of conclusions drawn on sketchy data are not uncommon in the history of science. Aristotle also thought that all celestial objects should be spheres and all orbits must be circles, because spheres and circles are the most perfect geometrical forms. Unfortunately, many centuries later the church and philosophers adopted Aristotle's ideas as truth (dogma), and they became obstacles in the development of astronomy.

Some of the later Greeks, particularly those who moved to Alexandria, then the seat of learning, did some excellent reasoning. They believed that earth was a sphere because of the shape of its shadow on the moon during an eclipse. They also observed that a ship moving toward the open ocean appears to go downhill, and the tip of the mast is the last thing seen. Aristarchus, in the third century B.C., reasoned that, because earth's shadow seen during an eclipse is larger than the moon, earth must be larger than the moon.

At about the same time, Eratosthenes actually measured the size of earth. His method is shown in Figure 16-3. Eratosthenes did not know the distance between the two cities used in his measurement very accurately, so his measurement was about 15 per cent too small (we are not sure of the size of the units that he used). In any case, his result was remarkable for his time. Interestingly, later repetitions of this measurement also gave results that were too small, and this is one reason that Columbus, many centuries later, thought that he had sailed all the way around the earth to Asia.

About 135 B.C. Hipparchus measured the distance to the moon as 30 earth diameters, very close to the actual distance. The method is shown in Figure 16-4. This method is similar to measuring the height of a flagpole by measuring the length of its shadow and the angle between the end of the shadow and the top, and then making a scale drawing.

These few examples should be enough to show us how much astronomical information can be obtained by careful observation and

F 16-3

F 16-4

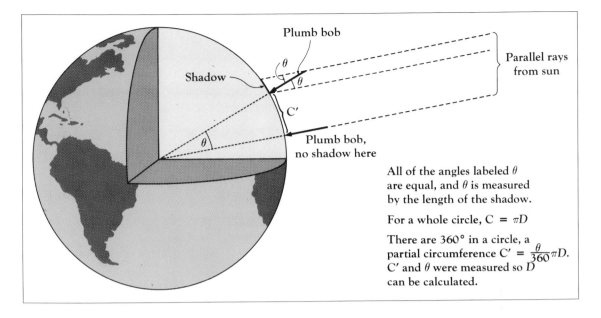

Plumb bob

θ

Shadow

θ

Parallel rays from sun

C'

θ

Plumb bob, no shadow here

All of the angles labeled θ are equal, and θ is measured by the length of the shadow.

For a whole circle, $C = \pi D$

There are 360° in a circle, a partial circumference $C' = \frac{\theta}{360} \pi D$. C' and θ were measured so D can be calculated.

Figure 16-3
Eratosthenes' method of measuring the size of earth.

good reasoning. Not all of these ideas were original with the Greeks or Alexandrians, because in many cases they extended the knowledge of earlier peoples. The Greeks also charted the wanderings of the planets among the stars, noting that they moved both east and west.

An important goal of science is to organize data in such a way that future events can be predicted. The planets wander through the stars, each seemingly moving in its own fashion. In about 150 A.D., Ptolemy organized the data collected over the centuries into a system that enabled prediction of future positions of the planets. His system was ingenious and was used until the Renaissance. Ptolemy assumed that earth is at the center of the universe and that the sun, moon, and planets orbit earth. To explain the observed back-and-forth move-

Figure 16-4
Sighting the moon from two points a known distance apart enables one to calculate the distance to the moon by trigonometry or by making a scale drawing.

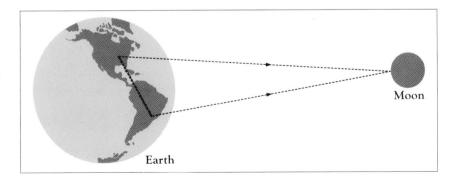

Moon

Earth

ments of the planets, he had to resort to orbits consisting of circles rolling inside circles, which he termed **epicycles** (see Figure 16-6B).

To understand the Ptolemaic system, we should compare his explanations with our modern understanding of the movements of planets. Figure 16-5 compares the **Ptolemaic,** or **geocentric** (*earth-at-the-center*), **system** with the **heliocentric** (*sun-at-the-center*) **system.** Notice

F 16-5

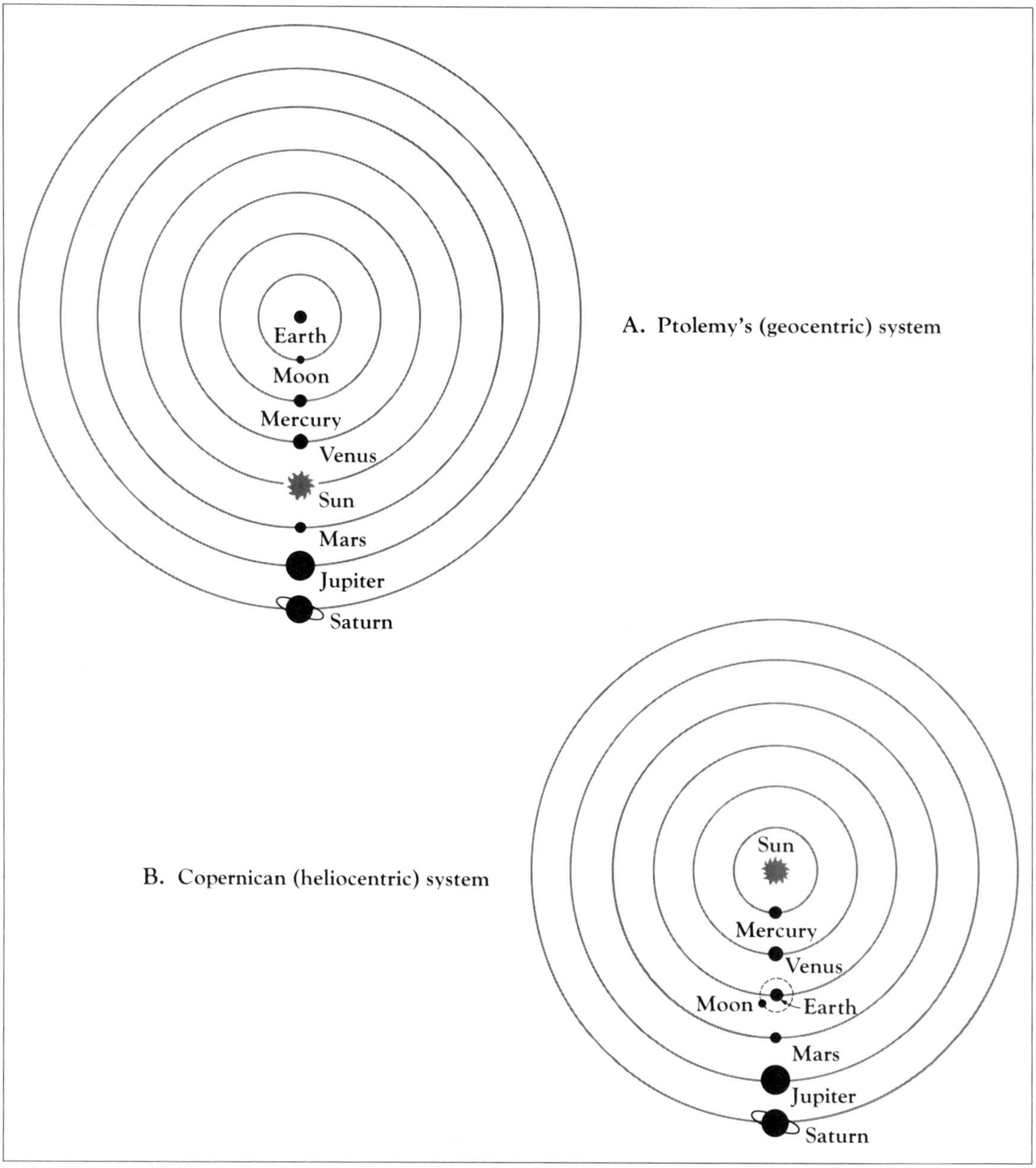

A. Ptolemy's (geocentric) system

B. Copernican (heliocentric) system

Figure 16-5 Comparison of the geocentric and heliocentric systems.

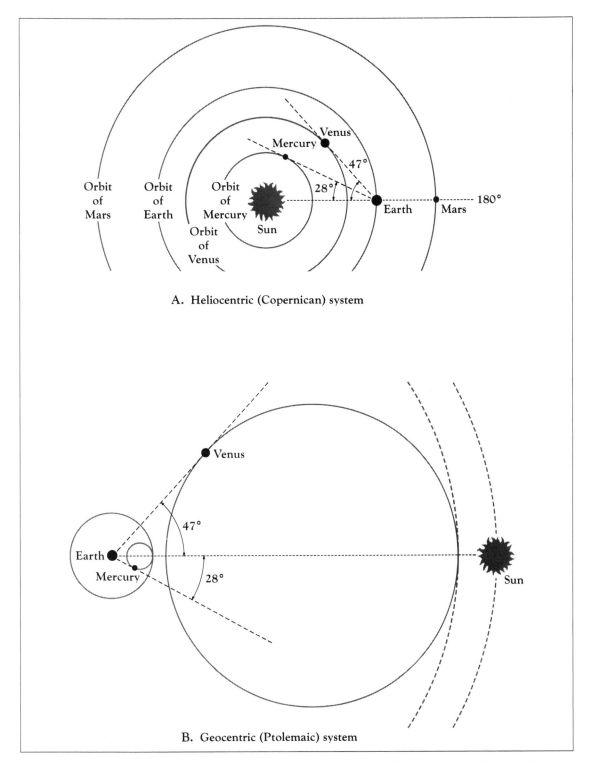

A. Heliocentric (Copernican) system

B. Geocentric (Ptolemaic) system

Figure 16-6 Explanation of the motions of Mercury and Venus in the geocentric and heliocentric systems.

F 16-6

Figure 16-7
The retrograde movement of a planet more distant than earth from the sun explained in both the heliocentric and the geocentric systems.

that the planets are not in the same sequence in the two systems and that only those planets known to the ancients are shown. First we will consider the planets that are actually closer to the sun than earth and then those farther away.

As viewed from Earth, Mercury and Venus are never far from the sun. In Figure 16-6A it is clear that Venus can only be about 47 degrees from the sun and Mercury 28 degrees. For this reason, Venus and Mercury can only be seen a few hours before sunrise or after

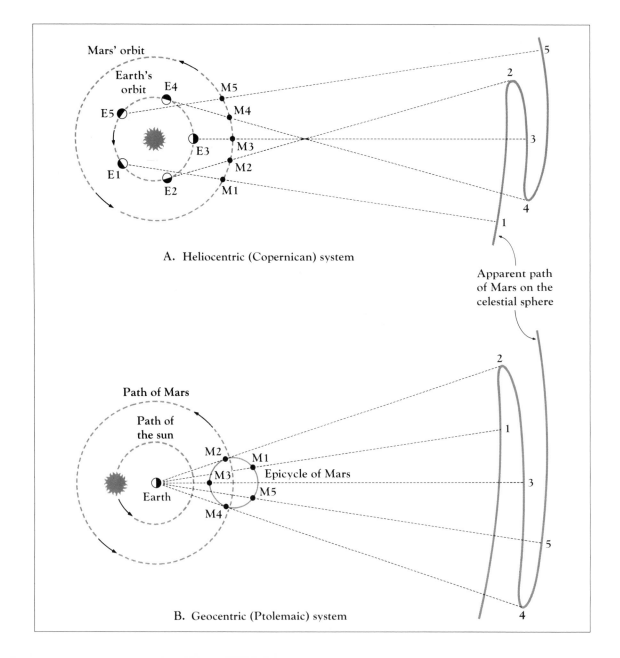

A. Heliocentric (Copernican) system

Apparent path of Mars on the celestial sphere

B. Geocentric (Ptolemaic) system

sunset, depending on where they are in their orbits. The explanation of this observation in the Ptolemaic theory requires that Mercury and Venus move in circular epicycles, the centers of which must move around Earth at the same rate that the sun moves around Earth (Figure 16-6B). Such motions are much more complex than those in the heliocentric system.

The apparent motions of planets more distant from the sun than Earth are still more complex. Mars will be used as an example, and Figure 16-7A shows that such more distant planets can be seen any- F 16–7 where on the ecliptic in the night sky and, unlike Mercury and Venus, can at times be seen at midnight. Mars generally moves eastward among the stars, but at times it slows, stops, and then moves westward for a time; then it slows, stops, and continues its eastward movement. Such *backward motion is termed* **retrograde motion.** In the Ptolemaic theory, retrograde motion is explained by epicycles as shown in Figure 16-7B. The current and probably true explanation is shown in Figure 16-7A. In this figure, the positions of Earth and Mars are shown at a number of times; that is, when Earth is at point E1, Mars is at M1; and when Earth is at E2, Mars is at M2. Notice that, because Mars is farther from the sun than is Earth, it travels slower around its orbit. Thus, if Mars is viewed from Earth at E1, it appears at point 1 among the stars; from E2 it appears at point 2. Because Earth travels faster than Mars, eventually Mars appears to change direction and move backward (retrograde), and then changes direction again and resumes its forward motion.

The geocentric theory of Ptolemy served a very useful purpose until the Renaissance. It was refined by adding epicycles and other complications to make it fit observations better. Its principal use was probably in casting horoscopes, but it was also used by many of the early navigators.

THE BEGINNINGS OF MODERN ASTRONOMY

Our modern understanding of the movements in the solar system began with the overthrow of Ptolemy's geocentric theory and was firmly established with Newton's discovery that gravity is the force that keeps the planets in orbit. The recognition that the sun, not earth, is at the center of the solar system was one of the great revolutions in thinking. It began with Nicolaus Copernicus, who was born in 1473. After noting small discrepancies in the then-current version of Ptolemy's system, he began trying to improve its accuracy. He soon discovered that placing the sun at the center of the solar system was a great simplification. His early work was published in 1512, but not widely circulated, and his main book was published in 1543. Although this book was originally titled *On Revolutions*, he did not wish to be the center of controversy, and so he did not publish it until the year of his death. Relegating earth to a minor place in the known universe was a revolution in theology as well as astronomy, and the Copernican Revolution was not immediately accepted by the church or by all intellectuals interested in astronomy.

Tycho Brahe (1546–1601), a Dane, set the stage for the next advance. As a young man, he, too, noted inaccuracies in the then-current version of the Ptolemaic system. His main contribution was to spend 20 years collecting accurate data on the positions of planets. He, of course, worked in the era before telescopes and used large protractors, called quadrants, in his measurements. He believed in a geocentric universe, although he did observe both a supernova, a star that becomes very bright for a short time, and a comet. He was able to show, using the same test as Aristotle (see Figure 16-2), that both of these objects were more distant than the moon. Thus he demonstrated that new objects could appear in the supposedly unchanging heavens. His observatory was built on an island near Copenhagen, and he was aided by the king of Denmark. He was a very strong personality, who wore a silver nose as a result of a duel in his youth. Late in his life he left Denmark after his royal support was withdrawn. He then moved to Prague, where he took on a student and assistant named Johannes Kepler.

Kepler (1571–1630) also started out believing in the Ptolemaic theory and thought that he would be able to refine it using Brahe's accurate data. He was a mathematician, and his study of Brahe's data led to his three laws of planetary motion. His approach was to analyze the data looking for regularities in them, and it was very similar to one of the ways that computers are used today. He spent almost 18 years at this study, which would require only a few minutes on a modern computer. He began by studying Brahe's observations of the retrograde motion of Mars and discovered that the orbit is not a circle but is an ellipse with the sun at one of the foci. This is Kepler's first law, and it was announced in 1609. (One way to draw an ellipse, the parts of an ellipse, and how an ellipse differs from a circle are shown in Figure 16-8.) Kepler's first law was the first departure from the idea of F 16-8 circular motion of the planets.

Kepler's second law was also announced in 1609. He discovered not only that orbits are not circles but also that a planet does not travel its orbit at a constant speed. The sun is not at the center of the planet's elliptical orbit, but is at one of the two foci. A planet travels faster when near the sun than when distant from it. Kepler's second law is that the change in speed is such that during equal time spans, the area swept out by a line between the planet and the sun is equal (Figure 16-9). F 16-9

Kepler announced his third law nine years later, in 1618. This law relates the orbital period of a planet to its distance from the sun. The

Figure 16-8
The ellipse.

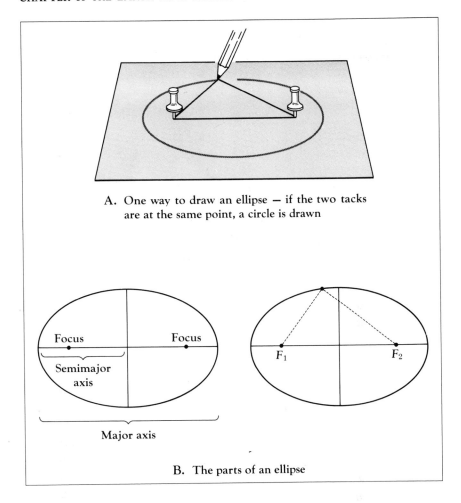

A. One way to draw an ellipse — if the two tacks are at the same point, a circle is drawn

Focus   Focus

Semimajor axis

Major axis

F_1 F_2

B. The parts of an ellipse

formal statement is that the ratio of the cube of the semi-major axis of a planet's orbit (*a*) to the square of its orbital period (*p*) is the same for each planet. The orbits of most planets are nearly circular, so the semi-major axis is approximately the radius of the orbit. An example will show its use. The average distance between earth and the sun is termed the **Astronomical Unit** (A.U.), and earth's orbital period is, of course, the year. If we use these units, the ratio for earth becomes

$$\frac{a^3}{p^2} = \frac{(1 \text{ A.U.})^3}{(1 \text{ yr})^2} = \frac{1}{1} = 1$$

For any other planet, then,

$$\frac{a^3}{p^2} = 1 \quad \text{or} \quad a^3 = p^2$$

Thus, as long as we use these same units, if we measure a planet's orbital period by observation, we can easily calculate its distance from the sun. Kepler's laws describe the motions of planets, but they

Figure 16-9
Kepler's second law.
A planet travels faster
when near the sun so
that equal areas are
swept in equal times.
In the figure, the
areas are all equal,
and the distance
moved by the planet
in equal times is
shown.

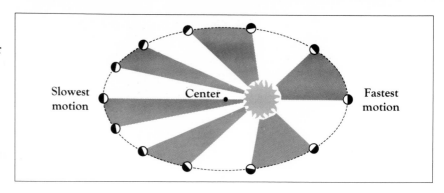

do not explain why the planets move as they do. An important point
not always appreciated is that, until Kepler's laws were applied to astronomical tables, the tables based on the Copernican system were no
more accurate than Ptolemaic tables. The initial advantage of the
Copernican system was that it was simpler, not that it was more accurate.

The next person to take up the Copernican cause is one of the
most important of the early scientists, Galileo Galilei (1564–1642), a
contemporary of Kepler. He was an Italian professor, and he differed
from his contemporary academicians in two important regards. He
was the first to perform experiments, and he was the first to publish in
a common language (Italian) instead of academic Latin. He developed
the telescope and was the first to use it in astronomy. With his telescope he made several discoveries that shook the Ptolemaic system.
He was the first to see Jupiter's moons, and this showed that not all
objects orbit Earth. He observed mountains on the moon, showing
that it, at least, was not a perfect sphere as Aristotle had said all celestial bodies must be. Saturn also did not appear spherical because his
telescope was not good enough to resolve its rings. He saw spots on
the sun, again suggesting that it was not perfect. He also saw that
Venus goes through phases like the moon's and that the full disk is
seen at times. As Figure 16-10 shows, in the Ptolemaic system Venus F 16-10
should always appear as a crescent. He announced these and other
discoveries in a book in 1610. In 1632 he published another book,
this time in Italian, advocating the Copernican system. He was tried
by the Inquisition, forced to recant, and put under house arrest for
his final years. He had, however, made his point by popularizing his
views. During the nine years he was under house arrest, he experi-
mented in the field of mechanics and published a book on this subject.

Isaac Newton (1642–1727) was born the year Galileo died. His
work in mechanics was based in part on Galileo's last studies. He,

along with Galileo, must be considered one of the great scientists of all time. Newton invented calculus when he needed this branch of mathematics for his studies of motion. He published his work in 1687. Newton deduced the three laws of motion:

1. A body with no force acting on it is either stationary or moves with a constant velocity.
2. A body acted on by a force is accelerated.
3. For every force on a body, there is an equal and opposite force.

He also stated his **law of gravity** in his book, and it is this principle

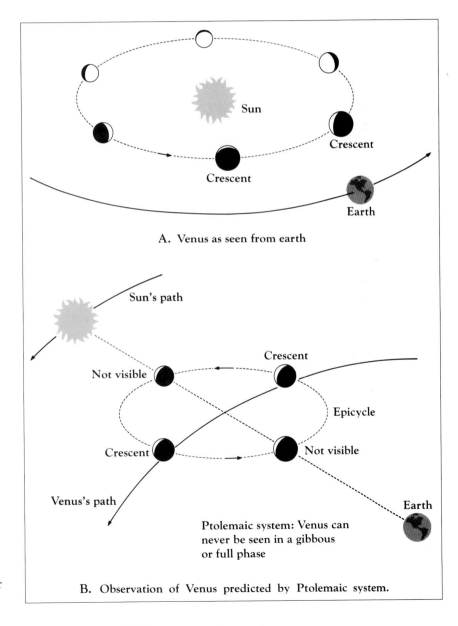

A. Venus as seen from earth

B. Observation of Venus predicted by Ptolemaic system.

Figure 16-10
Viewed from earth, Venus' size and illumination change, as shown in **A.** The Ptolemaic system predicts that Venus should always appear as a crescent, **B.**

that explains why the planets move in their orbits. The law of gravity is a basic law of physics and is very different from Kepler's laws. In fact, Kepler's laws can be easily derived from Newton's work. During Newton's lifetime, his explanation of the motions of the solar system was accepted, and at last the Copernican system was established.

Newton's law of gravity states that every piece of mass in the universe attracts every other mass with a force (F) that is proportional to the product of the masses ($M_1 M_2$) and inversely proportional to the square of the distance (d) between the masses; in algebraic terms,

$$F = G\frac{M_1 M_2}{d^2}$$

Thus, a person's weight is really the mutual gravitational force between the person and earth. Therefore, if you travel to a higher elevation, you are farther from the center of the earth, so the gravitational attraction is less and you weigh less.

To see how Newton's laws explain the motion of a planet, we will consider the behavior of earth. Like all other planets, earth is traveling in a elliptical orbit around the sun. Its motion can be considered as a combination of a tendency to travel in a straight line and a tendency to fall into the sun. If the gravitational force due to the mutual attraction between earth and the sun could be cut off, earth would have no force acting on it, and it would travel in a straight line at a constant velocity (Newton's first law). If earth's forward motion could be stopped, the gravitational attraction would cause it to fall into the sun, just as Newton's apple fell to earth. In the same way, to put a satellite into orbit around earth, we must give it a forward velocity fast enough that the gravitational force pulls it just enough to keep it in orbit (Figure 16-11). When we are closer to the sun, our orbital F 16-1 speed increases to balance the larger gravitational pull of the sun.

In 1851 Jean Foucault, a French physicist, gave a direct demonstration of the earth's rotation. If a pendulum is free to move in any

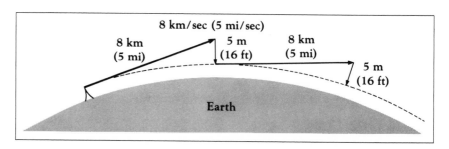

Figure 16-11 The earth's curvature is about 5 meters (16 feet) in 8 kilometers (5 miles), and a body will fall about 5 meters in 1 second. Therefore, if a body is projected at 8 kilometers per second, it will go in orbit around the earth.

Figure 16-12
Foucault pendulum. As the earth turns beneath it, the pendulum swings in the same direction; thus, the pendulum appears to move relative to the earth.

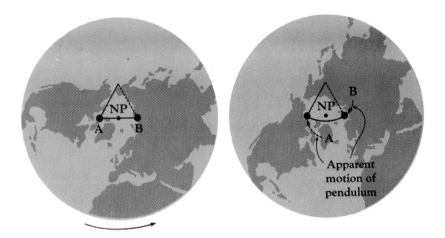

direction, it will, because of inertia, continue to swing in the same direction. Because the earth is spinning under the pendulum, the pendulum's direction of motion will change relative to the earth beneath it. Figure 16-12 shows a pendulum at the north pole, and its apparent motion relative to the earth below it is indicated. Similar apparent motion is shown everywhere on the earth except at the equator.

F 16-12

EARTH'S MOTIONS

Earth's main motions are its daily spin on its axis and its yearly orbit around the sun. These motions have great importance in creating the environment at the earth's surface and in establishing our units of time and our calendar.

Revolution

Earth's orbital trip, called the **revolution,** is the more complex of these motions. Earth's orbit is an ellipse with the sun at one of the foci; it is very close to a circle. Viewed from above the north pole, earth moves counterclockwise around the sun. At its *closest approach to the sun, called* **perihelion,** the earth is about 146 million kilometers (91 million miles) from the sun; *the earth's farthest point from the sun is called* **aphelion,** and the distance is about 150 million kilometers (93 million miles). Perihelion occurs in January, and aphelion in July.

Precession

Earth has another, much slower, motion related to its orbit. The motion is *a slow wobble of earth's axis called* **precession** (see Figure 16-13). This wobble is very slow, requiring about 26,000 years to complete one cycle. In spite of its slowness, precession was recognized by the

F 16-13

Figure 16-13
Earth's precession.
The earth, like a
spinning top, has a
wobble. (After
Goldsmith, 1981.)

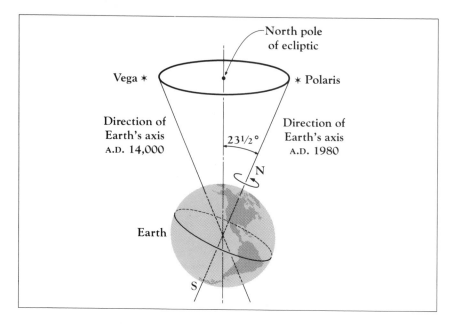

ancient Greeks because their observations did not agree with earlier records. One obvious effect of this wobble is that the earth's axis has not always pointed at Polaris, our present north star. Another result is that the constellation on the zodiac that the sun is in at equinox also changes. Because the equinox is a convenient reference, this slow movement is called the *precession of the equinox.*

One very interesting aspect of precession is that in classical times, when the calendar was established, the year began at the spring equinox. We still see this in the names of the months September, October, November, and December, meaning seventh, eighth, ninth, and tenth months. At that time, at the spring equinox the sun was in the constellation Aries. To astrologers, the year still begins with Aries. During the intervening centuries, because of precession the equinox has moved through the constellation Pisces to the edge of Aquarius. This is what is meant by the phrase "Age of Aquarius." Astrology, the belief that the positions of stars and planets affect our lives, is an ancient idea that has persisted to the present with very little change. It has changed so little that a person born in the month following the spring equinox is said to be an "Aries" because the sun was in the constellation Aries at the time of that person's birth; however, because of precession the sun was really in the constellation Aquarius. Astrology has changed so little that it has not kept up with simple observational astronomy.

Calendars

The time for earth to complete one revolution around the sun is our definition of the **year** and is the basis of our calendar. A normal year is 365 days, and each fourth year is a leap year with 366 days. Thus, our average year is 365¼ days. The actual period of earth's revolution, generally measured from one spring equinox to the next spring equinox, is 365 days, 5 hours, 48 minutes, and 46 seconds (365.242 days). Therefore, our calendar year is 11 minutes and 14 seconds longer than earth's orbital period. Given a few centuries, this difference would become a problem, as a look at the history of calendars will show.

Early calendars probably had 365 days in the year, with extra days added at times to keep the year in harmony with astronomical observations. This system worked reasonably well until Roman times, but by 46 B.C. the calendar and the seasons were badly out of agreement. Julius Caesar declared that the year 46 B.C. would be 445 days long and that henceforth each fourth year would be a leap year. In this way he got the calendar and the seasons into agreement, although the year 46 B.C. is still called the "year of confusion." This Julian calendar was used for many centuries, and the error between it and earth's actual orbital period accumulated.

Our present calendar, the Gregorian calendar, was devised by a commission appointed by Pope Gregory XIII. This calendar keeps leap year every fourth year as in the Julian calendar but omits leap year in century years not divisible by 400. In this way, the last century year that was a leap year was 1600 and the next one will be 2000. This change makes the average Gregorian year only 26 seconds too long, so it will take a long time to accumulate any appreciable error.

Devising the Gregorian calendar was difficult, but putting it into effect was even more difficult. When Pope Gregory proclaimed it in 1582, it was necessary to drop 10 days from that year to bring the calendar into agreement with earth's orbit. The Roman Catholic countries quickly adopted the Gregorian calendar, but most other countries were slow to adopt it. It was not adopted by England until 1752, when 11 days had to be dropped. The day after September 2nd was the 14th. This change caused riots and was widely misunderstood. Some felt that they had lost 11 days' wages, and rent laws were necessary to prevent abuses. The Gregorian calendar was not adopted by Russia until 1917, during the Russian revolution. By then it was necessary to omit 13 days. Needless to say, these changes are a source of confusion to historians.

Rotation and Time

The basic unit of time is the **second,** and the second is defined in terms of earth's daily rotation about its axis. This seems simple enough, but there are some complications. We live by solar time. *The* **solar day** *can be defined,* for example, *as the period from one noon until the next noon.* The stars can also be used to measure time. *The* **sidereal day**

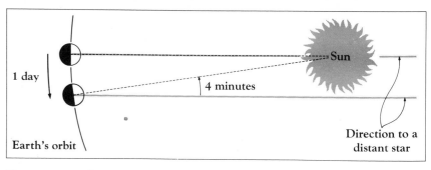

Figure 16-14 The sidereal day is about 4 minutes shorter than the solar day. In this figure, the solar day is from noon to noon, and the sidereal day is the interval between successive times that a star is seen at the same position.

can be defined as the period between successive times that a given star is at the same point in the sky. The relationship between these two days is shown in Figure 16-14. The sidereal day is somewhat shorter than the solar **F 16-14** day because of earth's movement in its orbit. Kepler's second law tells us that earth's velocity around its orbit is not constant, so the period of a solar day varies. Therefore, the solar day on which our civil, or clock, time is based is an average solar day averaged over a year. This average solar day is 24 hours, and the second is defined as

$$\frac{1}{60 \times 60 \times 24} \quad \text{or} \quad \frac{1}{86,400}$$

of an average solar day. The sidereal day is 23 hours, 56 minutes; said another way, a star rises about 4 minutes earlier each night.

Earth's rotation is at times either speeded or slowed by storms and earthquakes. In addition, the overall rate of rotation is slowing very gradually due to tidal friction. For these reasons, our second is now defined using the vibration frequency of cesium atoms.

THE MOON'S MOTIONS

As the moon makes its approximately one-month orbit of earth, its appearance from earth changes. These changes are both in the moon's shape and in the time of day that it is visible. These changes **F 16–15** are called **phases of the moon** (Figure 16-15). At **new moon,** the moon is between the sun and the earth. At this phase its only illumination, visible from earth, is from sunlight reflected from earth. Such reflected light is called **earthshine.** In the next week, the moon moves to **first quarter,** at which point it is seen from earth as a half-circle. Between new moon and first quarter, less than a half-circle is seen, and this is called **crescent moon.** During the next week, more and more of the moon is illuminated, and this is called **gibbous moon.** When the entire circle is illuminated, the phase is **full moon.** After full moon, the progression is reversed, and we have a week of gibbous moon again until **third-quarter** phase. One more week of crescent moon, and we are back to new moon. Note in Figure 16-15 that one can tell which phase the moon is in by which side of the moon is illuminated.

One can also tell which phase the moon is in by the time of day when it is visible. Figure 16-16 shows the new moon. Notice the direc-   **F 16-16** tion of the earth's rotation, and notice that the new moon is first seen at sunrise, so it rises with the sun and sets with the sun. In the same way, the first-quarter moon rises at noon and sets at midnight. The

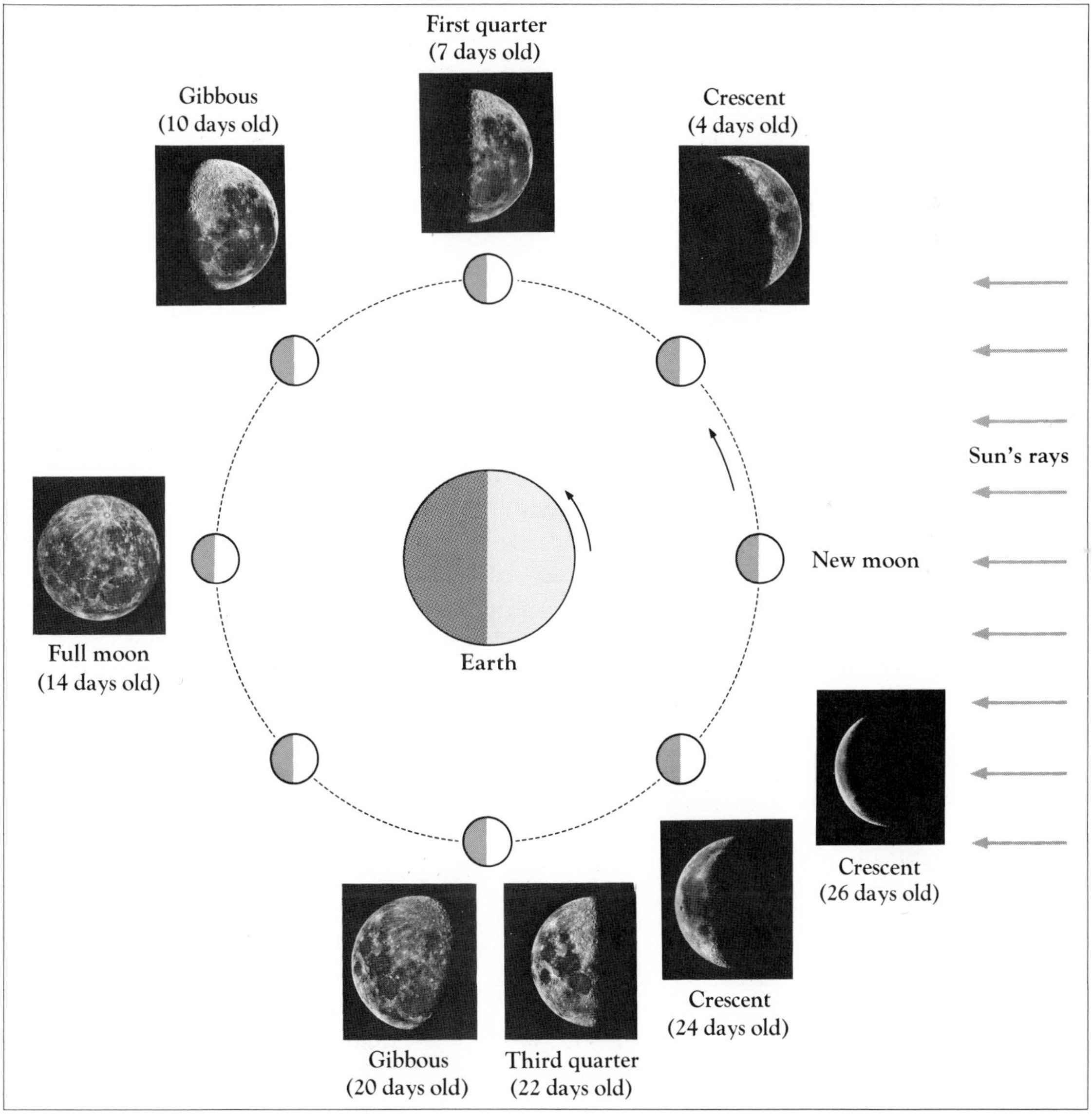

Figure 16-15 The phases of the moon. The photos show the moon as it would be seen from earth. (Photos from Lick Observatory.)

Figure 16-16
The new moon. An observer on earth would first see the new moon at point **A,** which is at sunrise. Point **B** is at noon, and point **C** is sunset, when the new moon is last seen.

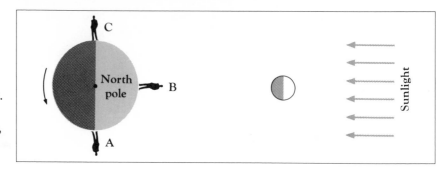

full moon rises at sunset and is last seen at sunrise. The third-quarter moon rises at midnight and sets at noon. You can easily test yourself on whether you understand moon phases by deducing the phase from its illumination and the time of day it is seen. Check yourself by looking at a calendar that has this information.

An *eclipse* occurs when the earth's shadow falls on the moon or when the moon is directly in line between the sun and the earth. Eclipses are rare events because the moon's orbit around the earth is tilted about five degrees from the plane of the earth's orbit around the sun, so normally the shadows of the two bodies miss each other. An **eclipse of the moon** can occur only at full moon (Figure 16-17), F 16-17 and an **eclipse of the sun** can occur only at new moon (Figure 16-18). F 16-18 Because the earth is much larger than the moon, its shadow is bigger than the moon; therefore, an eclipse of the moon can be seen by everyone on the dark side of the earth. During an eclipse of the sun, the dark part of the moon's shadow (**umbra**) covers only a small part of the daylight side of the earth, so only people in the path of this small area see a total eclipse. Those in the **penumbra,** or light part of the moon's shadow, see a partial eclipse (Figure 16-19).   F 16-19

The period of the moon's revolution around the earth depends on how it is timed. The period from, for example, full moon to full

Figure 16-17
A lunar eclipse can occur only at full moon. If only part of the moon passes through the earth's shadow, a partial eclipse results.

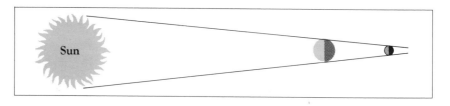

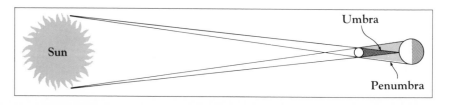

Figure 16-18
An eclipse of the sun occurs at new moon.

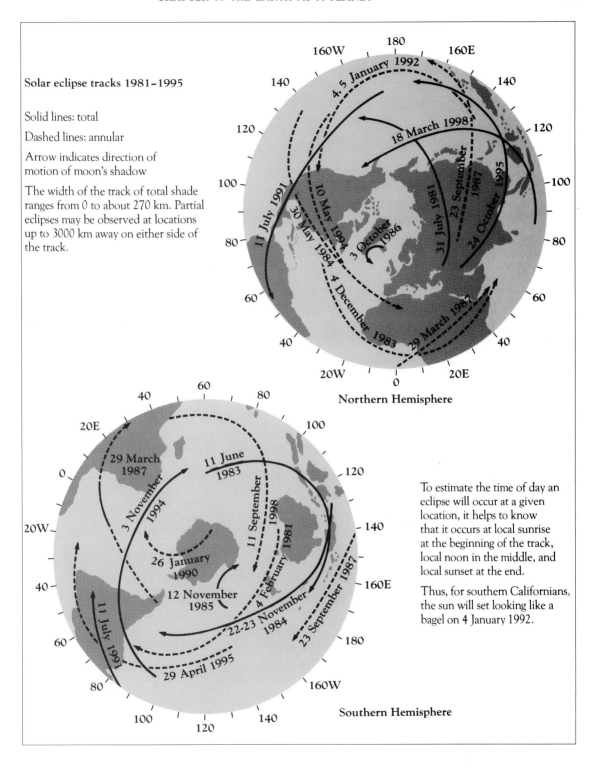

Solar eclipse tracks 1981–1995

Solid lines: total

Dashed lines: annular

Arrow indicates direction of motion of moon's shadow

The width of the track of total shade ranges from 0 to about 270 km. Partial eclipses may be observed at locations up to 3000 km away on either side of the track.

To estimate the time of day an eclipse will occur at a given location, it helps to know that it occurs at local sunrise at the beginning of the track, local noon in the middle, and local sunset at the end.

Thus, for southern Californians, the sun will set looking like a bagel on 4 January 1992.

Figure 16-19 Paths of solar eclipses through 1995. (After Goldsmith, 1981.)

Figure 16-20
The moon orbits the earth in 27.32 days—the *sidereal month* (point **A** to point **B**). Full moon to full moon is 29.53 days, called the *synodic month* (point **A** to point **C**).

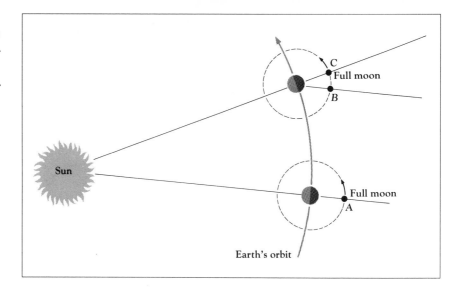

moon is 29.53 days and is called the **synodic month.** Because of the earth's revolution, this is a little over 2 days longer than it takes for the moon to complete one orbit of the earth when measured relative to the stars, the **sidereal month** of 27.32 days (Figure 16-20).

The moon's periods of rotation and revolution are the same, as shown in Figure 17-3, p. 426.

F 16-20

LIGHT, SPECTRA, AND TELESCOPES

Except for the moon rocks brought back by astronauts and the few meteorites that have fallen on the earth, astronomers can study only the radiation from celestial bodies. One usually thinks of an astronomer peering into a telescope, but this is only one of the astronomer's tools. Astronomers study radiation of many wavelengths, extending from X rays and gamma rays to radio waves, and not just visible light. From their studies they learn the temperature, size, velocity, and even the composition of objects so distant that the radiation, traveling at the speed of light, required many years to reach their instruments. To understand how they learn so much from the study of radiation, we must first understand the nature of light and other forms of radiation.

Light and Other Electromagnetic Radiation

Light, X rays, and radio waves are all forms of **electromagnetic radiation.** They travel through the near-vacuum of space at the speed of light. All types of electromagnetic radiation have electric and magnetic fields that vary rapidly at the same rate. The differences among the various forms of electromagnetic waves is the rate at which the electric and magnetic fields vary. *This rate of variation is called the* **frequency.** *The* **wavelength** *is the distance between successive crests of the wave* as the wave moves. The wavelengths of the various types of electromagnetic radiation are shown in Figure 11-10, p. 279. *A display of electromagnetic radiation spread out by frequency or wavelength,* such as shown in Figure 11-10, *is called a* **spectrum** (plural **spectra**).

Notice the tremendous range in wavelength in this spectrum of all types of electromagnetic radiation. The wavelengths of some radio waves are measured in kilometers, and radio wavelengths extend down to about one centimeter. Astronomy is most concerned with wavelengths from a few centimeters down to gamma rays that have wavelengths of less than 0.00000001 centimeter. To avoid such awkward numbers, another unit, the **micrometer** (*micron* is a nonstandard term for this unit), which is one-millionth of a meter, is used (see Appendix A).

We see the different wavelengths of visible light as different colors. Violet has a wavelength of about 0.4 micrometer, and red light has a wavelength of about 0.65 micrometer. The wavelengths smaller and larger are called **ultraviolet** and **infrared,** respectively, and our eyes are not sensitive to them.

The speed of light is 300,000 kilometers per second (186,000 miles per second) in a vacuum. Distances in astronomy are sometimes measured in **light years.** In a year, light will travel $300,000 \times 60 \times 60 \times 24 \times 365 = 9,460,800,000,000$ kilometers (5,865,696,000,000 miles). The speed of light is very slightly slower in space because space is not a perfect vacuum, and it is measurably slower in other media, such as glass. It is this slowing of light by the glass in a lens or prism that bends the light.

Spectra

If a beam of white light passes through a prism, as shown in Figure 16-21, it is bent, but the different colors, or wavelengths, of the white light are bent different amounts. The prism separates the wavelengths within the white light and so produces the spectrum of the white light. The blue light is slowed most, so it is bent most; the red light is slowed least, so it is bent least. The prism is not the only device used by astronomers in their study of celestial light. All spectra are not so simple as this example, and the study of spectra can reveal much about the nature of the radiating object.

The easiest spectrum to study is that of a hot, solid object, such as the filament of an ordinary electric light bulb. The spectrum shows all of the colors from red to violet, as shown in Figure 16-21. Such a

F 16–21

Figure 16-21
White light passing through a prism is separated into its various wavelengths. This is an example of a continuous spectrum.

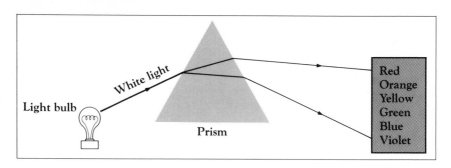

spectrum is called a **continuous spectrum** because *all colors, or wavelengths, are present in a continuous display.* Information about the nature of the hot, solid emitter can be obtained from study of the continuous spectrum. If we arrange an experiment in which we can vary the temperature of the light bulb by increasing or decreasing its voltage, we will find that the spectrum changes as the temperature of the bulb changes. As the temperature increases, the intensity of the blue light increases much more than does the intensity of the red light. Of course, as the temperature of the bulb increases, it will radiate more energy because it has more energy to radiate. A few minutes of experimenting with this apparatus would convince you that one can tell the temperature of the emitter by comparing the intensity of the blue light with the intensity of the red light. Indeed, this method is used to determine the temperature inside a furnace where it is inconvenient (or impossible) to place a thermometer.

Actually, you probably knew that a hot object radiates more blue light than a cooler one, but you may not have thought about it in this way. If you put a poker in a fire and leave it there, it will, of course, get hot. As it heats up, it will first become red hot. This occurs because most of its radiation is at the red end of the spectrum. As it gets hotter, the red becomes more orange or yellowish and, finally, if the fire is very hot, the poker will become white hot. It is radiating a white color because all of the colors of the spectrum, including at this temperature the blue, are about the same intensity; and, as we know, white light is a mixture of all colors.

If, instead of looking at the spectrum of a hot, solid object, we look at the spectrum of a hot, or otherwise energized, gas, we will see a totally different type of spectrum. This time we see only a number of bright lines (Figure 16-22). This is called a **bright-line,** or **emission,** F 16-2. **spectrum.** The important point about bright-line spectra is that the gas of every element (or molecule) emits a different series of bright lines. Thus, if the bright lines of every element are cataloged—and they have been—one can determine the composition of the gas from study of its bright-line spectrum. Spectrographic analysis of this sort is done routinely by chemists because it is a convenient and cheap means of determining the composition of an unknown substance.

Figure 16-22
The light from an excited gas gives a bright-line spectrum. The light from a neon sign would give this type of spectrum.

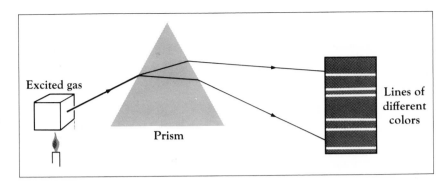

Figure 16-23
An absorption, or dark-line, spectrum is formed when light from an incandescent source is passed through an excited gas. The spectrum of most stars is this type.

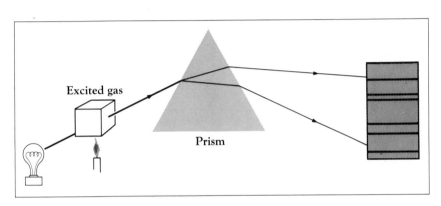

The third, and last, type of spectrum is really the combination of the continuous and the bright-line spectra. If the radiation from a hot, solid object, which would be a continuous spectrum, passes through a cooler gas, the gas will absorb certain wavelengths from the continuous spectrum. If we look at such a spectrum, we will see an otherwise continuous spectrum with some dark lines. This is called an **absorption,** or **dark-line, spectrum** (Figure 16-23). The dark lines occur at those wavelengths that the gas would radiate if it were energized. Thus, the absorption spectrum reveals the object's temperature from study of the continuous spectrum and the gas' composition from the wavelengths of the dark lines. Most stars emit absorption spectra, so we can determine the temperature of their surfaces and the composition of the gases that surround them (their "atmospheres").

F 16–23

Doppler Effect

The spectrum of a star can also tell us whether the star is moving toward us or away from us, and how fast. If a passing train blows its whistle, the pitch of the whistle gets higher as the train approaches and lower as the train moves away. This change is called the **Doppler effect,** and it *is a change in wavelength that occurs because of motion of either the emitter or the observer* (Figure 16-24). If a source of waves is

F 16–24

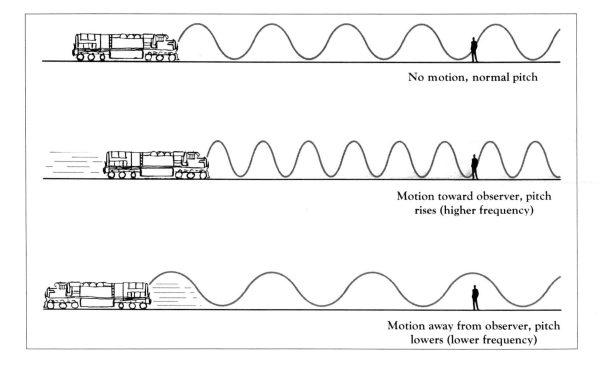

No motion, normal pitch

Motion toward observer, pitch
rises (higher frequency)

Motion away from observer, pitch
lowers (lower frequency)

Figure 16-24
The Doppler effect is
the change in
wavelength that
occurs when either
the emitter or the
observer is moving.

moving toward the observer, each successive wave front will reach the observer sooner because the source has moved closer before the next wave is emitted. If, however, the emitter is moving away from the observer, the waves will be farther apart when they reach the observer. It should also be clear that the same effect will be noted whether it is the source, the observer, or both that are moving. The amount of change in the wavelength is proportional to the relative velocity of the source and the observer.

The Doppler effect is seen in a star's spectrum by the shift of the dark lines in the spectrum. If the lines are shifted toward the blue end, the star is moving toward us; and if the lines are shifted toward the red, the star is moving away from us. The amount of the shift tells us the velocity toward or away from us. *The velocity directly toward or away from us is called the* **radial velocity,** and this is the motion that causes the Doppler effect.

Lenses, Mirrors, and Optical Telescopes

A simple astronomical telescope consists of two lenses, an **objective lens** and an **eyepiece** (Figure 16-25). A lens is a transparent medium, generally glass, shaped so that it brings light rays to a focus (Figure   F 16-25

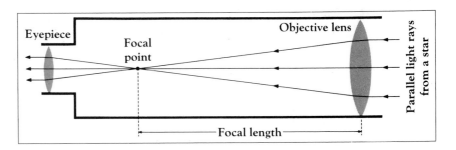

Figure 16-25
A simple telescope.

16-26). The **focal length** of a lens is the distance between the lens and   F 16–26
the place where the image is formed by parallel light rays entering the
lens. The objective lens of a telescope should be as large as possible so
that it can gather the most light possible. The purpose of the eyepiece
is to enable one to see the image formed by the objective lens. The
magnification of the telescope is determined by the focal length of
the eyepiece, because eyepieces, being small, can be changed; the big,
expensive objective lens cannot. Generally, several eyepieces of differing focal lengths are available for any telescope. The **magnification
of the telescope** is determined by dividing the focal length of the objective lens by the focal length of the eyepiece.

The purpose of an astronomical telescope is to gather light, so its
most important feature is the area of its objective lens. This is why
bigger and bigger astronomical telescopes are built. The magnification is important only when studying nearby objects such as planets.
Stars, except the sun, are so distant that they appear as points, even at
the highest magnification; so for stellar astronomy, the light-gather-
ing power is much more important than the magnification.

The bending of light rays by a lens or prism is called **refraction.** The
telescopes discussed so far are **refracting telescopes.** Refraction occurs
because light travels more slowly in the glass lens than it does in air.
As we noted in discussing spectra, the different wavelengths, or colors, of light travel through glass at different speeds. This is how a
prism separates light into its various wavelengths and so produces a
spectrum. The same thing happens in a lens, and the light of different
colors comes to focus at different distances from the lens (Figure
16-27). This is called **chromatic aberration** and is an inherent defect   F 16–27
in lenses. In many optical instruments, chromatic aberration is overcome by making lenses of several pieces of glass, each of which has
somewhat different properties, so that the images of the various colors
form at nearly the same place. This is not generally practical in the
large lenses used in optical telescopes.

Another problem with large refracting telescopes is the difficulty of
making the large piece of glass necessary for the objective lens with-

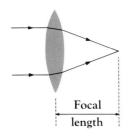

Figure 16-26
The focal length is
the distance between
the lens and the place
where parallel light
rays entering the lens
come to focus and
form an image.

Figure 16-27
Chromatic aberration.

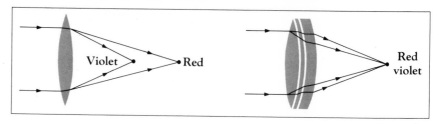

out bubbles or other defects. The weight of a large lens is also a problem. The surface of the lens is polished to form a very accurate shape. Because of its weight, however, the lens sags differently when it is pointed near the horizon than it does when it is close to vertical. Remember that the lens can be supported only at its rim. For these reasons, the practical limit for refracting telescopes is about 1 meter in diameter. The largest refracting telescope in the world is the 40-inch (1 meter) telescope, completed in 1893, at the University of Chicago's Yerkes Observatory.

The largest astronomical telescopes are **reflecting telescopes,** which use a curved mirror to focus the light. A mirror with the shape of a parabola will bring parallel light of any wavelength to focus at a point (Figure 16-28). The mirror can be supported at the bottom to prevent sagging. Some methods of getting the image away from the focal point of the mirror are also shown in the figure. Two of the large reflecting telescopes are the "200-inch" (5-meter) telescope at Palomar Mountain in California and the 6-meter (236-inch) reflector at Mount Pastukhov in the Soviet Union, completed in 1976. Even reflecting telescopes have practical limits because of the size and weight of their mirrors. A few even larger reflectors use a number of smaller mirrors to form the image.

F 16-28

The earth's atmosphere also limits observation from the surface by its turbulence. It is the turbulence of the atmosphere that makes the stars appear to twinkle. Before a new observatory is built, studies are made to be sure that turbulence is minimal at the proposed site. For this reason, and to avoid the night glow of cities, many observatories are located in remote places, especially on mountaintops, where there is less atmosphere above the observatory. Optical telescopes on orbiting satellites also obviously avoid problems with the atmosphere.

Telescopes for Invisible Radiation

Much of the radiation we receive from space is not in the visible wavelengths and so is not detected by optical telescopes. Radio wavelengths are an example, and very elaborate **radio telescopes** have been built. Radio telescopes are large antennnas that collect weak radio

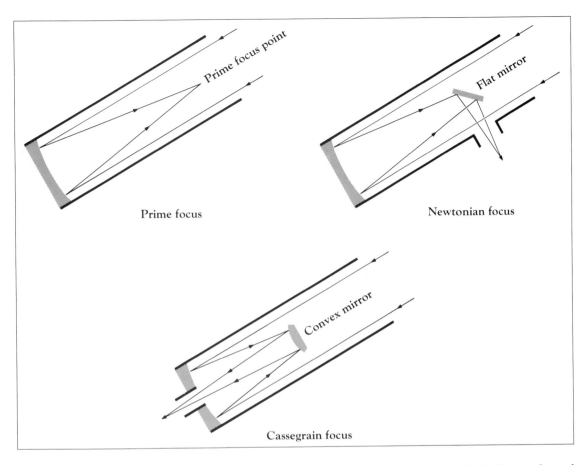

Prime focus point

Prime focus

Flat mirror

Newtonian focus

Convex mirror

Cassegrain focus

Figure 16·28
The reflecting telescope.

energy from space, and this energy is then amplified, detected, and recorded. Because radio waves are much longer than light waves, radio telescopes must be much larger than optical telescopes. Some of the antenna systems are made up of antennas similar to those used for television, but many such antennas are used to form the radio telescope. Others consist of one or more dish antennas (Figure 16-29). **F 16–29** The largest fully movable dish is 76.25 meters (250 feet) in diameter and is at Jodrell Bank, England. The dish at Green Bank, West Virginia, is 91.5 meters (300 feet) in diameter, but it is only partially steerable. The largest radio telescope in the world is at Arecibo, Puerto Rico; the fixed mirror there is made of 19 acres of steel mesh built in a valley (Figure 16-30). Because of the earth's spin and orbit, **F 16–30** even a fixed radio telescope points at a large part of space in the course of a year. Radio telescopes can, of course, be used day or night, in clear or cloudy weather.

Some celestial radiation is invisible to us because it cannot pass through the earth's atmosphere. Ultraviolet radiation, gamma rays, and X rays are examples. Instruments carried aloft by balloons have

Figure 16-29
The 300-foot (100-meter) radio telescope at Green Bank, West Virginia. (Photo from NRAO.)

Figure 16-30 The 1000-foot (300-meter) radio telescope at Arecibo, Puerto Rico. The reflector is built on the valley, and the receiver is suspended above. (Photo from Arecibo Observatory, NAIC, Cornell University.)

been used to detect such radiation, and, later, rockets were used to carry instruments even higher. Such flights were necessarily short, and now orbiting satellites are also used.

Some planets in the solar system, such as Venus, are surrounded by thick atmospheres that prevent direct optical observation from earth. Radar beams can penetrate such an atmosphere and be reflected back to earth. In this way, a radar image of the topography of the planet's surface can be made.

SUMMARY

- The apparent path of the sun in front of the stars is called the *ecliptic*.
- The *zodiac* is a zone 9 degrees on each side of the ecliptic in which the moon and the planets travel across the sky.
- Ptolemy, in 150 A.D., assumed that the earth is at the center of the universe and explained the motions of planets with a complex system of orbits and epicycles.
- Copernicus, in 1543, devised a much simpler system with the sun at the center.
- Tycho Brahe, in the pretelescope era, collected accurate observational data that enabled Kepler to discover these three laws of planetary motion:

  1. The orbits of planets are ellipses, not circles, and the sun is at one of the foci of the ellipse.
  2. Planets travel faster when they are near the sun, so that they sweep out equal areas in equal times.
  3. The relationship between a planet's distance from the sun and its orbital period is that the cube of the distance from the sun divided by the square of its period is a constant that is the same for all planets.

- Galileo was the first to use a telescope in astronomy, and his observations helped to establish the heliocentric theory.
- Newton's laws of motion and law of gravity explained why the planets orbit the sun.
- The orbital trip of a planet is called its *revolution*.
- At *perihelion* (closest approach to the sun), the earth is about 146 million kilometers (91 million miles) from the sun.
- At *aphelion* (farthest from the sun), the earth is about 150 million kilometers (93 million miles) from the sun.
- The earth's *precession* is a slow cyclical wobble requiring about 26,000 years to complete.
- The period of the earth's revolution is 365 days, 5 hours, 48 minutes, and 46 seconds, and this has resulted in various attempts to devise an accurate calendar.
- The *solar day* is the interval from one noon until the next noon.
- The *sidereal day* is the interval between successive times that a given star is at the same point in the sky and is about four minutes shorter than the solar day.

- The *second,* the basic unit of time, was defined on the basis of the average solar day, but is now defined using the vibration frequency of cesium atoms.

- The *moon's phases* are shown in Figure 16-15.

- An *eclipse of the moon* occurs when the earth's shadow falls on the moon. It can only occur at full moon.

- An *eclipse of the sun* occurs when the moon covers the sun and can only occur at new moon.

- The *synodic month* is the interval between two full moons and is 29.53 days.

- The *sidereal month* is the time for the moon to complete one orbit of the earth measured relative to the stars and is 27.32 days.

- The moon's periods of revolution and rotation are the same, and this is why only one side of the moon always faces the earth.

- The speed of light in space is about 300,000 kilometers per second (186,000 miles per second). In another medium, such as glass, it is slowed somewhat.

- A beam of light passing through a prism is separated into its various wavelengths, or colors, because each wavelength travels at a different speed in the prism. Such a display of wavelengths is called a *spectrum.*

- A hot solid object emits a *continuous spectrum;* that is, the spectrum has all colors. The temperature of the emitter can be determined from a continuous spectrum.

- An excited gas emits a *bright-line,* or *emission, spectrum.* Every element or molecule emits a different set of bright lines, so the lines are the fingerprints of the elements and can be used to determine the composition of the gas.

- If the light from a source of a continuous spectrum passes through an excited gas (the source of a bright-line spectrum), an *absorption,* or *dark-line, spectrum* results. The dark lines are at the wavelengths that lines would have appeared from the gas alone. An absorption spectrum can reveal both the composition and the temperature of its source. Most stars emit absorption spectra.

- The *Doppler effect* is the change in wavelength that occurs because of motion of either the source or the observer.

- The purpose of astronomical telescopes is to gather light, and this is why they are large.

- Telescopes that use lenses are called *refracting telescopes,* and they are limited by their weight to relatively small lenses.

- *Reflecting telescopes* use mirrors and can be much larger; they do not have *chromatic aberration.*

- *Radio telescopes* detect other wavelengths of radiation from space.

KEY TERMS

ecliptic

zodiac

epicycles

Ptolemaic system, or geocentric system

heliocentric system

retrograde motion

Astronomical Unit (A.U.)

law of gravity

revolution

perihelion

aphelion

precession

year

second

solar day

sidereal day

phases of the moon

new moon

earthshine

first-quarter moon

crescent moon

gibbous moon

full moon

third-quarter moon

eclipse of the moon

eclipse of the sun

umbra

penumbra

synodic month

sidereal month

electromagnetic radiation

frequency

wavelength

spectrum (spectra)

micrometer

ultraviolet radiation

infrared radiation

light year

continuous spectrum

bright-line, or emission, spectrum

absorption, or dark-line, spectrum

Doppler effect

radial velocity

objective lens

eyepiece

focal length

magnification of telescope

refraction

refracting telescope

chromatic aberration

reflecting telescope

radio telescope

REVIEW QUESTIONS

1. In the 16th century the Copernican model of planetary position and movement began to replace the Ptolemaic model, which had been in use since the second century A.D.
   (a) In one or two sentences state the basic differences between the Ptolemaic and Copernican systems.
   (b) How does the Copernican model explain the observation that Venus and Mercury appear only a few hours after sunset or before sunrise? (Explain in words or by making a sketch.)
   (c) How is retrograde motion explained by the Copernican system?
   (d) What was the initial advantage of the Copernican system?
2. State Kepler's three laws.
3. Mars has a period of 1.88 years; Jupiter's period is 11.86; Saturn's is 29.46.
   (a) Are any of these three planets closer to the sun than earth is?
   (b) Use Kepler's third law to calculate solar distance for each of the three planets.

4. Describe some ways in which Galileo's contributions to astronomy substantiate the following statements: As scientific equipment becomes more sophisticated, scientists are able to make better observations. As observations improve, scientific theories are often altered or refined.
5. Answer the following questions about earth's revolution:
   (a) What is the shape of the orbit?
   (b) Where within this orbit is the sun located?
   (c) Is the revolution clockwise or counterclockwise?
   (d) What is earth's closest approach to the sun called? When does close approach occur?
   (e) What is earth's farthest distance from the sun called? When does it happen?
6. Has Polaris always been our north star? Why or why not?

7. On what astronomical motion is our year based? What special features of the Gregorian calendar keep the calendar year and the astronomical year closely synchronized?

8. Until recently, the second was based on what motion? Why is the second now based on an atomic vibrational frequency?

9. Use descriptions of solar and sidereal days to explain why a star rises about four minutes earlier each night.

10. In which phase is the moon when the following observations are made:
    (a) A crescent with the convex side on the left; observed coming home from a party at 3:00 A.M.
    (b) Half-circle setting at midnight.
    (c) Half-circle seen on your way to an 8:00 A.M. class.
    (d) Gibbous with the fuller side on the right; seen in early evening.

11. In what phase must the moon be for a lunar eclipse to occur? Why is everyone on the dark side of earth able to see an eclipse of the moon? Why is it that lunar eclipses are not monthly occurrences?

12. In what phase must the moon be for a solar eclipse to occur? Why are total solar eclipses seen only along a very narrow zone of earth's surface?

13. What accounts for the difference between a synodic month and a sidereal month?

14. What characteristics do all forms of electromagnetic radiation have in common? Define a spectrum.

15. Why is light bent when it passes from the air into a prism? Explain how a prism produces the spectrum of white light.

16. Describe how a continuous spectrum looks. Discuss some changes you might see in the continuous spectrum if you changed the temperature of the solid emitter.

17. What produces a bright-line spectrum? What does a bright-line spectrum look like? Explain why scientists are able to discover the chemical composition of a substance by studying its bright-line spectrum.

18. How is an absorption spectrum produced? How are astronomers able to determine the surface temperature of a star and the composition of its "atmosphere"?

19. What causes the Doppler effect? Explain how the Doppler effect is used to determine the velocity of a star moving away from us.

20. Answer the following questions about refracting telescopes:
    (a) Why is the objective lens the most important feature of the telescope?
    (b) What determines magnification? When is magnification important?
    (c) Discuss the two major drawbacks of refracting telescopes.

21. What advantage does a reflecting telescope have over a refracting telescope?

22. State some ways in which atmosphere (both our own and other planets') has limited the study of celestial bodies. Also describe how astronomers have gotten around some of these limitations.

SUGGESTED READINGS

Aveni, A. F., S. L. Gibbs, and H. Hartung. "The Caracol Tower at Chichen-Itza: An Ancient Astronomical Observatory?" *Science* (June 6, 1975), Vol. 188, No. 4192, pp. 977–985.

Brandt, J. C., and S. P. Maran. *The New Astronomy and Space Science Reader.* San Francisco: W. H. Freeman, 1977.

Cox, R. E. "What's Inside Your Telescope?" *Astronomy.* "Part 1, Refractors" (March 1978), Vol. 6, No. 3, pp. 50–55; "Part 2, Reflectors" (April 1978), Vol. 6, No. 4, pp. 31–35; "Part 3, Catadioptrics" (July 1978), Vol. 6, No. 7, pp. 42–46.

Drake, Stillman, and James MacLachlan. "Galileo's Discovery of the Parabolic Trajectory," *Scientific American* (March 1975), Vol. 232, No. 3, pp. 102–110.

Eddy, J. A. "Mystery of the Medicine Wheels," *National Geographic* (January 1977), Vol. 151, No. 1, pp. 140–146.

Gingerich, Owen. "Musings on Antique Astronomy," *American Scientist* (March 1967), Vol. 55, No. 1, pp. 88–95.

Gingerich, Owen. "Tycho Brahe and the Great Comet of 1577," *Sky and Telescope* (December 1977), Vol. 54, No. 6, pp. 452–458.

Goldreich, Peter. "Tides and the Earth-Moon System," *Scientific American* (April 1972), Vol. 226, No. 4, pp. 42–52.

Goldsmith, Donald. *The Evolving Universe: An Introduction to Astronomy.* Menlo Park, Ca.: Benjamin/Cummings, 1981, 539 pp.

Hicks, R. D., III. "Astronomy in the Ancient Americas," *Sky and Telescope* (June 1976), Vol. 51, No. 6, pp. 372–377.

"A History of Astronomy in the United States," *Astronomy* (July 1976), Vol. 4., No. 7 (bicentennial issue).

North, J. D. "The Astrolabe," *Scientific American* (January 1974), Vol. 230, No. 1, pp. 96–106.

Ottewell, Guy. *Astronomical Calendar,* yearly. Sponsored by the Department of Physics, Furman University, in cooperation with the Astronomical League.

Philip, A. G. D. "A Visit to the Soviet Union's 6-Meter Reflector," *Sky and Telescope* (May 1974), Vol. 47, No. 5, pp. 290–295.

Stencel, Robert, William Blair, and Susan Conat-Stencel. "Astronomical Spectroscopy," *Astronomy* (June 1978), Vol. 6, No. 6, pp. 6–19.

Vehrenberg, Hans. *Atlas of Deep Sky Splendors.* Cambridge, Mass.: Sky Publishing, 1971.

Zirker, J. B. "Total Eclipses of the Sun," *Science* (December 19, 1980), Vol. 210, No. 4476, pp. 1313–1319.

THE PLANETS

Inner, or Terrestrial, Planets

EARTH

MOON

MERCURY

VENUS

MARS

Outer, or Jovian, Planets

JUPITER

SATURN

URANUS AND NEPTUNE

PLUTO

SATELLITES OF THE JOVIAN PLANETS

Jupiter's Satellites

Saturn's Satellites

Satellites of Uranus and Neptune

ASTEROIDS

COMETS

METEORS AND METEORITES

ORIGIN OF THE SOLAR SYSTEM

CHAPTER 17

THE SOLAR SYSTEM

Meteorite impact splashed lobes of material out of crater Yuty (18-kilometer diameter) on Mars. (Photo from NASA.)

In this chapter we will consider most of the objects that comprise our solar system. The architecture of the planets and their moons are described and compared with earth's. Asteroids, comets, meteors, and meteorites, as well as the origin of the solar system, are also considered. The principal member of the solar system—the sun, a star—is described in the following chapter.

THE PLANETS

The planets are of two distinct types. The **inner, or terrestrial, planets**—Mercury, Venus, Earth, and Mars—are more or less similar to Earth in size and composition. The **outer, great, or Jovian, planets**—Jupiter, Saturn, Uranus, and Neptune—are all very much larger, and they are made mainly of hydrogen and helium, with lesser amounts of ammonia, methane, and water ice. Pluto, normally the most distant planet, is small, and we know very little about it (Table 17-1, T 17-? Figure 17-1). Because Pluto's orbit is elliptical, it will be closer to the F 17-? sun than Neptune until 1999. Figure 17-2 shows an easy way to re- F 17-? member the order of the planets.

Inner, or Terrestrial, Planets

A useful approach to understanding the differences among the planets is to compare them with earth; this also leads us to a better understanding of the earth itself. The terrestrial planets are Mercury, Venus, Earth, and Mars, but here we will also consider our moon because it is similar to the others and we know more about it. All the terrestrial planets are similar in that they have two distinct types of crust, highlands and lowlands.

EARTH Earth is the largest of the terrestrial planets, with an equatorial radius of 6378 kilometers (3954 miles). This is slightly more (21 kilometers or 13 miles) than the polar radius because of the earth's spin. Earth's axis of rotation is tilted 23½ degrees with respect to its orbital plane, so we have seasons. The average temperature at the surface is 22°C (72°F). The atmosphere is composed of 78 per cent nitrogen and 21 per cent oxygen. The overall density of the earth is slightly more than that of Mercury and Venus and much more than the others. Earth has a thin *crust* composed of the volcanic rock basalt under the ocean-basin "lowlands" and of granitic rocks under the "highland" continents (see Figures 5-10, 5-11, and 5-13, pp. 134, 135, and 137). Below the crust is a thick layer called the *mantle*, composed of iron and magnesium silicate rocks. The *core* is composed of nickel and iron and is partially liquid. This liquid metallic core is believed to be the source of our magnetic field, the strongest field of the terrestrial planets.

Table 17–1 THE PLANETS

	Equatorial Radius (Kilometers)	Equatorial Radius (Miles)	Density (Grams per cubic centimeter)	Number of Satellites	Atmosphere	Rotation Period	Revolution Period	Average Distance from Sun (Millions of kilometers)	Average Distance from Sun (Millions of miles)
The Terrestrial, or Inner, Planets in Order of Size									
Moon	1738	1078	3.3	—	None	27.3 days	27.3 days	384,000 km (from earth)	239,000 mi (from earth)
Mercury	2439	1517	5.4	0	None	58.6 days	88 days	58	36
Mars	3398	2114	3.9	2	Carbon dioxide. Pressure is 0.7% of earth.	24 hours 37 minutes	687 days	227	141
Venus	6050	3751	5.3	0	Carbon dioxide. Pressure is 90 times earth.	243 days	225 days	108	67
Earth	6378	3954	5.5	1	78% nitrogen 21% oxygen	1 day	365.25 days	150	93
The Jovian, or Great, Planets in Order of Size and Distance from the Sun									
Jupiter	71,400	44,411	1.3	15	Mainly hydrogen and helium.	10 hours	11.9 years	777	483
Saturn	60,000	37,300	0.7	15	Mainly hydrogen and helium.	10 hours	29.5 years	1,426	887
Uranus	25,900	16,110	1.3	5	Mainly hydrogen and helium.	23 hours	84 years	2,869	1,783
Neptune	24,750	15,400	1.7	2	Mainly hydrogen and helium.	22 hours	165 years	4,494	2,793
Pluto, the Most Distant Planet, and Ceres, the Largest Asteroid									
Pluto	1500?	930?	0.7	1	Unknown	6.4 days	248 years	5,899	3,666
Ceres	502	311	—	—	None	0.38 days	4.6 years	416	258

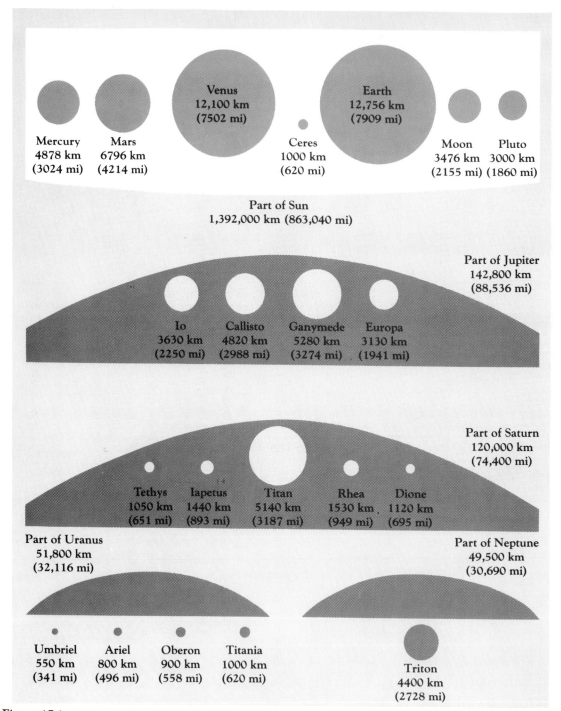

Figure 17·1
Relative sizes of the planets and their larger satellites. All are drawn to the same scale.

Besides the obvious feature of water in the form of oceans, glaciers, and clouds, the feature of the earth seen on no other terrestrial planet is our *folded mountain ranges*. Earth is an active planet, as evidenced by earthquakes, volcanoes, and old deformed rocks. The oldest rocks are more than 4000 million (4 billion) years old, and

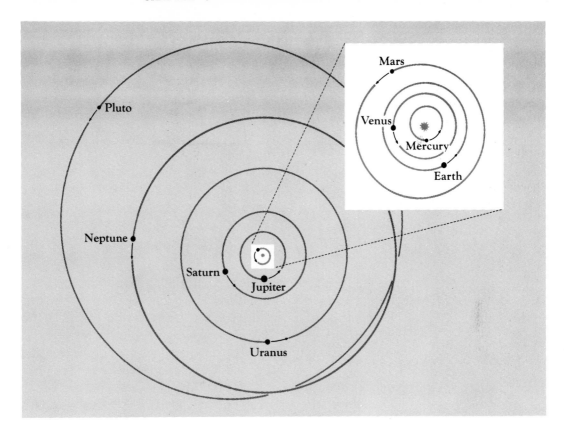

Figure 17-2
The orbits of the planets. The sequence from the sun outward can be remembered from Mary's Violet Eyes Make John Stay Up Nights Permanently.

their structures indicate that processes similar to those occurring to-day have always occurred on earth. The dynamic process occurring on earth is *plate tectonics,* and the energy source for plate tectonics is believed to be decay of radioactive elements within the earth and energy released as the liquid core crystallizes.

Because earth's liquid core may be its main energy source, one would think that similar-sized planets would probably have similar cores, so they would tend to be most like earth. Venus is the planet closest to earth in size, but unfortunately its surface is poorly known because it has a dense cloud cover. We will begin our comparison of earth to other planets with our moon because we know most about it.

MOON From earth we can only see one side of the moon because the same side of the moon is always toward the earth. Thus, everything we know about the far side of the moon we have learned from orbit-ing spacecraft. We actually see 59 per cent of the moon's surface be-cause the moon has a slight wobble called **libration.** One side of the moon is always toward earth because the periods of rotation and rev-olution of the moon are the same; therefore, the moon spins once on its axis each time it makes one orbit around earth. You can demon-strate that such motion will keep one face toward earth by turning a

Figure 17·3
The moon rotates once on its axis during each revolution around the earth, and this is why we can see only one-half of the moon. If Lincoln on the cent always faces the earth (the quarter), the cent must turn once on its axis during each orbit around the quarter.

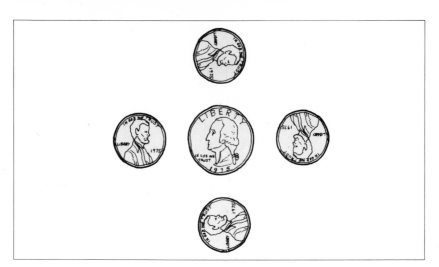

cent around a quarter (Figure 17-3). If Lincoln always faces the quar- F 17-3
ter, he will also see all four walls of the room each orbit.

The moon's orbit is elliptical; its distance from earth is about 356,000 kilometers (221,000 miles) at closest approach, about 407,000 kilometers (253,000 miles) when farthest away, and averages about 384,000 kilometers (239,000 miles). This change in distance causes the apparent size of the moon to change about 10 per cent.

The moon's radius is 1738 kilometers (1078 miles), and earth's is 6378 kilometers (3954 miles). Thus the size difference is about four times, but this difference is small enough that the earth-moon system is considered a double-planet system by many. The overall density of the moon is about 3.3 grams per cubic centimeter—that is, 3.3 times that of water. Earth's density is 5.5 grams per cubic centimeter, so the structure and composition of our moon is different from earth's. Because of the moon's smaller size and lesser mass and density, it lacks the gravity to hold an atmosphere. Temperature is the other factor in a planet's holding an atmosphere. High temperature causes atmospheric gases to move with high velocities and so enables atoms or molecules to escape the gravitational field.

From earth, with naked eye or telescope, dark and light colored areas are seen on the moon. The dark-colored areas were named **mare** (plural, **maria**) by the ancients, who thought that they resembled the oceans on earth. In the same way, the light-colored areas were called **terra** (plural, **terrae**) because they resembled continents. We call the light areas **highlands** because they are higher in elevation than the maria.

The most obvious feature of the whole moon's surface is that it is cratered by the impacts of materials probably similar to the meteorites that cause **impact craters** on earth. Some of the moon's craters have rays of material splashed radially out of them by the impact

(Figure 17-4). Earth has many fewer craters than the moon for two F 17-4 reasons. Meteorites must pass through earth's atmosphere, and all but the biggest are burned up by friction; also, erosion and other geologic processes soon obliterate a crater so that it is difficult to recognize. The moon has no atmosphere and no wind or water to cause erosion. The amount of cratering on any surface should be an index of the age of that surface, with older surfaces being more cratered than younger ones. Using this index, the lighter-colored, highland areas are older than the maria.

Volcanic features are also common on the moon's surface. Some of the craters may be volcanic, and so may some of the moon's domes, especially those topped by craters. **Rilles** are sinuous valleys that may be collapsed lava tubes (Figure 17-5). On two occasions in 1963, a red F 17-5 glow was seen at the crater Aristarchus, and *Apollo* astronauts noted gas discharge from this region. This is the only instance of any type of volcanic activity actually seen on the moon. The moon's mountains are all caused either by cratering or by volcanic activity; they are in no way like the folded and faulted mountains on earth.

Apollo astronauts have returned rocks from the moon's surface, and study and radioactive dating of these samples enables us to decipher some of the moon's history. The moon rocks have all been gathered from the surface or near-surface. They are all made up of fragments of many different rocks; such rocks are called **breccia.** These rocks are the product of impacts of meteorites of all sizes, and

Figure 17-4
The moon, showing light-colored rays of splashed material from many of its craters. The large dark areas, maria, are also probably of impact origin (see Figure 17-7.) (Photo from NASA.)

Figure 17-5
Rille on the moon.
The origin of these
features is not clear,
but they may be
collapsed lava tubes.
(Photo from NASA.)

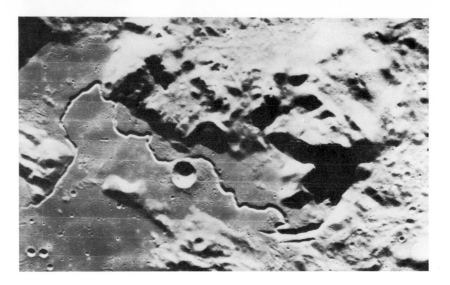

they are fractured and melted; many of the fragments are themselves older breccia (Figure 17-6). Thus the surface of the moon records a F 17-6 long history of meteorite impacts. Although the astronauts could only collect near-surface rocks, the cratering has brought deep rocks to the surface.

Most of the moon rocks that have been collected are similar to **basalt,** a dark volcanic rock containing appreciable amounts of iron and magnesium, common on earth and underlying its oceans. Although the moon basalts are very similar to earth rocks, the minor differences between them may be quite significant. The moon rocks contain less of the volatile elements such as hydrogen, helium, mercury, and lead, and more of the refractory elements such as aluminum and titanium than do earth rocks. Such compositional differences imply that at one time the moon was heated to high temperature.

The early moon flights landed on the relatively smooth maria, and returned mainly basalt samples. Later flights landed in rougher, more complex areas, and a few of the rocks came from the highlands. These rocks are composed mainly of plagioclase, one of the feldspar minerals (the most common mineral group on earth); such plagioclase rocks are called **anorthosite.**

Radioactive dating of moon rocks reveals that the anorthosites of the highlands are about 4000 million (4 billion) years old, slightly older than the oldest rocks on earth and probably the time that the moon's crust formed. The basalts of the maria are between 3700 and 3200 million (3.7 and 3.2 billion) years old. The maria are generally large—up to 1200 kilometers (740 miles)—circular areas that appear to be huge impact craters. This suggests that huge meteorites, up to about 100 kilometers (60 miles) in diameter, may have crashed into the newly formed crust and either caused the melting that produced the maria basalt or simply broke through the newly formed crust and released the already molten basalt below (Figure 17-7).   F 17-7

The *Apollo* missions left a seismometer to record moonquakes. This instrument revealed that most moonquakes occur in a monthly cycle when the earth, the moon, and the sun are in line; flexing by tidal forces causes the moonquakes. These moonquakes are very gentle, and most would be in the range of 1 to 2 on the Richter earthquake scale. Quakes this small are detectable only with instruments. The moonquakes do reveal some of the internal structure of the moon. Like earthquakes, moonquakes have both P waves and S waves. The S, or shear, waves cannot pass through the moon's deep interior, revealing that it is soft (Figure 17-8). The attenuation of the S waves   F 17-8 reveals a soft zone at a depth of 1000 to 1400 kilometers (620 to 870 miles) below the surface. Seismic study also reveals a thin crust on the moon, about 65 kilometers (40 miles) thick on the earth side of the moon, and about twice as thick on the far side.

The present magnetic field of the moon is very small, suggesting that, unlike earth, the moon does not have a liquid conducting core. However, the very old rocks from the moon retain some of the magnetism they acquired when they formed about 4500 million (4.5 billion) years ago. Calculations based on this fossil magnetism suggest that, at an early stage, the moon had a small liquid-iron core about 300 kilometers (186 miles) in radius. Cooling since then has solidified the core.

Figure 17-6
Breccia from the moon. Many fragments can be seen, and the black circular areas are small impact craters that are glass lined. (Photo from NASA.)

Figure 17-7
The maria may have formed by huge impacts that melted the rocks or broke through a relatively thin crust above still-liquid material.

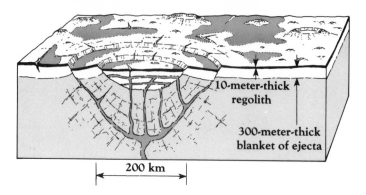

10-meter-thick regolith

300-meter-thick blanket of ejecta

200 km

From this evidence we can deduce a history of the moon (Figure 17-9). As noted above, the oldest rocks from the moon are about 4500 million years old; this is probably about the time both the earth and moon formed. Soon after forming, the moon probably melted. At this time, the liquid core, the iron-magnesium silicate mantle, and the anorthosite crust formed as the heavy elements moved toward the center and the light elements floated to the surface. The crustal rocks from the moon are anorthosite 4000 to 4500 million years old, the date of crustal formation on both the earth and the moon. From that

F 17-9

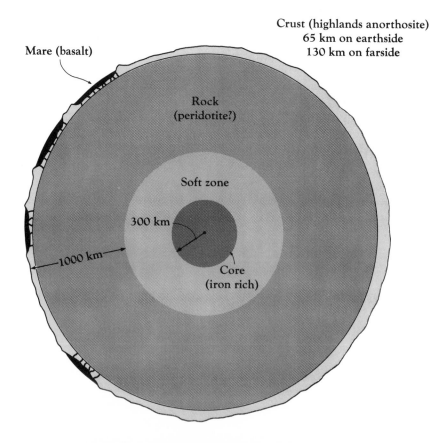

Crust (highlands anorthosite)
65 km on earthside
130 km on farside

Mare (basalt)

Rock (peridotite?)

Soft zone

300 km

1000 km

Core (iron rich)

Figure 17-8
The moon's internal structure.

Figure 17-9
The history of the moon. By about 3000 million years ago, the interior had solidified, so there was no source of internal energy, and the moon had become a dead planet.

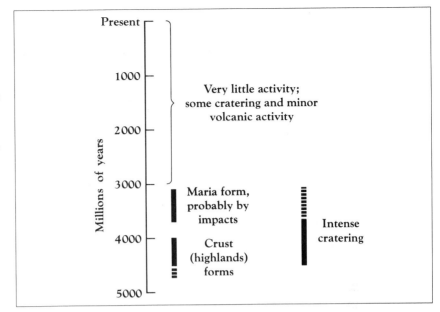

point onward, the histories of the earth and the moon are different. In the period between 3700 and 3200 million years ago, the maria formed, probably as a result of the impacts of very large meteorites. The maria are much less cratered than the highlands, so the rate of cratering must have slowed about 4000 million years ago; the highly cratered highlands may record the late stages of the infall of the material that formed the moon. After about 3000 million years ago, little has occurred on the moon except minor volcanic activity and meteorite impacts. This probably means that the moon's interior had, by that time, solidified and that there was no longer any source of internal energy. Thus, the moon is a dead planet.

MERCURY Our information about Mercury is much newer and less complete than that about the much closer moon. It is the closest planet to the sun, making telescopic observations very difficult.

Mercury is about 50 per cent larger than the moon. Its radius is 2439 kilometers (1517 miles), and its average distance from the sun is about 58 million kilometers (36 million miles). Its orbit is the least circular of all of the planets except Pluto's; and, with the same exception, its orbit is most inclined at 7 degrees from the ecliptic. Mercury's orbital period—its year—is 88 days; and its period of rotation, its day, is 58.6 days. Thus, it spins on its axis one and one-half times during each orbit of the sun. Mercury has no moons. It has no atmosphere because of its small mass and high surface temperatures. The surface temperatures reach up to 325°C (620°F) on the side toward the sun and drop to about −175°C (−280°F) on the dark side.

In 1974, the spacecraft *Mariner 10* passed within 6000 kilometers

Figure 17-10
The edge of Caloris
basin on Mercury.
Craters and wrinkle
ridges are prominent.
(Photo from JPL.)

Figure 17-11
Larger image of part
of Figure 17-10,
showing details of
wrinkle ridge on
Mercury. (Photo from
JPL.)

(3700 miles) of Mercury, went around the sun to pass it closely again, and finally repeated the trip around the sun to pass a mere 327 kilometers (203 miles) from Mercury, sending us images. The surface of Mercury, much like that of the moon, has highly cratered highlands and large, less cratered basins (Figure 17-10). The basins appear to be floored with volcanic flows and are believed to have been formed by huge meteorite impacts soon after the crust formed. The basins resemble the maria on the moon except that they have more cracks and wrinkle ridges (Figure 17-11). The terrain at the points on the opposite side of Mercury from the large basins is unusual, with hills and ridges that cut across both smooth and cratered areas (Figure 17-12). This terrain is believed to have been formed by the shock waves of

F 17-10

F 17-11

F 17-12

Figure 17-12
In this area that is opposite a big basin on Mercury, the hills and ridges cut across both craters and smooth areas. This terrain may have been formed by the shock waves of the impact that created the large basin. (Photo from JPL.)

Figure 17-13
The scarps or cliffs in this view of Mercury are several hundred kilometers long. They may be evidence of earlier internal processes. (The *Mariner* photography has been provided by the National Space Science Data Center.)

the basin-forming impact. Mercury also has many cliffs over 485 kilometers (300 miles) long and up to 3 kilometers (1.8 miles) high, believed to be fault scarps (Figure 17-13).

Mercury's internal structure is not known in any detail. Its density is 5.4 grams per cubic centimeter, almost the same as earth's, and it has a magnetic field about 1 per cent as strong as earth's. These points indicate that Mercury has a large iron core at least 1800 kilometers (1120 miles) in radius that may be fluid. The layer surrounding the core is probably iron-magnesium silicates as on the moon and earth (Figure 17-14).

Without datable rock samples, it is difficult to deduce a history of Mercury. We assume that it formed about 4500 million (4.5 billion) years ago, as did the other planets, and that the crust formed about

Figure 17-14
Possible structure of
Mercury.

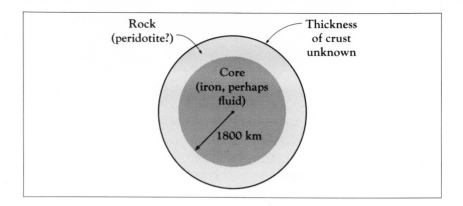

4000 million years ago, as it did on moon and earth. The crustal high-lands, the basin formation, and the relative intensity of the cratering appear to record a history similar to the moon's. The last events were the formation of the long, high cliffs. They may be due to shrinking as Mercury cooled, or they may be evidence of internal forces.

VENUS Venus has been called earth's sister because they are both about the same size and mass. Because Venus is closer to the sun, it is seen only near sunrise and sunset. Venus is very different from earth, but the differences are not apparent because Venus is veiled with a thick cloud cover that has been penetrated only by radar beams and a few spacecraft. The picture of Venus that has emerged is one of a hellish inferno, whose surface is hot enough to melt lead.

From earth, Venus appears yellowish white, and its disk is feature-less; only a little more detail is seen in ultraviolet photographs of the cloud top. Venus is about 108 million kilometers (67 million miles) from the sun and so at times is only 42 million kilometers (26 million miles) from us. Its equatorial radius is 6050 kilometers (3750 miles), very close to earth's, which is 6378 kilometers (3954 miles). The Venus year is 225 earth days. Its period of rotation on its axis, first measured by radar in 1962, is 243 days. The direction of rotation is opposite to that of all of the other planets except Uranus. *Motion or spin in the opposite direction is called* **retrograde.** Viewed from above the sun, the planets all move counterclockwise in their orbits, and most also rotate counterclockwise on their axes. The reason for Venus' re-verse, or retrograde, spin is not known, but it could have been caused by a collision or by tidal forces due to earth. Venus' rotation axis is almost perpendicular to its orbital plane. Venus has no moons.

In the 1960s temperature measurements using radio and far in-frared frequencies revealed that the surface of Venus, far below the clouds, is about 475°C (890°F). This temperature was confirmed in 1970 by the first spacecraft to land successfully on another planet. This Soviet lander lasted only 23 minutes, but in this time it sent back data on temperature and the composition of Venus' atmosphere. The atmosphere of Venus is composed mainly of carbon dioxide, and the atmospheric pressure at the surface is about 90 times greater than earth's.

This atmosphere causes Venus' high surface temperature through the **greenhouse effect,** discussed in Chapter 11. Certain gases, and carbon dioxide is one, allow the short-wavelength energy from the sun to pass through them, but these gases are opaque to the longer-wavelength radiation from the surface of the planet. Thus the sun's heat is trapped by the atmosphere, and the temperature rises to the point at which the radiation from the surface can penetrate the atmosphere. It should be clear that all planets must radiate back into space the same amount of energy that they receive from the sun; otherwise, they would be either heating up or cooling down.

We now know that Venus' atmosphere is composed of 97 per cent carbon dioxide, about 2 per cent nitrogen, and minor amounts of carbon monoxide, water vapor, hydrochloric acid, and hydrofluoric acid. The clouds, about 80 kilometers (50 miles) thick, are composed largely of droplets of sulfuric acid. It is logical to inquire why Venus' atmosphere is so different from earth's, even though the two planets are nearly the same size and so might be similar. The actual amounts of carbon dioxide on both planets are about the same, but earth's carbon dioxide has dissolved in the oceans and precipitated as calcium carbonate (the mineral *calcite*) in the rock limestone. Perhaps Venus lost water at an early stage because it is closer to the sun and hotter; thus the atmospheres of the two planets developed differently.

Our knowledge of the surface of Venus comes almost entirely from radar studies, carried out both from earth and from orbiters. A very few landers have sent back images showing angular and rounded boulders as well as dust. The hot, reactive atmosphere probably causes weathering of surface rocks, and the high wind velocities recorded by some spacecraft suggest that at least some erosion could occur. The surface of Venus is different from the other terrestrial planets, although it has some similar features (Figure 17-15).

F 17-15

Venus has a gently rolling surface with some abrupt highlands. About 60 per cent of the planet is a rolling plain whose elevation only varies within one kilometer (3280 feet); Soviet soft-landers suggest that it is composed of granite. This plain has many craters 400 to 600 kilometers (250 to 370 miles) in diameter that are only 200 to 700 meters (660 to 2300 feet) deep. This is shallower than the craters on other planets, perhaps because the hot surface may be plastic. There are, of course, no small craters because small meteorites are destroyed by friction in Venus' thick atmosphere.

About 16 per cent of Venus, an area about the size of the North Atlantic Ocean, is a smooth, uncratered area about 2.9 kilometers

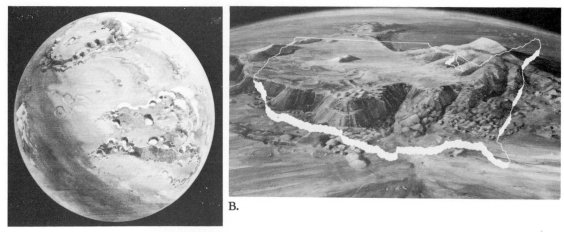

Figure 17-15
Artist's conceptions of the surface of Venus. **A.** Whole disk. Aphrodite Terra in foreground, Ishtar Terra at top, and Beta Regio at left. **B.** Ishtar Terra. **C.** Beta Regio. **D.** Rift Valley at east end of Aphrodite Terra. (Photos from NASA.)

(1.8 miles) below the elevation of the plains. This surface is probably formed by young basalt.

The highlands, composed of two continent-like features, are only about 8 per cent of the surface of Venus. The larger of these, Terra Aphrodite, about the size of northern Africa, is about one kilometer above the plains, and has two areas of mountains whose elevations reach 9 kilometers (5.6 miles) and 4.3 kilometers (2.7 miles). Terra Aphrodite has a rougher surface than the other "continent" and so is presumed to be older. The other highland, the size of Australia, is Terra Ishtar (named for the Assyrian goddess of love and war). It is a smooth plateau about 3.3 kilometers (2.1 miles) higher than the plains, with steep-sloped mountains ranging between 5.7 and 7 kilometers (3.5 and 4.4 miles) high. The highest elevation on the planet, on Ishtar, is Maxwell Montes, about 10.8 kilometers (6.7 miles) high. Maxwell has a depression near the top and so is probably a large shield volcano.

Beta Regio is an area with two features that, like Maxwell Montes, appear to be giant shield volcanoes similar to those on Mars and larger than the Hawaiian Island chain. Beta is 4 kilometers (2.5 miles) high and 1000 kilometers (620 miles) across, and it is in line with other possible volcanic features. The rocks, probably basalt, were tentatively identified as such by a Soviet lander.

Surface deformation possibly indicative of internal energy is suggested at two places on Venus. Alpha Regio is an area with parallel mountains, rising about 1.8 kilometers (1.1 miles) above the plains, that appear to be like the faulted basin-and-range mountains of southwestern United States. The other place is unlike anything on earth. Two parallel ridges, about 1400 kilometers (870 miles) long and 2 kilometers (1.2 miles) high, divide a deep valley 90 kilometers (56 miles) across. This valley, the deepest point on Venus at 2.9 kilometers (1.8 miles) below the plains, is similar to rift valleys on earth but much larger.

Because Venus is about the same size as earth, we might expect to find similar geologic features and rock types, but it is different. Venus has a very small magnetic field compared with earth's; although this may indicate that it does not have a fluid core, its very slow rotation might also be the reason. Venus might have a much smaller core than earth; this, too, could explain its small magnetic field and apparent lack of internal energy.

MARS Mars is the only one of the terrestrial planets farther from the sun than earth, and so it is the only one that at times can be seen all night. Telescopic observation revealed polar ice caps that form and melt seasonally, and seasonal color changes from greenish gray to red-brown. These observations, together with faint markings, led to the idea that life might exist on Mars, but so far none has been found.

A.

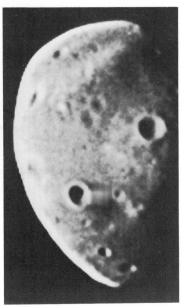

B.

Figure 17-16
Mars' moons have irregular shapes and are cratered and grooved, suggesting that they may be captured asteroids. **A.** Closeup of Phobos. **B.** Deimos, 12 by 8 kilometers (7 by 5 miles). (Photos from NASA.)

Mars is about 227 million kilometers (141 million miles) from the sun, giving it an orbital period of 687 earth days. Its period of rotation, or day, is 24 hours 37 minutes. At closest approach to earth, it is 56 million kilometers (35 million miles) away; only Venus approaches closer. Mars, with an equatorial radius of 3398 kilometers (2114 miles), is about half as large as earth.

Mars has two moons, or satellites, named Phobos (*fear*) and Deimos (*dread*), two of the god Mars' attendants in Greek mythology. Both are irregularly shaped and have craters and grooves (Figure 17-16), suggesting that these moons may be fragments of larger bodies F 17-16 or at least that they have been involved in many collisions. The origin of the grooves is not known, but they may be caused by tidal forces. One possibility is that the moons are asteroids from the nearby belt (described later) that were captured by Mars.

The composition of Mars' atmosphere is more like Venus' than Earth's but is very much thinner than either. Like Venus', its main constituent is carbon dioxide, but it lacks Venus' acids (see Table 17-1). The atmospheric pressure on the surface of Mars is only 0.7 per cent of that on earth. The polar ice caps appear to be mainly frozen carbon dioxide, possibly with some water ice (Figure 17-17). Some F 17-17 surface features suggest that water was once present on Mars, and it may now be found as permafrost in the Martian "soil." The amounts of gases like argon, krypton, and xenon in the Martian atmosphere suggest that it may once have had a surface pressure a quarter to a half as much as earth's.

Figure 17·17
Mars' polar ice cap is shown in this distant view. The circular feature is Olympus Mons, the huge volcano. (Photo from JPL.)

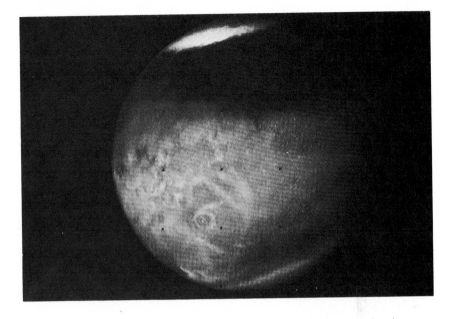

The surface temperatures on Mars are cooler than earth's because it is farther from the sun and has a thin atmosphere. The maximum temperature at the equator is about 27°C (80°F), the minimum about −85°C (−120°F). The daily range is almost as great because of the thin atmosphere. At the poles, the temperature generally stays colder than −75°C (−100°F).

Thus Mars is a cold desert. The surface is covered with rocks (Figure 17-18), and they and the finer material in which they rest are all a red-brown color, probably a coating of rust or iron oxide; even the atmosphere close to the surface is this color because of dust. Big dust or sand storms have been detected by orbiting spacecraft, but only minor amounts of sediment have settled on the landers. Even the thin atmosphere is capable of high velocities and is able to erode and transport sediment, as indicated by dune fields (Figure 17-19). Ero-

F 17-18

F 17-19

Figure 17·18
The surface of Mars is covered with rocks and probably dust and sand. (Photo from NASA.)

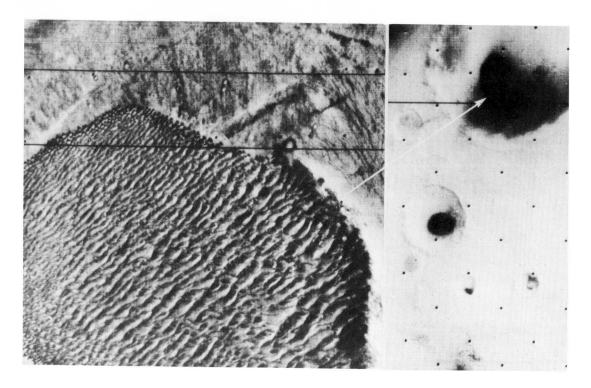

sion, presumably by wind, has rounded some of the craters and other surface features so that relative age can be determined.

The volcanoes of Mars are very impressive. The largest, Olympus Mons (Figure 17-20), is 500 kilometers (300 miles) in diameter and 13 **F 17-20** kilometers (8 miles) high. It is one of almost a dozen huge, probably basaltic shield volcanoes clustered in two areas; most of these volcanoes are larger than any on earth. Most of the surface rocks of Mars are probably basalt, and much of the surface is cratered; younger lava flows cut across some of the cratered surfaces.

The other evidence of internal energy on Mars is a system of parallel valleys (Figure 17-21) that extends over 4800 kilometers (3000 **F 17-21** miles). The valleys are about 1.6 kilometers (1 mile) wide. Their origin is not known, but because they are relatively straight, they are believed to be evidence of faulting. It is difficult to relate such faults to a process like earth's plate tectonics, but they may be an early stage. The volcanoes and valleys of Mars are more similar to structures on Venus than to those on earth.

Other valleys on Mars suggest erosion by running water (Figure 17-22). There is presently not enough pressure in Mars' atmosphere **F 17-22** to allow liquid or running water, and at least some of these patterns could be caused by wind erosion. Some of the river-like features start in areas where the surface is jumbled. These areas are similar in appearance to places on earth where permafrost or subsurface ice has

Figure 17-20
Olympus Mons on Mars is the largest volcano in the solar system. It is 500 kilometers (300 miles) wide and 13 kilometers (8 miles) high. (Photo from JPL.)

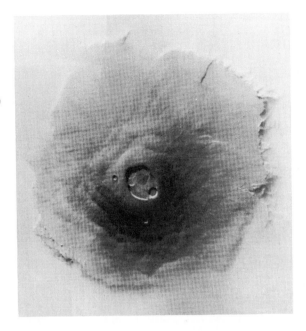

Figure 17-21
This valley is part of one of a number of parallel valleys on Mars that extend more than 4800 kilometers (3000 miles). These valleys may have been caused by faulting. (Photo from NASA.)

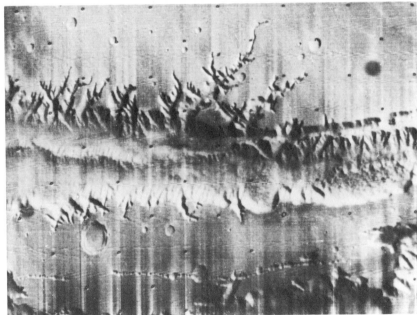

melted. Perhaps in the past Mars had surface water, and some might now be trapped in the "soil" as ice. A meteorite impact could melt the ice and cause the flooding. At present some clouds of water vapor have been noted over the big volcanoes, which may be releasing gases into the atmosphere.

Spacecraft have landed on Mars and searched for life but have found none. The evidence of possible running water gives hope of

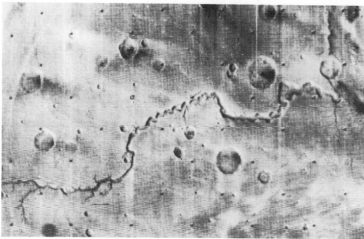

A.

B.

Figure 17-22
Evidence of erosion by running water on Mars. Water is no longer present on Mars, but some ice may be trapped in the "soil." **A.** Sinuous, river-like feature. **B.** Teardrop shapes suggest running water. (Photos from NASA.)

finding some life buried in the "soil." Our hopes of finding life on Venus were dashed by the discovery of the high surface temperatures. The moon with no atmosphere seemed unlikely to have life, and none was found. The search still goes on, and some are speculating that certain of the moons of the great planets may possibly have life.

Mars has a density of only 3.9 grams per cubic centimeter, compared with earth's 5.5 grams per cubic centimeter. It has a small magnetic field and so is believed to have a fluid iron core. Between the core and the thin crust is probably a mantle composed of iron-magnesium silicates of which little is known. A crystallizing liquid core and radioactive decay can be the sources of internal energy, and the big volcanoes and the fault valleys are probably evidence of internal processes. We see no features like the plate tectonics of earth. Perhaps the volcanoes are so large because the surface on which they accumulate is stationary; on earth, the moving plates cause a string of volcanoes like the Hawaiian Islands instead of a single huge volcano.

Outer, or Jovian, Planets

Four outer planets—Jupiter, Saturn, Uranus, and Neptune—are all very much larger than the inner, terrestrial planets. The composition and structure of the Jovian planets and their moons are very different from those of the inner planets, although a number of the Jovian moons are similar in size to the terrestrial planets and may give some insight into their history. The fifth outer planet, Pluto, also resembles the terrestrial planets in size, but we know little about it.

JUPITER Jupiter is the largest of the planets and contains about 71 per cent of the total mass of all of the planets. It has a polar radius of

67,000 kilometers (41,674 miles), and, because of its rapid rotation, an equatorial radius of 71,400 kilometers (44,411 miles), making it more than ten times larger than earth. Jupiter rotates on its axis in about a 10-hour period. The clouds near its equator rotate with a period of 9 hours 50 minutes, and nearer the poles the period is 9 hours 56 minutes, showing that the outer visible surface is gaseous. Jupiter is 777 million kilometers (483 million miles) from the sun, and it revolves once around its orbit every 11.9 years.

Jupiter's atmosphere is composed mainly of hydrogen and helium, with minor amounts of water, ammonia, and methane. The average temperature at the top of the clouds, constant day and night, is near −140°C (−220°F). The surface markings are a number of stripes, parallel to the equator, of dark shades of red-brown and light shades of white to yellowish (Figure 17-23). The light stripes are colder, sug- F 17–23 gesting that they are higher than the dark stripes; this, in turn, suggests convection. Such convection, caused by internal heat, and Jupiter's rapid rotation would cause the parallel stripes in a process similar to the formation of trade winds on earth. Lightning near the

Figure 17-23
Jupiter. The stripes
are white on yellow
and red-brown.
(Photo from NASA.)

cloud tops has been recorded by spacecraft. The colors of the stripes are probably caused by differences in the chemical compounds because the gases mentioned above are all colorless. Organic compounds and sulfides have been suggested, but little is known of the chemistry of Jupiter's atmosphere. Perhaps at some level, Jupiter's atmosphere may be the same composition and temperature as earth's primitive atmosphere, in which life is believed to have originated.

Perhaps the most striking detail of Jupiter's cloud tops is the Great Red Spot (Figure 17-24), a feature known since 1665. Its width is F 17-2 about 14,000 kilometers (8700 miles), and its length varies between 30,000 and 40,000 kilometers (19,000 and 25,000 miles). It is believed to be a permanent hurricane-like storm of great size. Similar smaller storms occur but are not permanent features.

Jupiter radiates about twice the amount of thermal energy it receives from the sun, implying that it is warmer at some depth below the clouds. The source of this energy is probably gravitational, caused by very slow contraction. Jupiter's interior, unlike a star, is nowhere near hot enough for nuclear reactions to occur.

Jupiter also radiates radio energy, believed to be related to its magnetic field, the strongest of any of the planets. An electron moving through such a magnetic field will emit radio energy. In addition, when Io, one of Jupiter's moons, is in certain positions, we receive radio bursts; this radio source is not understood. Jupiter's magnetic

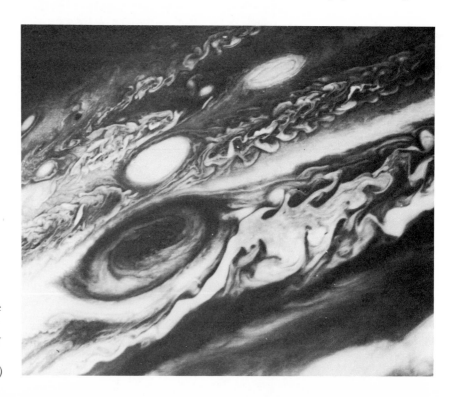

Figure 17-24
Jupiter's Great Red Spot, a permanent hurricane-like storm tens of thousands of kilometers long. (The *Voyager* photography has been provided by the National Space Science Data Center.)

Figure 17-25
Jupiter's internal
structure.

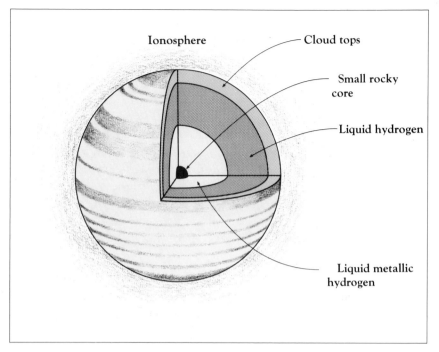

field traps particles from the sun, causing radiation belts like our Van Allen belts; and particles moving along magnetic lines cause auroras near Jupiter's poles like our northern and southern lights. A strong magnetic field implies that Jupiter has a large conducting fluid layer as its cause.

Jupiter's density is 1.3 grams per cubic centimeter, only 1.3 times that of water, and because of its great size and mass, the internal pressures are very high. It is believed to be composed mainly of hydrogen with a small core of iron and silicate rock (Figure 17-25). The temperature of this core is thought to be about 30,000°C (54,000°F). The hydrogen surrounding the core is not in the more familiar molecular form, with the electrons held tightly by the molecules, but is in the **liquid metallic** form, in which it is tightly compressed so that the electrons can move freely through the liquid. In this state hydrogen can conduct electricity like a metal, and this may be the layer where Jupiter's magnetic field forms. At about 46,000 kilometers (28,500 miles) from the center, the hydrogen changes from liquid metallic form to **liquid molecular.** The liquid molecular hydrogen layer is about 24,000 kilometers (15,000 miles) thick, and above it is the atmosphere described above.

Fifteen satellites of Jupiter have been discovered so far, some by spacecraft. Some of these, the size of terrestrial planets, will be described later. Jupiter also has a ring, discovered by *Voyager 1* in 1979. Jupiter's ring, much smaller than Saturn's, begins about 55,000 kilometers (34,000 miles) above the clouds and is about 6000 kilometers

F 17-25

(3700 miles) wide and one kilometer (0.6 mile) thick. The ring is made of small particles, probably up to a few centimeters in diameter. Some planets have rings because a moon cannot exist close to a planet; if a large moon formed too close to a planet, the tidal forces would break up the moon. The distance from the planet in which a large moon cannot exist is called **Roche's limit.**

SATURN Saturn is next in size after Jupiter as well as next farther out from the sun. Saturn, almost ten times larger than earth, has a radius of 60,000 kilometers (37,300 miles), and is 1,426 million kilometers (887 million miles) from the sun. Saturn requires about 29.5 earth years to complete one orbit of the sun. Saturn's atmosphere, like Jupiter's, is mainly hydrogen and helium, with minor amounts of water, ammonia, and methane. It has less ammonia and more methane than Jupiter because its lower temperature (about $-170°C$, or $-270°F$, near the top of its atmosphere) causes the ammonia to freeze out at lower levels. The surface markings are yellow and tan stripes; and although there is nothing like the Great Red Spot of Jupiter, there is a small red spot as well as occasional bright disturbances that last for weeks. Saturn is the least dense of all of the planets and so is believed to have a higher proportion of hydrogen than the others. Its average density is only 0.7 gram per cubic centimeter, which means that it could float in water. It is believed to have a small rocky core surrounded by a layer of water, methane, and ammonia ices. Above the ice layer is a thin layer of liquid metallic hydrogen; most of the planet is probably liquid molecular hydrogen (Figure 17-26).

F 17-26

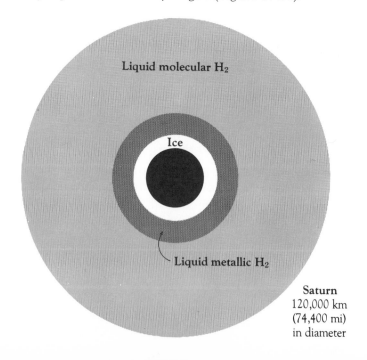

Figure 17-26
Saturn's internal structure.

Liquid molecular H_2

Ice

Liquid metallic H_2

Saturn
120,000 km
(74,400 mi)
in diameter

Figure 17-27
Saturn's rings, photographed by *Voyager 2*, show dark "spokes" that have not yet been explained. (Photo from NASA.)

Saturn has become famous for its ring system. The rings were long thought to be unique, but it is now known that other great planets also have rings. Galileo is generally credited with the discovery of Saturn's rings, but his telescope was poor, and he drew it as a triple planet. Forty-five years later, Christian Huygens first reported the rings. The rings begin about 10,000 kilometers (6200 miles) above the atmosphere and extend about 65,000 kilometers (40,000 miles) radially outward. They are estimated to be about 2 kilometers (1.2 miles) thick and are composed of fragments up to about 10 centimeters (4 inches) in diameter. *Voyager 2* revealed thousands of rings in the system. (Figure 17-27). The plane of the rings is tilted 27 degrees to the orbital plane, so we see the rings from above, below, and edge-on as Saturn travels around its orbit. The fragments may be left over from Saturn's formation or may be from a satellite or asteroid that was torn apart by the tidal forces inside Roche's limit. Saturn also has an extensive family of satellites described later.

URANUS AND NEPTUNE Uranus was discovered by accident by William Herschel in 1781 as a byproduct of a star-mapping project. Its orbit was not as predicted, leading to the discovery of Neptune. The orbital discrepancy prompted J. C. Adams in England and Urbain Leverrier in France to calculate the position of the unknown planet whose gravity was affecting Uranus' orbit. Adams completed his work in 1845 and Leverrier in 1846, but neither could get his country's astronomers to look for the planet. Finally, in 1846, Leverrier sent his data to a German astronomer, who found the new planet within hours. This caused a scandal in England because the older professors there would not take the calculations of a recent graduate seriously; thus England lost the glory of discovering a new planet.

Uranus and Neptune are similar. Uranus has a radius of 25,900 kilometers (16,110 miles), and Neptune a radius of 24,750 kilometers (15,400 miles); thus both are a bit more than six times earth's size. Uranus is about 2869 million kilometers (1783 million miles) from the sun, and its orbital period is 84 years. Neptune is 4494 million kilometers (2793 million miles) from the sun; its orbital period is 165 years. Uranus' day, or period of rotation, is about 23 hours; Neptune's is about 22 hours. Uranus' rotational axis is tilted 98 degrees from its orbital plane, so its rotation is retrograde, or opposite to that of all of the other planets except Venus. Neptune's axis is inclined 29 degrees.

The atmospheres of both planets, mainly hydrogen and helium with some methane, seem to be much less thick than those of the other great planets. Their temperatures are almost the same: Uranus is −210°C (−350°F); Neptune, 3°C colder. Both are believed to have rocky cores about 8000 kilometers (4976 miles) in radius, surrounded by an 8000-kilometer (4976-mile-) thick layer of water, methane, and ammonia ice; the remainder of each is composed of liquid molecular hydrogen (Figure 17-28). The density of Uranus is F 17-28 1.3 grams per cubic centimeter; Neptune is somewhat denser, at 1.7 grams per cubic centimeter.

Uranus has faint rings, discovered in 1977 when the rings passed between us and a star. So far, nine small rings have been discovered. Uranus has five moons, and Neptune has two.

PLUTO Pluto is the most distant planet yet discovered, although Planet X has been announced several times but never confirmed. Because of Pluto's great distance and small size, we know very little

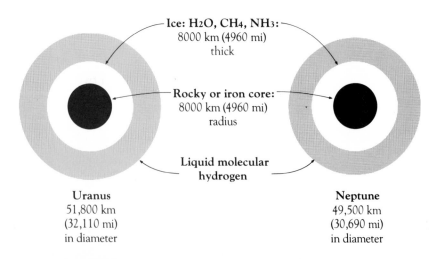

Figure 17-28
Internal structure of Neptune and Uranus.

Ice: H₂O, CH₄, NH₃:
8000 km (4960 mi)
thick

Rocky or iron core:
8000 km (4960 mi)
radius

Liquid molecular
hydrogen

Uranus
51,800 km
(32,110 mi)
in diameter

Neptune
49,500 km
(30,690 mi)
in diameter

about it; it may resemble the satellites of the Jovian planets, which are described in the next section. Pluto was discovered in 1930 as the result of a year-long search for a planet that was known to disturb the orbits of Uranus and Neptune. Pluto is estimated to be smaller than our moon. Its average distance from the sun is 5899 million kilometers (3666 million miles), and it requires 248 years to complete one orbit. Pluto's orbit is inclined 17 degrees to the rest of the solar system, much more than any other planet's. Its orbit is also more elliptical than any other; thus, Pluto is now closer to the sun than is Neptune and will remain so through 1999. Its rotational period is probably 6.4 days, the same as the orbital period of its only satellite, a moon discovered in 1978, whose period is 6 days, 9 hours, 17 minutes. The surface of Pluto is very cold, $-235°C$ ($-390°F$), and may be covered with methane frost. Pluto's density is estimated to be about 0.7 gram per cubic centimeter, so it must be composed largely of hydrogen.

SATELLITES OF THE JOVIAN PLANETS

Satellites in the solar system seem to be of two general types. The larger ones, such as our moon, are similar to the terrestrial planets. We know much less about the smaller ones; the few we have been able to observe closely seem to be like the moons of Mars. Our attention will be focused on the larger moons in the hope of getting more insight into the development of the terrestrial planets, especially earth.

Jupiter's Satellites

Fifteen satellites of Jupiter have so far been discovered. Four are large enough to be considered small planets; they are called the **Galilean moons** because they were discovered by Galileo. In 1979 *Voyagers 1* and 2 passed close to Jupiter and returned excellent images of these moons. In addition, two small moons were discovered on these images.

Jupiter's outer moons are in two groups of four. The closer group is about 12 million kilometers (7.5 million miles) from Jupiter, and their elliptical orbits are inclined about 27° to Jupiter's equator. They range in diameter from about 8 to 170 kilometers (5 to 106 miles). The other group is more uniform in size, ranging between about 17 and 27 kilometers (11 and 17 miles). Their orbits are retrograde (clockwise), inclined near 60 degrees, more elliptical, and about 23 million kilometers (14 million miles) from Jupiter.

The inner moons move in nearly circular orbits, almost parallel with Jupiter's equator. Of the three small, innermost satellites, only *Amalthea*, an elongated body about 260 kilometers (160 miles) long

Figure 17-29
Amalthea, one of
Jupiter's small inner
moons. (The *Voyager*
photography has been
provided by the
National Space
Science Data Center.)

and 140 kilometers (90 miles) in diameter, was known before the *Voyager* missions (Figure 17-29). Amalthea and the four large moons all keep one face toward Jupiter as they orbit it, just as our moon keeps one face toward earth.   F 17-29

The four Galilean moons are different from any other objects we have met, and each is different from the others. *Io*, nearest to Jupiter, is similar in size (1816-kilometer radius) and density to our moon (Table 17-2). Io is, however, much different from our moon in most other respects. It is the only object, except earth, in the whole solar system on which we have observed volcanic activity (Figure 17-30). *Voyager 1* recorded eight active volcanoes; seven of these were still active four months later when *Voyager 2* passed by. Io's surface has no craters, showing that its many volcanoes have obliterated all evidence of impacts (Figure 17-31). The surface colors are red-orange, black, and white, making it resemble a large pizza pie. Spectrographic studies show that sulfur and sulfur dioxide probably cause these colors. The volcanoes may be erupting these materials, or, as a few have suggested, these substances may be coloring ordinary lava. Io's composition may be similar to our moon and Mars because of its similar density. The heating that causes the volcanic activity is probably not internal but due to tidal flexing. The gravitational attraction of the other Galilean moons, especially nearby Europa, causes Io to move slightly in the strong gravitational field of Jupiter; this tidal flexing heats Io. Apparently Io has lost all of its volatile materials except sulfur.   T 17-2   F 17-30   F 17-31

The other three Galilean satellites have not experienced this much tidal flexing. Also, the temperature near Jupiter is much colder than it is near the inner planets, so these Galilean moons have been able to retain water at their surfaces. *Europa*, slightly smaller than our moon, has a density of 3.03 grams per cubic centimeter and so is be-

Table 17-2   SUMMARY OF JUPITER'S LARGER SATELLITES

	Radius		Density *Grams per cubic centimeter*	Revolution Period	Distance from Jupiter	
	Kilometers	*Miles*		*Days*	*Kilometers*	*Miles*
Amalthea	130*	80*	3–4	0.49	181,000	112,475
Io	1816	1128	3.53	1.77	422,000	262,230
Europa	1563	971	3.03	3.55	671,000	416,960
Ganymede	2638	1639	1.93	7.16	1,070,000	664,900
Callisto	2410	1498	1.79	16.69	1,880,000	1,168,230

*Irregular body 260 by 140 kilometers (160 by 87 miles) in diameter.

Figure 17-30
Volcanic eruption on Io. Io is the only other object in the solar system on which we have seen active volcanoes. (The *Voyager* photography has been provided by the National Space Science Data Center.)

Figure 17-31
Io has circular structures that are sulfur volcanoes. Io's colors and markings resemble a pizza. (The *Voyager* photography has been provided by the National Space Science Data Center.)

lieved to be composed of silicate rocks with a water and ice layer about 100 kilometers (60 miles) thick at the surface. This ice layer makes Europa the brightest of Jupiter's moons. Its most prominent feature is a tangle of lines over 1000 kilometers (600 miles) long, 200 to 300 kilometers (125 to 185 miles) wide, and only about 100 meters (300 feet) high (Figure 17-32). These lines are believed to be cracks, of F 17-32 unknown origin, that have been healed by water moving into them and freezing. Europa has only three small features that resemble craters, suggesting that at the time of intense crater formation early in the formation of the solar system, its surface was not yet frozen.

Ganymede, the next farthest from Jupiter of the large moons, is slightly larger than Mercury. It has a density of 1.93 grams per cubic centimeter and so is probably composed of equal amounts of rock and water. Its surface must have frozen earlier than Europa's because many craters are preserved. The young rayed craters are white, suggesting that the impacts have exposed new ice (Figure 17-33). Proba- F 17-33 bly the oldest surface is a heavily cratered dark area. Because the craters are not equally clearly defined, this surface may have begun as slush and later frozen hard. This surface has some large, faint, concentric rings, perhaps formed by a large impact before the crust was completely frozen; Callisto has a similar feature. This old dark surface is cut by a younger, light-colored grooved terrain (Figure 17-34), F 17-34 consisting of up to 20 parallel ridges and troughs, each 5 to 15 kilometers (3 to 9 miles) wide and up to several hundred kilometers long.

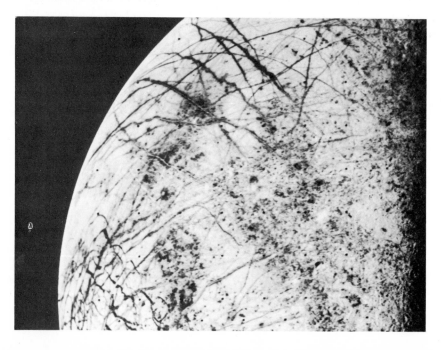

Figure 17-32
The surface of Europa shows a tangle of lines. (The *Voyager* photography has been provided by the National Space Science Data Center.)

Figure 17-33
Ganymede's surface shows white rayed craters and grooved terrain. (The *Voyager* photography has been provided by the National Space Science Data Center.)

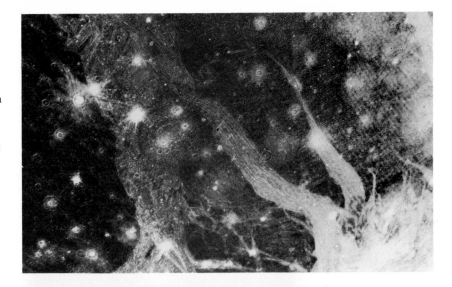

Figure 17-34
Ganymede's older dark surface appears to be cut by the light-colored grooved terrain. (The *Voyager* photography has been provided by the National Space Science Data Center.)

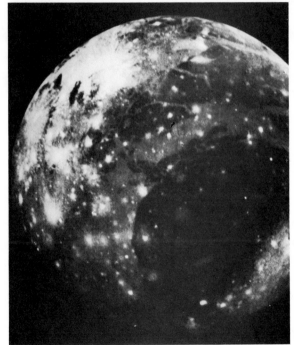

At most places these curving grooves end abruptly where they meet and are not offset where they cross; but at a few places they are apparently offset (Figure 17-33), suggesting internal energy and faulting. Ganymede is the most complex of Jupiter's moons, and the origin of the grooved terrain is unknown. Perhaps the grooves are similar to the cracks on Europa. The number of craters on the grooved terrain is variable, suggesting that it formed over a long time.

Figure 17-35
The surface of Callisto is covered with craters. (The *Voyager 2* photography has been provided by the National Space Science Data Center.)

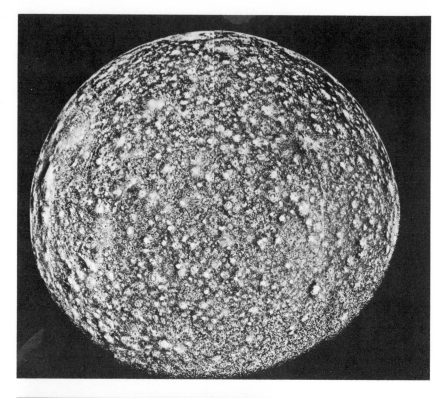

Figure 17-36
Concentric circles on Callisto (near the bottom of this image) are probably scars of a large impact. The rings appear older than the craters that cut them. (The *Voyager 2* photography has been provided by the National Space Science Data Center.)

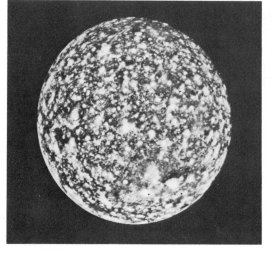

Callisto, the most remote large moon of Jupiter, is about the same size as Mercury. It is the least dense of the Galilean satellites and so probably has much water. The surface is almost saturated with craters, implying that it froze earlier than the others (Figure 17-35). The craters, unlike those on our moon, do not have much relief; this smooth surface implies that at the time of cratering, the lower crust was soft or slushy. The most prominent feature is a series of concentric circles (Figure 17-36). These rings, 50 to 200 kilometers (30 to

F 17-35

F 17-36

120 miles) apart, start at a diameter of about 600 kilometers (360 miles) and extend to nearly 2600 kilometers (1600 miles). Only a few craters are inside the inner ring. The rings are younger than many craters and are, themselves, cut by some craters. This feature is interpreted as the scar of a large impact that occurred when the lower crust was still mushy.

Saturn's Satellites

Saturn has 15 known satellites, but only Titan is large. (Table 17-3). T 17-3 Except for the two most distant, Iapetus and Phoebe, they all move in nearly circular orbits almost parallel to Saturn's equator. *Phoebe* moves in retrograde in a highly inclined orbit. *Iapetus* has a less inclined orbit, and it is notable because it has one bright side, probably because that side is covered with frost or snow. The densities are known for a few of the small inner moons; they are between 1.1 and 1.5 grams per cubic centimeter, suggesting that they are composed of more water than rock (Figure 17-37).   F 17-37

Titan is slightly bigger than Mercury, but its density is only 1.9 grams per cubic centimeter, so its composition is much different. It is probably composed of about half rock and half ice. It is the only satellite in the solar system with an appreciable atmosphere, composed mainly of nitrogen, less than ten per cent methane, and lesser amounts of other hydrocarbons. The total atmospheric pressure is about 1.6 earth's. The atmosphere has reddish clouds, perhaps similar to some types of photochemical smog on earth. Although Titan has much less mass than Mercury, it can hold an atmosphere because it is much colder; the surface temperature is about −180°C (−290°F).

Figure 17-37
Two of Saturn's 15 moons. **A.** Enceladus has craters and grooves like Jupiter's Ganymede. **B.** Tethys has heavily cratered and less cratered areas, as well as large trenches. (The *Voyager 2* photography has been provided by the National Space Science Data Center.)

A. Enceladus

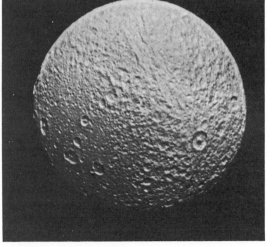

B. Tethys

Table 17·3   THE SATELLITES OF SATURN IN ORDER OF DISTANCE FROM SATURN (THE UNNAMED MOONS WERE DISCOVERED IN 1980)

	Radius		Density *Grams per cubic centimeter*	Revolution Period *Days*	Distance from Saturn	
	Kilometers	*Miles*			*Kilometers*	*Miles*
——	15	9	——	0.60	137,000	85,320
——	110	68	——	0.61	139,000	86,620
——	100	62	——	0.63	142,000	88,000
——	45 (90 × 40 km diam)	28 (56 × 25 mi diam)	——	0.69	151,422	94,100
——	50 (100 × 90 km diam)	30 (62 × 56 mi diam)	——	0.69	151,472	94,125
Mimas	195	120	1.2	0.96	188,000	116,960
Enceladus	250	155	1.1	1.39	240,000	149,250
Tethys	525	325	1.0	1.91	297,000	184,280
——	80	50	——	2.74	378,600	235,260
Dione	560	350	1.4	2.76	379,074	235,560
Rhea	765	475	1.3	4.53	528,000	327,990
Titan	2570	1600	1.9	15.94	1,221,000	759,000
Hyperion	(150 × 100)	(90 × 60)	——	21.74	1,502,000	933,500
Iapetus	720	450	1.1	79.24	3,559,000	2,211,800
Phoebe	80	50	——	406.49	10,583,000	6,576,400

Satellites of Uranus and Neptune

Uranus has five satellites that all move in nearly circular orbits parallel to Uranus' equator. Uranus itself has a 98 degree tilt of its axis, so either Uranus or its satellites were all tilted together, or Uranus captured its moons after it was tilted. Very little is known about these small, distant satellites (see Figure 17-1).

Neptune has only two moons. *Nereid,* the smaller, very distant moon, moves in a very elliptical, inclined orbit. *Triton,* also not very well known in spite of being about the same size as Mercury, may turn out to be the largest of the satellites. Triton moves in the retrograde (backward or clockwise) direction in a circular, inclined orbit.

ASTEROIDS

The **asteroids** are a large number of small bodies, most of whose orbits are between Mars and Jupiter. About 100,000 of them have been seen, but orbits have been determined for only just under 2000. The largest, Ceres—with a radius of 500 kilometers (300 miles)—is about one-quarter the size of our moon. The asteroids range in size down to one kilometer (0.6 mile) or less in diameter. They are planet-like bodies, whose spectra show that they are mainly rocky; some contain metallic iron as well. The rocky ones appear to be basaltic and so are like the moon, Mars, and earth. As they move in their orbits, the amount of sunlight that they reflect varies, showing that they have irregular shapes.

Their orbits, generally somewhat elliptical and inclined compared with planets', are much more circular and less inclined than orbits of comets. Most of them are in the **main asteroid belt** between Mars and Jupiter. Another group, called the **Apollos** for one of its prominent members, crosses earth's orbit. About 25 Apollos are known, and in 1968 Icarus came within 6 million kilometers (4 million miles) of earth. The fate of the Apollos is to crash into earth or some other planet, causing impact craters. It is estimated that they remain in orbit only a few hundred million years before crashing, implying that some asteroids that come close to Jupiter are thrown into Apollo orbits by Jupiter's gravity. Another group of asteroids crosses Mars' orbit; they may be the cause of craters on Mars. Two other swarms of asteroids are found in Jupiter's orbit.

Determining the origin of the asteroids is a problem. Their irregular shapes suggest that they are fragments of one or more larger bodies. All of the asteroids together are not enough to form a planet. They are probably material left from the origin of the solar system. If only 10 or 20 small bodies were there initially, mutual collisions in the last 4000 million (4 billion) or more years would form the number we see now. Perhaps we have already seen what asteroids look like when we viewed the moon of Mars (Figure 17-16). Asteroids and comets are probably the source of meteorites, discussed later.

COMETS

A few comets are spectacular objects; some reappear every few years. Although feared for centuries, they are now believed to be chunks of frozen gas only a few kilometers wide. *The head of a* **comet** *has a bright point called the* **nucleus;** *the nucleus is surrounded by a bright diffuse glow called the* **coma.** The coma may be 100,000 kilometers (62,000 miles) or more in diameter, and the tail may be tens of millions of kilometers long (Figure 17-38). Spectrographic study shows that comets are F 17-38 composed of water, carbon dioxide, ammonia, and methane, as well as dust; thus comets are "dirty snowballs." The light that we see from a comet is partly reflected sunlight and partly light emitted by the gases that are excited by the sun's radiation. The tails are of two types, and some comets have both. The ionized gases from the coma are energized by the sun's radiation and form a tail pointing away from the sun in a straight line (Figure 17-39). The dust particles in F 17-39 the tail may have a curved path. A comet's tail is very tenuous; it would be an excellent vacuum in the laboratory. In 1910, earth passed through the tail of Halley's Comet with no ill effects despite dire predictions.

An unusually bright comet is visible with the naked eye for a number of days or weeks, and then fades, although many can be followed for many months with telescopes. Comets are visible only when they are within about 3 Astronomical Units of the sun. They follow Kepler's laws and speed up as they near the sun. They approach the sun from all angles and directions, and their generally very elliptical orbits then carry them far away from the sun. Some return with periods as short as 3.3 years; others have periods of at least 80,000 years.

That comets are made of small chunks of dirty ice is suggested by many things other than their chemical composition. They are seen only when near the sun. A few have been seen to break up into several pieces. Each passage near the sun eliminates some of the ice, and it has been estimated that after 100 to 1000 passes they are destroyed. Earth passes through the orbits of old comets several times each year, and the remaining dust causes meteors. Comets are similar in composition to Jovian planets and their satellites, so it is thought that they may be material left from the formation of those objects.

Figure 17-38
Comet West, March, 1976. (Photo from Lick Observatory.)

Figure 17-39
The orbits of comets are generally inclined, and the tail points away from the sun.

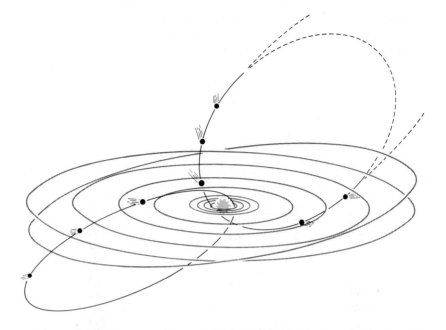

METEORS AND METEORITES

Meteors *are streaks of light in the sky,* commonly called "shooting stars." Their origin is believed to be small bits of interplanetary material that are burned by friction when they pass through earth's atmosphere. On a typical evening about three per hour are seen, but after midnight the sky is in the direction of the earth's advance, and up to 15 per hour can be expected. **Meteor showers,** times when the number of meteors exceeds the average, occur when earth passes through the orbit of a burned-out comet; dust from the comet is their source. Table 17-4 lists the bigger meteor showers, named for the area of the sky from whch they appear to come. T 17-4

Meteorites *are fragments that are not burned up by friction and so fall to earth.* Fragments larger than about a centimeter can become meteorites. Large meteorites may form impact craters; smaller ones have been known to break through roofs (Figure 17-40). About 10 to 20 F 17-40 are recovered each year, and about 3000 are now in museums and other scientific collections.

Meteorites are of three types—stony, iron, and stony-iron. Most of those found are iron meteorites because they are easily recognized. If only those seen to fall are considered, most are stony.

Meteorites have been very closely studied and are believed to be fragments of a few to a few tens of larger parent bodies. The size and arrangements of minerals show that meteorites were once parts of bigger bodies, believed to have been a few hundred kilometers in diameter, that cooled and crystallized slowly. The composition of meteorites is similar to the compositions suggested by spectral studies of the logical source of meteorites, the Apollo asteroids, whose orbits

Table 17-4   PRINCIPAL METEOR SHOWERS

Meteor Swarm	Normal Period of Visibility	Night of Maximum Visibility	Average Number of Meteors Visible per Hour
Quadrantids	Jan. 2–4	Jan. 3	30
Lyrids	Apr. 20–22	Apr. 22	8
Eta Aquarids	May 2–7	May 4	10
Delta Aquarids	July 20–Aug. 14	July 30	15
Perseids	July 29–Aug. 18	Aug. 12	40
Draconids	Oct. 10	Oct. 10	
Orionids	Oct. 17–24	Oct. 21	15
Taurids	Oct. 20–Nov. 25	Nov. 4	8
Leonids	Nov. 14–19	Nov. 16	6
Andromedids	Nov. 15–Dec. 6	Nov. 20	
Geminids	Dec. 8–15	Dec. 13	50
Ursids	Dec. 19–23	Dec. 22	12

From The *Evolving Universe: An Introduction to Astronomy* by Donald Goldsmith. © 1981 by Benjamin/Cummings Publishing Company, Inc., Menlo Park, California. Reprinted by permission.

cross earth's. Meteorites have radioactive ages of about 4600 million (4.6 billion) years, and, because they are thought to have formed at the same time as the rest of the solar system, this is used as the age of the solar system.

ORIGIN OF THE SOLAR SYSTEM

The sun and the planets are believed to have formed as part of the same event about 5000 million (5 billion) years ago. Our galaxy, the Milky Way, is much older; the origin of stars and galaxies will be described in the next chapter.

The story begins with a huge cloud of gas and dust in space. When this cloud became dense enough, it began to contract because of gravitational attraction. It can be shown both experimentally and theoretically that the initially random motions of the dust cloud would resolve into rotation in a single direction as the cloud condensed. The rotation would become faster as material moved toward the center, just as skaters spin faster when they bring their arms in toward their bodies (Figure 17-41). This process has two effects: the dust cloud is flattened into a disk, and the center rotates faster than the rim.

The gravitational compression causes the cloud to be heated. The pressure caused by the hot gas slows the gravitational collapse and allows the disk to form; if the disk did not form, all of the cloud

F 17-41

Figure 17-40
Wolf meteorite crater, the largest crater in Australia. (Photo from Australian Information Service.)

would become the sun. The center of the disk is heated by compression to 10 million°C, causing nuclear reactions to begin—the sun is born. The disk will become the planets, explaining why they are in a plane. The temperature of the disk at this stage is about 1700°C (3100°F); at this temperature the disk is entirely gas. As the disk cools, condensation occurs; the particles clump together under the influence of gravity and adhesion in low-velocity collisions, forming the particles in the disk. Some meteorites have structures that may have formed in this way.

The initial dust cloud was composed mainly of hydrogen and helium and a small percentage of the other elements. This is the composition of the sun, and most of the cloud went to form it. The condensation in the disk that formed the planets was controlled by

Figure 17-41
Origin of the solar system. (After Goldsmith, 1981.)

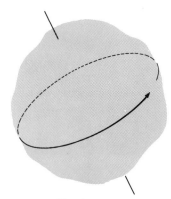

Cloud rotates
more rapidly as
it contracts

Cloud flattens to
pancakelike
configuration

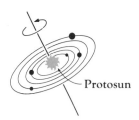

Protosun

Planets accrete at
their present distances
from sun

the temperature; the central part was hotter than the outer part. At 1700°C (3100°F) no condensation occurs, but when the temperature dropped to about 1300°C (2400°F) aluminum, titanium, and calcium oxides began to condense. At 1100°C (2000°F) iron and nickel grains formed; and at 1000°C (1830°F) silicate minerals formed. Apparently, temperatures in the central part of the disk were in this range because the inner planets have such compositions. The outer planets contain much water, ammonia, and methane ice; these materials condense at temperatures in the range of −170°C to 30°C (−275°F to 85°F).

The best estimates are that about 100 million years were required to form the sun and the initial planets. Studies of the moon strongly suggest that during the 500 million years after it formed, it was bombarded by meteorites or other objects. Similar events may have occurred on other planets, and this may have been the final stage in gathering up the parts of the disk. At this late stage, some of the objects may have been quite large; some of these may have become moons. Collisions between planets and some of these larger objects may have affected the axial tilt and rotation of some planets. After the planets formed, they went through a stage of melting. The heat was generated by radioactive elements and by the compression caused by condensation. At this stage, iron and nickel migrated to form the cores, and the lighter silicates formed the outer shells of the terrestrial planets.

Soon after the sun formed, it developed a magnetic field that interacted with the disk of dust and gas. The result of this interaction was to slow the sun's rotation and to transfer momentum to the disk, accounting for the present distribution of momentum between the planets and the sun. The sun's radiation also cleared much of the remaining dust and gas out of the solar system.

This theory of origin of the solar system suggests that planets may also have formed with stars other than our sun. If some of those planets have water and temperatures like earth's, they too could have life.

SUMMARY

- The *inner* or *terrestrial* planets are Mercury, Venus, Earth, and Mars; they are all more or less like Earth.

- The *outer, great,* or *Jovian* planets, Jupiter, Saturn, Uranus, and Neptune, are all much larger than the terrestrial planets and are composed mainly of hydrogen and helium.

- Earth, the largest terrestrial planet, is slightly more dense than the others. Its equatorial radius is slightly more than its polar radius because of the centrifugal effect of its daily spin. Earth's axis is tilted 23½°; this is the cause of seasons. Our atmosphere is composed of 78 per cent nitrogen and 21 per cent oxygen. The thin *crust* is composed of basalt under the oceans and granite under the continents. The *mantle* is a thick layer composed of iron and magnesium silicate rocks. The *core* is composed of iron and nickel, is partly melted, and is the source of our magnetic field, the strongest field of the terrestrial planets. Superficial surface processes are erosion and deposition of sedimentary rocks. The earth's internal energy causes *plate tectonics,* and the resulting deformation of surface rocks is unique to earth.

- Moon is large enough to be considered a small planet. Its density is much less than earth's, so its mass is not great enough to hold an atmosphere; thus erosion does not occur on the moon. Moon has dark areas called *maria* (*mare,* singular) composed of basaltic rocks, and light-colored areas called *terrae* (*terra,* singular) or highlands made of feldspar-rich rocks called *anorthosite*. Thus, like earth, it has two types of crust. *Craters* caused by meteorite impact are common, and the surface rocks are mainly *breccia* produced by meteorite impacts. Volcanic features are also common. Moon has a very small magnetic field, and *moonquakes* are caused by tidal forces rather than internal energy. The highlands formed about 4000 million years ago, and between 3700 and 3200 million years ago the maria formed, probably by huge meteorites that crashed through the thin crust. No rocks younger than 3000 million years have been brought to earth, suggesting that moon solidified about that time and has been a dead planet since.

- Mercury is about 50 per cent larger than the moon. Its orbit is inclined 7 degrees from the ecliptic, its year is 88 days, and its rotation period is 58.6 days. It has no atmosphere or moons. It is closest to the sun, and its surface near the sun reaches 325°C (620°F) and its dark side −175°C (−280°F). Its surface is cratered, and it has large basins like the moon's maria. Mercury's density is almost the same as earth's, and its magnetic field is one per cent of earth's. It has a few long cliffs, suggesting faulting.

- Venus is very close to earth in size and density. It has a very thick cloud cover that traps its heat, making the surface temperatures above the melting point of lead. Its atmosphere is mainly carbon dioxide. Its year is 225 earth days, and its rotation period is 243 days. Its clockwise direction of rotation (spin) is opposite or *retrograde* to the other planets. Radar images show that the surface has boulders and that Venus has some highland areas that may be made of deformed rocks like earth's mountains. It has a small magnetic field.

- Mars, farther from the sun than earth, has a year that is 687 earth days long. Its day is 24 hours, 37 minutes. The equatorial radius is about half earth's. Mars has two moons; their irregular shapes suggest that they are captured meteorites or asteroids. Its atmosphere is mainly carbon dioxide like Venus', but it is much thinner than earth's. Mars' axis is tilted, so it has seasons, and ice caps can be seen to form and melt. Dust storms are common. Evidence of running water, perhaps from buried ice, is seen at a few places. Mars has huge volcanoes, the largest in the solar system. Internal energy is also suggested by very long valleys that may be evidence of faulting.

- Jupiter is the largest planet, with a polar diameter of 134,000 kilometers (83,350 miles) and equatorial diameter of 142,800 kilometers (88,825 miles), the difference caused by its 10-hour rotation. The atmosphere is mainly hydrogen and helium, and is striped or banded in yellow and red-brown. Lightning occurs, and the *Great Red Spot* is probably a huge permanent storm. Jupiter has a strong magnetic field and so emits radio energy. It also emits thermal energy, probably from continuing gravitational collapse. Jupiter probably has a small solid core of iron and rock surrounded by a layer of liquid metallic hydrogen, then a layer of liquid molecular hydrogen, and finally by its atmosphere. Jupiter also has a ring.

- Saturn is similar to Jupiter but somewhat smaller. Its density is only 0.7 gram per cubic centimeter, about half that of Jupiter. It probably has a small rocky core surrounded by a layer of water, methane, and ammonia ices. Above this is a thin metallic hydrogen layer and a molecular hydrogen layer. Saturn has the most extensive ring system.

- Uranus and Neptune are similar to Saturn in structure but are slightly less than half as large. Their densities are more like Jupiter's.

- Pluto is the most distant planet, and its orbit is inclined 17 degrees to the rest of the solar system. It is probably smaller than our moon and probably is composed largely of hydrogen because its density is about 0.7 gram per cubic centimeter.

- Jupiter has a total of 15 satellites that have been discovered so far.

- Io is closest and is similar in size and density to our moon. It has a number of active volcanoes erupting sulfur and sulfur dioxide, giving it a red-orange, black, and white appearance like a pizza pie. The volcanic energy probably comes from tidal flexing. Its composition is probably similar to our moon's.

- Europa is slightly smaller than our moon but is bright because its surface is a thin layer of ice. Its surface is a tangle of lines, perhaps cracks in the ice layer healed by upwelling water that froze.

- Ganymede is slightly larger than Mercury, and its density of just less than 2 grams per cubic centimeter suggests that it is composed of about equal parts of ice and rock. Its surface has light and dark areas and some craters. The younger light surface is grooved, and the grooves end abruptly where they meet, suggesting faulting.

- Callisto is about the same size as Mercury and is probably composed of much ice because it is the least dense of Jupiter's moons. Its surface is saturated with craters, some very large.

- Saturn also has 15 known moons, but only one, Titan, is large. Titan, slightly larger than Mercury, is the only satellite with an atmosphere. The atmosphere is nitrogen and small amounts of hydrocarbons, giving it a red color like some types of smog.

- The *asteroids* are small bodies between Mars and Jupiter. About 100,000 are known; the largest is about one-quarter the size of our moon. They appear to be composed of rock and iron, and may be the source of meteorites.

- *Comets* are chunks of ice and dust (dirty snowballs) a few kilometers in diameter. They have highly elliptical orbits; when near the sun the frozen gases are released and ionized, forming the visible head and tail. The head has a bright point called the *nucleus* and a diffuse glow called the *coma*. The tail may be millions of kilometers long.

- *Meteors*, streaks of light in the sky, are caused by small bits of interplanetary material burned by friction as they pass through our atmosphere.

- *Meteor showers*, times when many meteors are seen, occur at a number of places on the earth's orbit and may be the dust remaining from burned-out comets.

- *Meteorites* are fragments from space that reach the earth's surface, ranging in size from dust to several meters. Some are stony, some iron, some stony-iron. Most seen to fall are stony, but most found on the surface are iron because they are most easily recognized. They probably formed at the same time as the rest of the solar system.

- The solar system is believed to have formed about 5000 million years ago. An initial dust cloud, contracting gravitationally, would begin to rotate and would be flattened. The gravitational compression would also cause heating. The center of the disk would become very hot—millions of degrees Celsius—nuclear reactions would begin, and so the sun was born. Other much smaller centers of accumulation would develop in the disk, becoming the planets. The sun's radiation then drove the remaining dust away.

KEY TERMS

inner, or terrestrial, planets

outer, great, or Jovian, planets

libration

mare (pl., maria)

terra (pl., terrae)

highland

impact crater

rille

breccia

basalt

anorthosite

retrograde motion or spin

greenhouse effect

liquid metallic hydrogen

liquid molecular hydrogen

Roche's limit

Galilean moons

asteroids

main asteroid belt

Apollo asteroid

comet

nucleus of comet

coma of comet

meteor

meteor shower

meteorite

REVIEW QUESTIONS

1. What two features are unique to the planet earth?

2. The liquid portion of the earth's core is believed to be responsible for two striking characteristics of our planet. What are these two characteristics, and why is the liquid core believed to be responsible for them?

3. Why are we able to see only one side of the moon from earth?

4. (a) State the reasons why the earth's moon lacks the gravity to hold an atmosphere.
   (b) What characteristic besides gravity determines whether a planet can hold an atmosphere?

5. The moon's surface is peppered with impact craters, whereas the earth's surface has few. Give two reasons for this difference.

6. Contrast the origins of the moon's mountains with the origins of earth's mountains.

7. (a) Define the term *breccia*. How many of the rocks gathered so far from the moon's surface are breccia? What does the abundance of breccia tell us about the moon's history?
   (b) What is a basalt? Where on the moon do rocks of this type predominate?
   (c) From what part of the moon does anorthosite originate?

8. The highlands of the moon are older than the maria. Cite two types of evidence for this difference in age.

9. In one or two sentences contrast the cause of moonquakes with the cause of earthquakes.

10. What has each of the following activities told us about the history and current state of the moon?
    (a) Radioactive dating of moon rocks.
    (b) Measurements of the moon's fossil magnetism and of the moon's present magnetic field.
    (c) Waves from moonquakes.
    Organize your answers into a chronological history of the moon.

11. Does Mercury have an atmosphere? Why?

12. (a) What features of Mercury indicate that it has a history similar to that of the moon's?
    (b) Why do Mercury's magnetic field and long, high cliffs indicate that it may still be internally active, as is the earth, rather than dead, as is the moon?
    (c) What is another possible cause of Mercury's long, high cliffs?

13. In what way are Venus' and Uranus' rotations different from those of all other planets?

14. Venus is only a little closer to the sun than earth is, yet Venus' surface temperatures are much higher than those of Mercury. Why?

15. (a) Describe Venus' atmosphere.
    (b) In what ways does earth's atmosphere differ from Venus'?
    (c) State a possible reason why Venus and earth developed different atmospheres.

16. Is it possible that weathering and erosion occur on Venus? Why or why not?

17. Describe Venus' surface in each of the following areas (omit measurements of elevation; include rock type, relative age, cratering, mountains):
    (a) plains
    (b) depression below the plains
    (c) highlands
18. What features of Venus' surface indicate that the planet might have remained internally active until relatively recent times?
19. List three possible reasons why Venus' magnetic field is so weak. Which of these reasons also explain the planet's current lack of internal energy?
20. (a) Compare Mars' atmosphere with Venus'.
    (b) Of what do Mars' polar ice caps appear to be made?
    (c) Compare Mars' temperature with Earth's.
21. Cite evidence of the following on Mars:
    (a) weathering
    (b) erosion and transport by wind
    (c) erosion by water
22. (a) Does Mars appear to have internal energy? Give reasons for your answer.
    (b) Is there evidence of plate tectonics on Mars?
    (c) Give a possible reason why Mars' shield volcanoes are so huge.
23. (a) Of what is Jupiter's atmosphere composed?
    (b) What is believed to be the cause of the planet's parallel stripes? What might be the cause of the stripes' colors?
    (c) What is the Great Red Spot believed to be?
24. (a) Jupiter radiates twice the thermal energy that it receives from the sun. What does this observation say about the temperatures below the planet's clouds? What is the probable source of this thermal energy?
    (b) Jupiter radiates radio energy. State one probable source of this energy.
25. Describe the surface and interior of Jupiter. Which layer is responsible for the planet's strong magnetic field?
26. Jupiter's rings are made of what? Why is it

that a moon cannot exist closer than a certain distance to a planet?
27. What is the chief difference between Saturn's atmosphere and Jupiter's? Explain the cause of the difference.
28. Why is Saturn believed to have more hydrogen than any other planet?
29. Describe the surface and interior of Saturn.
30. Describe Saturn's ring system. What are the suggested origins of the rings?
31. What did the discoveries of Neptune and Pluto have in common?
32. Describe the temperature, atmosphere, and structure of Uranus and Neptune.
33. Describe Pluto's orbit, size, and temperature. What is the possible composition of Pluto's surface and interior?
34. Summarize the chief differences between the terrestrial planets and the Jovian planets.
35. Concerning Jupiter's Galilean moons:
    (a) Which of the moons is volcanically active? What is the probable source of energy for the volcanism?
    (b) Which moon may be internally active or, at least, have remained internally active long after surface freezing? What observation suggests internal activity?
    (c) Which three moons have ice at their surfaces? State two reasons why they have been able to retain water.
    (d) Why does Io lack impact craters? Describe the cratering on the other three moons and explain what the cratering tells us about each moon's history.
    (e) Of what materials are the moons composed? Does their composition make them more like Jupiter or earth?
36. Describe Titan's unique feature.
37. (a) What are asteroids?
    (b) Where are most located?
    (c) Describe their size, shape, and composition. Why is it suspected that the moons of Mars were asteroids that were captured by the planet?
    (d) How might asteroids have originated?
38. (a) Why are comets called "dirty snowballs"?

(b) Give several types of evidence that comets are dirty ice.

(c) If a comet is made of ice, why is it seen as a bright light?

(d) What is believed to be the origin of comets? Why?

39. What are meteors? What is believed to be their source?

40. What are meteorites? Of what are they composed? From where are meteorites believed to have come?

41. Write or tell a history of the solar system that covers the following points:

(a) Formation of the disk from the cloud.

(b) Why the sun, inner planets, and outer planets have different compositions.

(c) Aggregation of particles into larger bodies.

(d) Remelting and formation of planets' interior structures.

(e) Development of the sun's magnetic field.

SUGGESTED READINGS

Arvidson, R. E., A. B. Binder, and K. L. Jones. "The Surface of Mars," *Scientific American* (March 1978), Vol. 238, No. 3, pp. 76–89.

Carr, M. H. "The Volcanoes of Mars," *Scientific American* (January 1976), Vol. 234, No. 1, pp. 32–43.

Carr, M. H. "The Geology of Mars," *American Scientist* (November–December 1980), Vol. 68, No. 6, pp. 626–635.

Dyal, Palmer, and C. W. Parkin. "The Magnetism of the Moon," *Scientific American* (August 1971), Vol. 225, No. 2, pp. 63–73.

Goldreich, Peter. "Tides and the Earth-Moon System," *Scientific American* (April 1972), Vol. 226, No. 4, pp. 42–52.

Goldsmith, Donald, *The Evolving Universe: An Introduction to Astronomy*. Menlo Park, Ca.: Benjamin/Cummings, 1981.

Hartmann, W. K. "Cratering in the Solar System," *Scientific American* (January 1977), Vol. 236, No. 1, pp. 84–99.

Head, J. W., and others. "Geologic Evolution of the Terrestrial Planets," *American Scientist* (January–February 1977), Vol. 65, No. 1, pp. 21–29.

Horowitz, N. H. "The Search for Life on Mars," *Scientific American* (November 1977), Vol. 237, No. 5, pp. 52–61.

Lawless, J. G., and others. "Organic Matter in Meteorites," *Scientific American* (June 1972), Vol. 226, No. 6, pp. 38–46.

Lewis, J. S. "The Chemistry of the Solar System," *Scientific American* (March 1974), Vol. 230, No. 3, pp. 51–65.

Mason, Brian. "The Lunar Rocks," *Scientific American* (October 1971), Vol. 225, No. 4, pp. 48–58.

Murray, B. C. "Mars from *Mariner 9*," *Scientific American* (January 1973), Vol. 228, No. 1, pp. 48–69.

Sagan, Carl. "The Solar System," *Scientific American* (September 1975), Vol. 233, No. 3, pp. 22–31. (The entire issue is on the solar system.)

Schmitt, H. H. "*Apollo 17* Report on the Valley of Taurus-Littrow," *Science* (November 16, 1973), Vol. 182, No. 4113, pp. 681–690.

Soderblom, L. A. "The Galilean Moons of Jupiter," *Scientific American* (January 1980), Vol. 242, No. 1, pp. 88–100.

Stone, E. C., and A. L. Lane. "*Voyager 1* Encounter with the Jovian System," *Science* (June 1, 1979), Vol. 204, No. 4396, pp. 945–1007.

Waldrop, M. M. "*Voyager 1* at Saturn," *Science* (December 5, 1980), Vol. 210, No. 4474, pp. 1107–1111.

Whipple, F. L. "The Nature of Comets," *Scientific American* (February 1974), Vol. 230, No. 2, pp. 49–57.

Wood, J. A. "The Lunar Soil," *Scientific American* (August 1970), Vol. 223, No. 2, pp. 14–23.

CHAPTER 18

An unusual group of galaxies. (Photo from Palomar Observatory, California Institute of Technology.)

SUN, STARS, AND UNIVERSE

THE SUN

The sun is our closest star; the next nearest is 4.3 light years distant. The sun, an average star, is 1,390,000 kilometers (863,000 miles) in diameter, and its average distance from earth is 150 million kilometers (93 million miles). The sun is the only star close enough to see as a disk; all the others are points of light even on big telescopes. It was a long time before people realized that the sun is a star, similar to those points of light; but now we learn much about other stars from study of the sun.

There are problems in observing the sun. Never look directly at it, even for a very short time, because the eye will be damaged or even blinded; pointing a telescope or binoculars at the sun is even worse. Even dark glass or film may allow enough radiation to pass through to hurt eyes. The easiest way to observe the sun is to view its image formed by a pin hole, as shown in Figure 18-1A; however, the image is small and shows little detail. A better way is to project the image from a telescope or binoculars onto a screen, as shown in Figure 18-1B. In some ways the sun is harder to study than more distant stars because it must be observed during the day, when its heating makes the atmosphere turbulent, reducing the sharpness of the image. For this reason, observatories built for study of the sun must be very carefully located in areas of reduced turbulence.

F 18-1

Structure of the Sun

Only the sun's outer layers, its "atmosphere," are observable from earth. Beneath this atmosphere is the main body of the sun, and near its center is the core, where its energy is produced. Our knowledge of the inner sun is based mainly on theoretical studies. It is easy to measure the energy that the sun radiates: it is the equivalent of the amount of energy produced by burning 7300 kilograms of coal per square meter (1500 pounds per square foot) of the sun's surface every hour. Of course, this is not the way that the sun produces its energy, but this measurement did lead astronomers to realize that the stars cannot be eternal and that they must change with time. It also led to the realization that the source of the sun's energy must be nuclear

Figure 18-1
The sun can be safely viewed by obtaining an image with a pinhole or a telescope, but it should never be viewed directly.

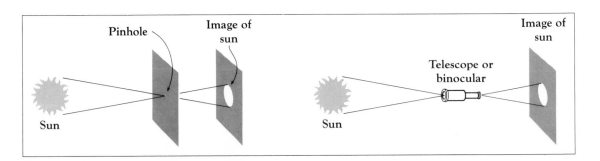

Figure 18-2
Structure of the sun.

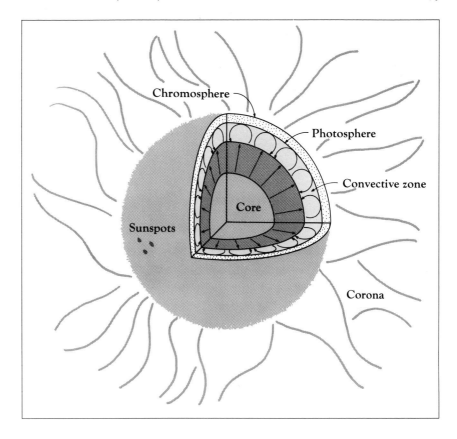

reactions. *In ordinary* **chemical reactions,** *such as burning coal, only the electrons that form a cloud around the nucleus of an atom are involved in the reaction. In a* **nuclear reaction,** the atoms are much closer together, and interactions among the nuclei occur.

The sun's spectrum reveals its composition. The spectrum is an absorption or dark-line spectrum like that of most stars. The main elements in the sun are hydrogen, 78.4 per cent; helium, 19.8 per cent; all other elements, 1.8 per cent. Interestingly, helium was discovered from study of the sun's spectrum in 1868 and was not discovered on earth until 1891. The source of the sun's energy is the transformation of hydrogen into helium. This and the other stellar nuclear reactions in which mass is changed into energy will be described later.

The sun's energy is produced in the core. *The radius of the* **core** *is about one-quarter of the radius of the whole sun,* and this small volume (1/64th of the sun's volume) contains about half of the sun's total mass. The temperature of the core is estimated to be 15 million °C, and the density is about 158 times greater than water. The energy produced in the core moves through the main body of the sun by radiation until it gets near the top of this zone, where the energy-transporting process changes to convection (Figure 18-2).

The layer of the sun that we see is called the **photosphere,** *a cloudlike*

F 18-2

Figure 18-3
Granules on the sun
are caused by
convection. (Photo
from Big Bear Solar
Observatory,
California Institute of
Technology.)

layer that is the base of the sun's atmosphere. This visible layer is only about 400 kilometers (250 miles) thick, so it appears like a surface. Its temperature is 5400°C (9800°F). Its surface has **granules,** which are the upper surfaces of the convection cells mentioned above (Figure 18-3). The granules are 1000 to 2000 kilometers (600 to 1200 F 18-3 miles) in diameter, and each one lasts only a few minutes. Larger patterns in the granules—called **supergranules,** about 30,000 kilometers (18,000 miles) in diameter—shoot **spicules** 10,000 kilometers (6000 miles) above the surface (Figure 18-4). Spicules are composed of in- F 18-4 candescent gas and last only a few minutes. The photosphere also has sunspots and is the source of flares and other features described in the next section.

The layer above the photosphere is the **chromosphere,** a pink-glowing layer visible only during an eclipse. It is about 2500 kilometers (1500 miles) thick, and its temperature is 10,000°C (17,500°F). Above the chromosphere is the **corona,** which is composed of very low-density gas. It extends for a few million kilometers from the surface, and it is only visible, without instruments, during an eclipse (Figure 18-5). The temperature of the corona is between 1 and 2 mil- F 18-5 lion °C.

Surface Activity of the Sun

Sunspots are dark areas on the sun. They appear dark because they are somewhat cooler—3700°C to 4200°C (6700°F to 7600°F), as compared with 5400°C (9800°F), the temperature of the photosphere.

Figure 18-4
Spicules on the solar
surface. (Photo from
Big Bear Solar
Observatory,
California Institute of
Technology.)

Figure 18-5
The sun's corona is seen during an eclipse. (Photo from Dearborn Observatory, Northwestern University.)

Figure 18-6
Sunspots and a close-up view. (Photos from Big Bear Solar Observatory, California Institute of Technology.)

When present, they are easy to see (Figure 18-6), and from observa- F 18-6
tion of sunspots the rotation of the sun was first noted. The sun's
rotation period is about a month (27.3 days), and, because the sun is
gaseous, the equatorial region rotates faster than the poles. Sunspots
range in size from about 1600 kilometers (100 miles) to 160,000 kilometers (100,000 miles) and may last for a few days to a few months.
Sunspots are apparently storms on the sun's surface, caused by the
sun's magnetic field. The magnetic field on the sun's surface can be
detected because a magnetic field causes certain spectral lines to split.

Sunspots have been observed for many years, and an 11-year cycle
has been noted (Figure 18-7). They tend to form as pairs sym- F 18-7
metrically above and below the sun's equator. At the beginning of a
sunspot cycle, there are very few sunspots, and they form at about 30
degrees above and below the equator. At the sunspot maximum, they
are at 20 degrees above and below the equator. At the end of a cycle,
the number of sunspots is at a minimum, and they are about 10 degrees above and below the equator. The new cycle will begin with
sunspots near 30 degrees. Interestingly, in the new cycle the sun's
magnetic field is reversed, so perhaps we should say that the sunspot
cycle is really 22 years. Why the sun's magnetic field reverses we do
not know. The earth's magnetic field also reverses, but at much
longer intervals and not regularly.

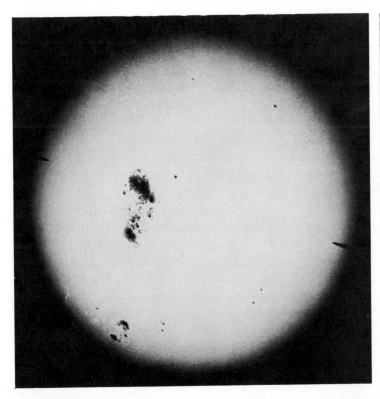

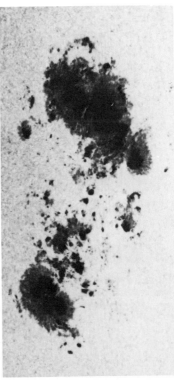

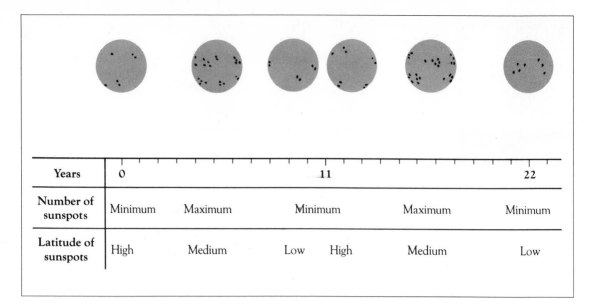

Years	0			11		22
Number of sunspots	Minimum	Maximum	Minimum		Maximum	Minimum
Latitude of sunspots	High	Medium	Low	High	Medium	Low

Figure 18-7
The sunspot cycle.

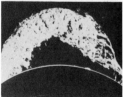

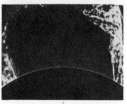

The region around sunspots is the site of other activities. **Prominences** are large clouds of gas shot up hundreds of thousands of kilometers above the sun's surface (Figure 18-8). Some are quiet flows of gas that may last for hours or days; others shoot out at speeds of 1000 kilometers per second (600 miles per second). **Flares,** the largest and most violent of these features, originate from large, active sunspots (Figure 18-9). Flares begin quickly and may last up to about four hours. Temperatures in a flare may reach 5 million °C, and, although the amount of visible light from the sun may only change 1 per cent during a flare, the amount of X rays may increase 100 times. Particles ejected by a flare reach the earth in a number of hours or days and may disrupt radio communications and produce the aurorae (northern or southern lights) described in the following paragraph.

The material ejected from sunspots causes the **solar wind,** an extension of the corona. The particles of the solar wind are traveling at velocities between 600 and 1000 kilometers per second (360 to 600 miles per second) when they reach the earth. When these high-energy particles reach the vicinity of the earth, they interact with the earth's magnetic field to cause the **Van Allen radiation belts** (Figure 18-10). Eventually they move toward the earth's magnetic poles; and when they pass through the atmosphere, they collide with the molecules in the air, causing them to emit spectral wavelengths, or colors. The resulting colored lights in the skies, generally seen only at high latitudes, are called the **aurora borealis** and the **aurora australis** (the northern and southern lights) (Figure 18-10B).

Figure 18-8 Four stages in a single solar prominence. (Photos from Yerkes Observatory.)

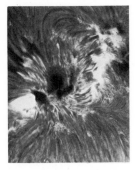

Figure 18-9
A large solar flare.
(Photo from Big Bear
Solar Observatory,
California Institute of
Technology.)

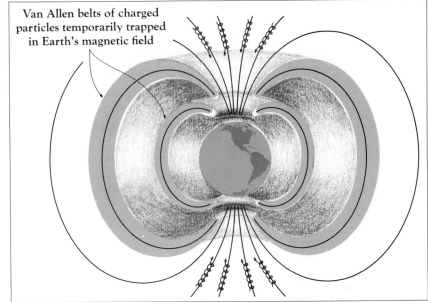

Van Allen belts of charged
particles temporarily trapped
in Earth's magnetic field

A.

Figure 18-10
A. Van Allen
radiation belts are
caused by charged
particles trapped by
the earth's magnetic
field. **B.** Aurora
borealis, northern
lights. (Photo from
NOAA.)

B.

MEASURING THE PROPERTIES OF STARS

Distance—Parallax

The realization of the immensity of the universe was a revelation of
astronomy. Distances to the planets were first measured in classical
times, and the results were refined later. Stellar distances, however,
had to wait until better instruments were built.

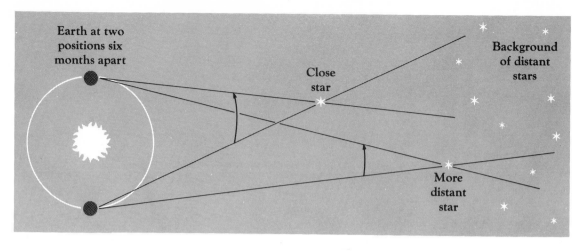

Earth at two
positions six
months apart

Close
star

Background
of distant
stars

More
distant
star

Figure 18-11
The parallax, or
angular shift, of a
close star is much
more than that of a
more distant star. The
actual angles are very
much smaller than
shown here.

Distance to close stars can be measured by the technique called
parallax. The basic idea underlying this method was mentioned in
Chapter 16, because it was one of the classical tests of whether or not
the earth revolved around the sun (see Figure 16-2, p. 388). If the
earth orbits the sun, the position of a nearby star should apparently
shift as the earth moves in its orbit. It was not until 1838 that this
shift was first measured because the shift is so small.

Parallax uses a right triangle with the radius of earth's orbit as a
base. An astronomer twice notes the position of a nearby star relative
to the more distant stars at intervals six months apart; photographic
plates are used now for these observations. The amount of shift of the
nearby star's position is measured as an angle—that is, how much the
telescope has to be turned to move the field of view the amount that
the nearby star shifted in six months. Half of this angle is the star's
parallax angle (Figure 18-11). The angle is very small and is used to F 18-11
define the distance unit used by most astronomers. *If the shift is* **one
second of arc,** *the distance to the star is one* **parsec** (Figure 18-12). Said F 18-12
another way, one parsec is the distance that has a PARallax of one
SECond of arc. As shown in Figure 18-12, if the angle is one half-
second, the distance is two parsecs, and so on. Each parsec is 3.26
light years. (Remember, a circle has 360 degrees; each degree is divided into 60 minutes, and each minute is divided into 60 seconds.)

The first parallax that was measured was one-third of a second,
giving a distance of three parsecs, or 9.8 light years. The parallax of
the closest star is about 0.76 second, or 1.32 parsecs, or 4.3 light years.
The limit of accurate parallax measurements is about one one-hun-
dredth (0.01) of a second, or 100 parsecs. Many astronomers place the
limit at half of this or even less. About 10,000 stars are close enough
to measure their distances by parallax. All of the other methods of
measuring the distance to a star are based on parallax measurements
made on these close stars.

Luminosity and Magnitude

The **luminosity** *of a star is a measure of how much energy it radiates.* If all stars were the same distance away, it would be easy to measure their luminosity. If two stars have the same luminosity, the closer one will appear brighter than the more distant one. The scale used by astronomers to measure brightness is one inherited from Hipparchus about 130 B.C. He called the brightest stars *first-magnitude stars* and the dimmest that can be seen with the naked eye *sixth-magnitude stars.* When accurate measurements of brightness were made in the 19th century, it was discovered that a difference of five magnitudes, as from first magnitude to sixth magnitude, was about a difference of 100 times in brightness. This information was used to define the **magnitude** scale. To make each step equal, we need a number that when multiplied by itself five times equals 100—that is, the fifth root of 100, or $\sqrt[5]{100}$. The number is 2.512. Therefore, each magnitude is 2.512 times brighter than the next magnitude. This means that the difference in brightness between magnitude 1 and 2 (or 4 and 5) is 2.512 times, and the brightness difference of 3 magnitudes is $2.512 \times 2.512 \times 2.512 = 15.85$ times. Note again that, the higher the magnitude number, the dimmer the star. A magnitude 0 star is 100 times brighter than a magnitude 5 star, and a star with a negative magnitude is even brighter.

The **apparent magnitude** *of a star is its magnitude as seen on earth. The* **absolute magnitude** *is defined as what its magnitude would be if it were moved to the standard distance of 10 parsecs (32.6 light years).* Obviously, the apparent magnitude of any star can be measured because it is merely the brightness of the star as we see it here on earth. If one knows the distance to a star, and, of course, its apparent magnitude, the absolute magnitude can be calculated quite easily. We know the distances to the close stars from measurement of their parallax, so we can calculate the luminosities of those stars. All we need to know is how brightness changes with distance.

Figure 18-12
The parallax angle is used to define the parsec, a distance used by astronomers. One parsec equals 3.26 light years. The closest star (except the sun) has a parallax angle of ¾ second of arc and therefore is at a distance of 4/3 = 1.33 parsecs, or 4.3 light years.

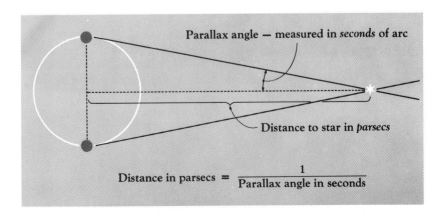

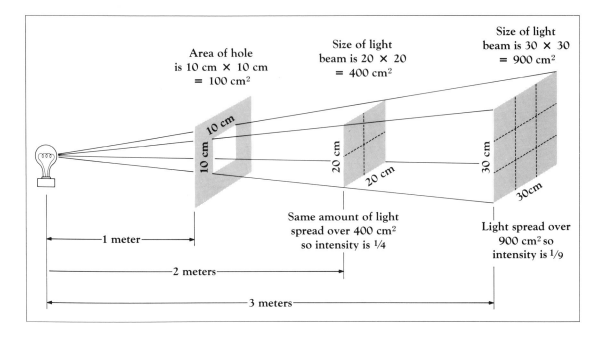

Area of hole
is 10 cm × 10 cm
= 100 cm²

Size of light
beam is 20 × 20
= 400 cm²

Size of light
beam is 30 × 30
= 900 cm²

10 cm

10 cm

20 cm

20 cm

30 cm

30cm

Same amount of light
spread over 400 cm²
so intensity is ¼

Light spread over
900 cm² so
intensity is ⅑

1 meter

2 meters

3 meters

Figure 18-13
The intensity of light is reduced with distance from the source, because the light is spread over a larger surface. The amount of reduction in intensity is 1 divided by the square of the distance, as shown in this example.

The brightness of light varies inversely with the square of the distance; that is, if a light source is moved twice as far away, the brightness drops to one-fourth of what it was. If the distance is halved, the brightness will increase four times. This relationship is illustrated in Figure 18-13. At a distance of one unit from the source, the light F 18-13 covers a square; at a distance of two units, the light covers four squares. The distance doubled, and the light that covered one square now covers four squares, so the brightness, or intensity, of the light is one-fourth. The **inverse square law** enables us to calculate the absolute magnitude of stars whose distances are known. We will use this information after we learn a little more about stars from their spectra.

Classification of Stars—Composition and Temperature

The spectrum of a star reveals the composition of the gases surrounding the star. The spectrum of a star is an absorption spectrum, and the dark lines are those of the gases that surround the star. The composition of most stars is much like that of the sun—mainly hydrogen and helium. There are, however, differences in the spectra of stars, and the differences have been used to classify stars into several **spectral types** (Figure 18-14). F 18-14

The spectrum of a star also reveals its temperature. Soon after the first attempts were made to classify stars on the basis of their spectra, it was noted that the spectral types corresponded closely to the temperatures. The spectral types were originally designated A, B, C, and so on, based on the strength of the lines of elements such as hydrogen. When put into sequence with the hottest stars first, the sequence is O B A F G K M. The sequence is easily remembered by the mnemonic device: Oh be a fine girl kiss me. This sentence was made up by the noted American astronomer Henry Norris Russell, whose work we will discuss next. About 99 per cent of the stars fit into this classification, and most of the rest fit into three lesser classes.

H-R DIAGRAM   In the early part of this century, the Danish astronomer Ejnar Hertzsprung and H. N. Russell plotted the spectral types of stars versus their luminosity or absolute magnitude. The resulting diagram is called the **Hertzsprung-Russell diagram,** or the **H-R diagram** (Figure 18-15). It shows the relationship between absolute magnitude and spectral type for the stars that form a diagonal band across the graph. This diagram will also be the starting point for later discussions of the evolution of stars. For now, our interest is the relationship between spectral type and absolute magnitude.

F 18-15

Figure 18-14
The spectral types. Most stars fall into one of these spectral types. (Photo from Palomar Observatory, California Institute of Technology.)

Distance—Spectral Type

The H-R diagram could be drawn only for those stars whose distance was known from parallax because they were the only stars whose luminosity or magnitude could be calculated. Once the relationship be-

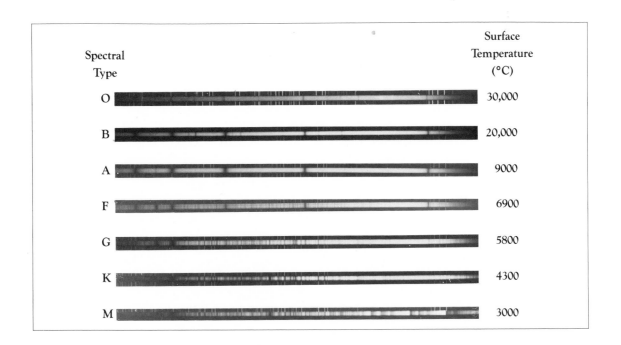

Spectral Type		Surface Temperature (°C)
O		30,000
B		20,000
A		9000
F		6900
G		5800
K		4300
M		3000

Figure 18-15
The Hertzsprung-
Russell (H-R)
diagram.

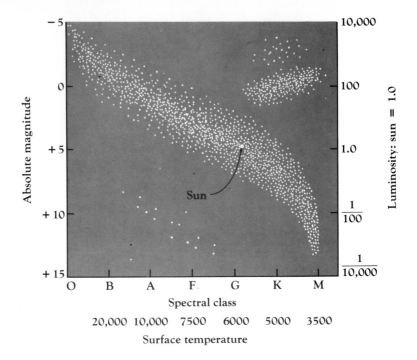

tween magnitude and spectral type was established, however, the magnitude of stars too far away to measure their distances by parallax could be estimated from their spectra. In this way, our yardstick for measuring the distance to stars was greatly enlarged, because the spectra of even distant stars are easily recorded on film with time exposures taken through telescopes.

The procedure to find distance is to determine luminosity from the spectral type, using the H-R diagram. For instance, if the star has a type G spectrum, read up from G on the horizontal axis on Figure 18-15 to the diagonal band of stars, and from there read across to the vertical axis, where the absolute magnitude 6 is read. Each spectral type is further subdivided, so that the uncertainty in magnitude is less than suggested in our example. Once the absolute magnitude is known, it is compared with the apparent magnitude and the distance is calculated.

Magnetic Field, Rotation, and Diameter

The magnetic field of a star can be determined from its spectrum. A magnetic field causes some of the lines in a spectrum to split into two or more closely spaced lines. The strength of the magnetic field can be estimated by the number of lines split and the amount of separation. Magnetic storms on the sun were described earlier, and some stars have magnetic fields thousands of times greater than the sun's.

The rotation of a star can also be determined from spectral studies. The Doppler effect causes a shift toward the blue in the spectrum of the side of the star moving toward the observer (see Figure 16-24, p. 410), and the side moving away is shifted toward the red. This can be seen on close stars (and planets), and the rate of rotation can be estimated from the amount of the shift. For more distant stars, the effect of this shift is to broaden the spectral lines received from the whole star, because it is not possible to record only the light from one side of the star. The rate of rotation, in this case, is estimated from the degree of broadening of the lines. In general, the bigger and more luminous the star, the faster the rotation.

The diameter of a very few, very large stars can be measured directly using widely spaced optical or radio telescopes. The diameters of most stars are calculated from their temperatures and absolute magnitudes. The temperature is determined from the continuous part of the star's spectrum, and the absolute magnitude from the spectral type. The rate of energy production is related to the star's temperature, with hot stars producing energy much faster than cooler stars. The absolute magnitude of a star is a measure of the total amount of energy produced by that star. Thus, for any given temperature, which determines the rate of energy production, a certain size is necessary to produce a given total amount of energy (absolute magnitude). Thus the size, or diameter, of a star can be calculated. The sizes of stars vary from smaller than the earth to larger than the diameter of Mars' orbit.

Mass—Binary Stars

More than half of the stars are **binary stars**—that is, *two stars that revolve around a mutual center of mass.* This is fortunate, because we can measure the mass of stars only if they are binary, or double, stars. There are various types of binaries. **Visual binaries** are double stars that can be distinguished with the naked eye or with telescopes (Figure 18-16A). **Eclipsing binaries** are binaries that pass in front of each F 18-16 other, causing the overall amount of light reaching us to change abruptly (Figure 18-16B). If the binary is distant or the stars are close together, we may not be able to see both stars, but spectra from both may be observable. If rotation can be detected from either a change in spectral type or from Doppler shift, the binary is called a **spectrographic binary.**

The importance of binary stars is that they can reveal the masses of the two stars if we can determine their period of revolution and their distance apart. The period of revolution can be found from observation. Their angular distance apart can be measured from observation

Figure 18-16
Binary stars.

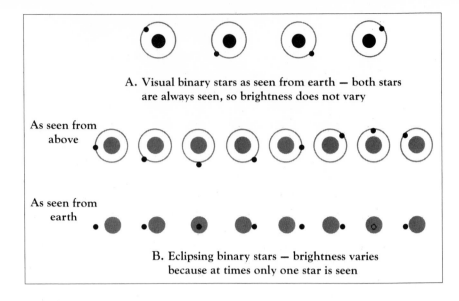

A. Visual binary stars as seen from earth — both stars
are always seen, so brightness does not vary

As seen from
above

As seen from
earth

B. Eclipsing binary stars — brightness varies
because at times only one star is seen

when they are most separated as observed from earth, and, if the distance to the binary is known from either parallax or absolute magnitude, their distance apart can be calculated. Newton showed that Kepler's third law can be rewritten as follows: the cube of the distance between the stars divided by the square of their period of revolution is equal to the sum of the masses of the two stars, or

$$\frac{D^3}{T^2} = M_1 + M_2$$

From this simple calculation, the sum of the masses of the binary pair is determined. The next step is to determine how the total mass is apportioned between the two stars. Any two bodies revolving due to their mutual gravity revolve around their center of gravity. The center of revolution of two stars can be determined by observation, so the distribution of mass between the two stars can be determined. For instance, if two stars revolve around a point one-third of the distance between them, the star nearest the point of revolution has two-thirds of the mass, and the more distant star, one-third.

Study of the masses of binary stars revealed that the mass of a star is very closely related to its spectral type. A plot of mass and absolute magnitude is very similar to the H-R diagram, which is a plot of spectral type and absolute magnitude. Thus, from spectral type we can estimate the mass of a nonbinary star.

The masses of stars range from one-tenth or less than that of our sun to 50 or more times the mass of the sun. There are many more small stars than large ones, but most of the light from stars comes from the few massive stars.

Distance—Cepheid Variable Stars

All of the methods of determining the distance to a star depend ultimately on parallax, the first method we covered and the only direct way to measure distance. In this discussion of a third method, we meet a new kind of star, the variable star, whose luminosity varies.

Cepheid variable *stars are stars whose luminosity varies with a regular period.* Cepheids have periods ranging from one day to 50 days, and their luminosity varies up to about one magnitude. The period of variation of a cepheid is related to its average magnitude. Once this relationship was determined, it became possible to know the magnitude of the cepheid by measuring its period of pulsation from brightest to dimmest. This means that, when studying a distant group of stars, one need only measure the period of a cepheid to find the distance, rather than obtain the spectrum of a single star. This method greatly extends our yardstick of measuring distance to other galaxies and the size of the universe itself.

The relationship between period and luminosity of cepheid variables was determined in 1912 by Henrietta S. Leavitt of the Harvard College Observatory. She studied the stars in the Magellanic Clouds, two galaxies visible from the southern hemisphere. The stars in these galaxies are so far away that she was able to assume that they are all about the same distance from us. She then determined the relationship between period and apparent magnitude for the two dozen or so cepheids in the Magellanic Clouds. All that was needed then was to determine the distance or absolute magnitude of any cepheid variable star anywhere in the heavens. This last difficult step was done in 1917 by Harlow Shapley.

EVOLUTION OF STARS

Stars must change, because their energy is produced by nuclear processes that change one element into another; in the process, mass is converted into energy. Thus, their compositions change and their masses are reduced. Even though tremendous amounts of energy are produced, the amount of mass reduction is very small. If stars undergo changes in composition, they must eventually use up their nuclear fuel and fade into darkness. But we still have stars, so stars must also be born. Once this truth was realized, astronomers looked at the physical properties of stars in an effort to decipher their life cycles. The H-R diagram mentioned earlier was one of the keys in this search.

H-R Diagram—Giants, Dwarfs, and Main-Sequence Stars

The original H-R diagram (Figure 18-17) plotted spectral type on the F 18-17 horizontal axis and absolute magnitude on the vertical axis. Because of the close relationship between spectral type and temperature, temperature can also be used on the horizontal axis. *On the H-R diagram, most stars plot in a diagonal band that is called the* **main sequence.** The main-sequence stars are the ones whose absolute magnitude can be determined from their spectral type, and these are the stars whose distance can be found from the spectral type–absolute magnitude method described previously. The main-sequence stars can be thought of as the ordinary stars consisting mainly of hydrogen, although there are tremendous differences among them. The stars that do not plot on the main sequence are even more interesting.

The stars that lie above the main sequence radiate more energy than main-sequence stars of the same spectral type or temperature. As noted, the diameter of a star can be determined from its temperature and absolute magnitude, or rate of energy production. The diameters of the stars above the main sequence are very much larger than the diameters of stars with similar spectral types on the main sequence. Such **giant stars** have K and M type spectra and so are reddish in color; hence, they are sometimes called red giants. Some giant stars are brighter than type O stars and are called **supergiants.**

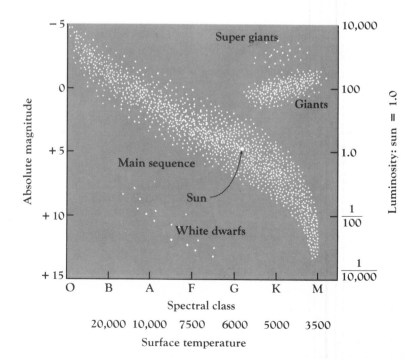

Figure 18-17
The Hertzsprung-Russell (H-R) diagram.

Some stars lie below the main-sequence stars on the H-R diagram. They radiate less energy than stars of the same spectral type on the main sequence. Such stars are called **dwarfs.** Most are neither red nor blue in color, and so they appear white and are sometimes called white dwarfs.

Very few stars are giants; most are main-sequence stars, and there are many more small main-sequence stars than large ones. Many of the bright stars we see at night, however, are giants or large main-sequence stars. Dwarf stars are probably numerous also, but because of their dimness they are difficult to see.

The H-R diagram enabled astronomers to distinguish several different groups of stars—main-sequence, giant, and dwarf. The next step in understanding stars came in the 1920s. Arthur Eddington, an Englishman, assumed that all main-sequence stars produced their energy in the same way, and so the only differences among them were in their sizes and masses. His approach was a fruitful one of applying basic physics to explain how internal processes in a star could produce the physical properties observable with telescopes. We now realize that main-sequence stars are converting hydrogen into helium and that they remain stable for relatively long times during this process. The rate of energy production in large stars is very much faster than that in small ones, so such stars remain on the main sequence a much shorter time. This is one reason for the small number of very large main-sequence stars. Giants and dwarfs produce their energy by different processes.

Internal Processes

A star begins as a condensing cloud of dust and gas. Gravitational attraction causes contraction. The contraction is a compression of the gas, so the gas is heated. Because of the star's great mass, the heating is great, especially at the core. The heating and compression bring the hydrogen atoms in such close contact that nuclear reactions begin in the core. The nuclear reactions produce energy, and the energy must be radiated out from the core. A **stable star** *is one in which the pressure of the outward-moving energy balances the gravitational contraction.*

In main-sequence stars, as we noted, hydrogen is converted into helium. When the hydrogen in the core of a main-sequence star is depleted, these reactions cease, and different nuclear reactions begin in which helium is converted into heavier elements. When the hydrogen-to-helium process stops, the star becomes unstable and contracts gravitationally because, without energy production in the core, there is no outward radiation pressure. The contraction compresses the core, raising the temperature there until the helium reactions begin. These reactions release great amounts of energy, and the resulting radiation causes the surface of the star to expand, forming a giant star. The helium reactions release less energy than the hydrogen reactions did, so the giant phase does not last so long as the main-sequence phase did. After the helium runs out, further contraction starts nuclear reactions that use somewhat heavier elements to form even heavier elements.

Life Cycles of Stars

We now have enough background to enable us to combine observation and theory into a glimpse into the private lives of stars. As we do so, we will discover many types of stars that may seem odd or even bizarre to us.

A **nebula**—a cloud of dust and/or gas—contracts, and a star is born when the temperature in the compressing core becomes high enough for nuclear reactions to begin. The observational evidence for this process is that the youngest stars are to the right of the main sequence on the H-R diagram at the cooler, smaller end of the plot. Some of these stars, called **cocoon stars,** have just formed and are still surrounded by dust; very soon after formation, the radiation drives the dust away. If the dust cloud is too small, the contraction may form a planet-like object or a black dwarf. As the nebula condenses, it becomes warmer and so moves toward the main sequence. The smaller the star or nebula, the longer it takes to reach the main sequence (and the longer it stays on the main sequence).

The star then moves onto the main sequence, where it stays most of its life. The rate of energy production is much greater in the large, hot, main-sequence stars, so there are fewer of them. A star the size of our sun will remain on the main sequence for about 9000 million (9 billion) years. The sun is estimated to be about 4500 to 5000 million (4.5 to 5 billion) years old, so there is no need for worry. The fact that there are any massive main-sequence stars shows that star formation is an ongoing process. An H-R diagram drawn for a group of stars believed to have all formed at the same time also shows many fewer massive stars than smaller, cooler stars. Eventually, a main-sequence star uses up its available hydrogen and moves off the main sequence.

On moving off the main sequence, a few of the smaller stars, less than 0.4 of the mass of the sun, simply contract and become white dwarfs. Most stars, however, contract until their cores become hot enough to begin transforming helium into heavier elements. This process creates energy so fast that the star must expand to get enough surface area to radiate the energy. In this way a giant star is formed. It moves to the right on the H-R diagram, and in the case of smaller stars it also moves upward. Stars with masses more than six times that of the sun become supergiants because they are the size of giants while still on the main sequence. Giant and supergiant stars use up their helium rapidly, so this phase does not last so long as does the main sequence. At the end of the giant and supergiant phase, the stars become unstable, and their behavior depends on their masses.

All the stars that were smaller than type O stars when they were on the main sequence become **variable stars** when they leave the main sequence. Close to 23,000 variable stars are known. Some have regular periods such as the Cepheid variables mentioned earlier, and many more are irregular. Apparently they become variable stars because, after their helium is consumed, they contract, causing the core to heat until the temperature is high enough for nuclear reactions to begin again. During contraction, they are small and radiate less energy; and when nuclear reactions begin, they expand and radiate more energy. These pulsations are slower for the more massive stars.

At the end of the variable phase, even more spectacular events may occur. A single star that was type G while on the main sequence will expand into a red giant. Later, as helium is used, it will expand rapidly and shed as much as half of its mass to form a nebula, or dust cloud. The core will then contract and become a white dwarf. If, however, this star was part of a binary pair of stars, it may become a **nova** (new star). *A nova is a star that becomes very much brighter for a short time.* Typically a hot but faint star becomes up to 12 magnitudes (60,000 times) brighter in a period of a few days and then, at first, declines in brightness almost as rapidly, although the total decline may take a few years. All that remains of the star is a rapidly expanding gaseous nebula. Apparently what happens is that, in a binary pair, the larger star evolves more rapidly and becomes a white dwarf. Later the less massive star may expand into a giant star, so that some of its material spills onto the dwarf, causing the dwarf to become unstable; an explosion results. Thirty to fifty nova explosions are estimated to occur in our galaxy each year.

The more massive stars die even more spectacular deaths. **Supernova** explosions result when stars with masses more than 30 times that of the sun collapse because energy production slows in their cores (Figure 18-18A). The collapsing outer layers prevent the energy F 18-18 that is still radiating from the core from escaping. The trapped energy eventually builds up, causing a massive explosion. In some cases, the core then becomes a white dwarf, but the most massive stars become more bizarre types of star. Supernovae are the most energetic stellar explosions known. They radiate more energy in a year than our sun radiates in 1000 million years. About 14 have been recorded in our galaxy in the last 2000 years. One was recorded by Chinese astronomers in 1054 in the region of the Crab nebula (Figure 18-18B).

Nova and supernova explosions are very important to us. They

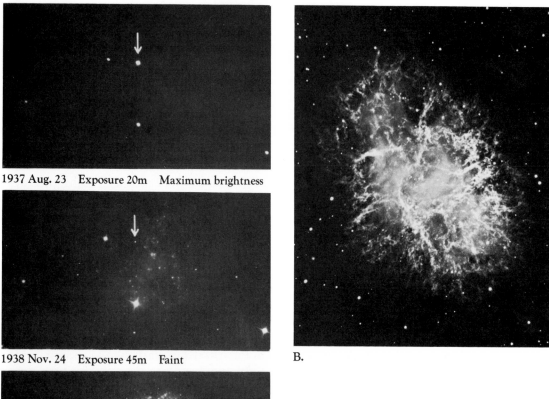

1937 Aug. 23   Exposure 20m   Maximum brightness

1938 Nov. 24   Exposure 45m   Faint

1942 Jan. 19   Exposure 85m   Too faint to observe

A.

B.

Figure 18-18
A. Distant supernova. Note the differences in exposure times in the three photos. B. The Crab nebula. This was formed by a supernova explosion in 1054 A.D. (Photo **A** from Mt. Wilson Observatory, California Institute of Technology; **B** from Lick Observatory.)

return heavy elements to interstellar space, and it is from nebulae, or dust clouds, containing these elements that our earth and solar system are made. Without these explosions the heavy elements would be trapped within stars. The elements of which earth is made were manufactured billions of years ago in stars that no longer exist. We are truly children of the stars.

White dwarfs are the final stages in the evolution of many stars. They are planet-sized, very dense stars. Because they are not bright, they are hard to observe, so only about 200, all relatively nearby, are known. Gravitational attraction pulls their atoms very close together, and the pressure is great enough that the electrons have been stripped from the atoms. Most of the weight of an atom is in its nucleus, so white dwarfs, composed of close-together atomic nuclei held apart by a sea of electrons, are very dense. A cubic inch of white-dwarf material would weigh a ton.

The most massive stars, those that are type O while on the main sequence, become supergiants, then supernovae, then they too collapse. Because they have much more mass, the collapse is more complete than in those stars that become white dwarfs. The pressure is so intense that the nuclei also collapse. The nucleus of an atom is composed of protons with a positive charge and neutrons with no charge. The electrons, which surround the nucleus, have a negative charge and are equal to the protons in number. When the nucleus collapses, the protons and electrons combine to form neutrons, so these stars become **neutron stars.** This is a very dense form of matter, and a cubic inch would weigh hundreds of millions of tons. A neutron star would have the mass of the sun but would be the size of a small asteroid.

Most neutron stars have not yet been seen by astronomers, but a few have been seen. In 1967 astronomers using radio telescopes found radio sources that radiated strong pulses of radio waves at approximately one-second intervals. At first, they thought that they were receiving a message from another galaxy, but as more sources were discovered, they realized that they were observing a natural phenomenon. These objects were named **pulsars,** and they are believed to be neutron stars. If a neutron star is spinning rapidly, its magnetic field could produce radio waves that would be beamed into space in much the same way as a lighthouse beams its beacon light into the night. The center of the Crab nebula is a visible pulsar, so pulsars and supernovae appear to be related.

What might happen if a very massive star blew up in a supernova explosion and then collapsed? If the star were big enough, a **black hole** might be formed. A black hole would be caused by a star so dense that nothing, not even radiation, could escape from its gravity field. Astronomers have shown the existence of black holes, and these black holes surely have been an inspiration to writers of science fiction.

DISCOVERY OF GALAXIES

The stars are not uniformly distributed through the universe. On a clear, dark night, the Milky Way, composed of seemingly countless stars, can be seen as a band of light across the sky. Beginning with Galileo, who first used the telescope, astronomers have easily resolved the Milky Way's stars. By the 18th century it was recognized that the **Milky Way** is a disk-shaped **galaxy,** that we are looking into an edge,

Figure 18-19
The North American nebula in the Milky Way. Both dark and light nebulae are seen. (Photo from Palomar Observatory, California Institute of Technology.)

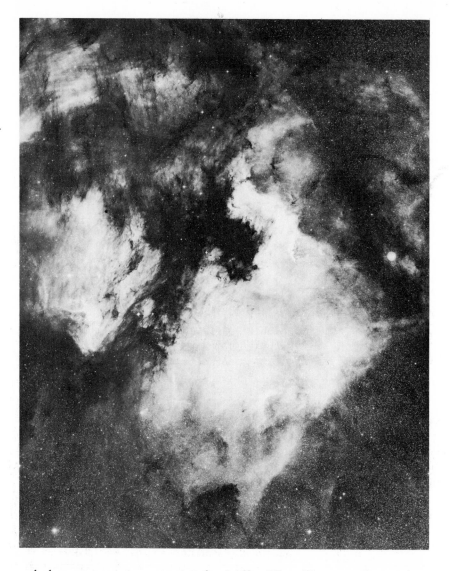

and that our sun is a star in the Milky Way. Photographs, such as Figure 18-19, show that parts of the Milky Way are obscured by dark dust clouds; other parts are luminous clouds. This clearly shows that dust is present in at least parts of the Milky Way, obscuring stars and causing problems in deducing the nature of the Milky Way from within.

In the early part of the 20th century, the methods described for finding the distance to stars were applied to groups of stars within the Milky Way, and a picture of its size and nature emerged. As we realized that our sun is a part of a galaxy, the question arose whether or not the Milky Way galaxy is unique. The term *island universe* had already been applied to galaxies by the philosopher Immanuel Kant in 1755. The proof of other galaxies had to wait until 1924.

From the 18th century on, astronomers found patches of fuzzy light as well as the pinpoints of light from stars. These fuzzy objects were carefully cataloged so that they would not be mistaken for newly appeared comets. As telescopes improved in the middle of the 19th century, some of these objects were seen to have a pinwheel, or spiral, shape. Many astronomers felt that they were galaxies like our own, but others believed that they were nebulae in our own galaxy. The debate went on until the completion in 1924 of the first of the big telescopes, the 100-inch (2.5-meter) telescope on Mount Wilson in southern California. Using this new telescope, Edwin Hubble was able to resolve individual stars in the spiral Andromeda galaxy (Figure 18-20) and determine its distance from its faint Cepheid vari- **F 18-20** ables. We had finally moved out of our own galaxy into the universe.

STAR CLUSTERS

Star clusters are much smaller groupings than galaxies, but their study was very important in understanding galaxies. Some clusters, such as the Pleiades (Seven Sisters) are visible to the naked eye. Serious study of clusters had to wait until the 1920s, when both large telescopes and the methods of measuring distance and other properties of stars previously described became available. Two types of clusters were recognized, open clusters and globular clusters, each with distinct characteristics and star types.

Figure 18-20
The Andromeda galaxy. (Photo from Lick Observatory.)

Open clusters are irregular, loose groupings, generally of 100 to 1000 stars. About 900 of them have been found in the plane of the Milky Way. They are composed mainly of young, hot, blue stars, with some heavy elements and dust.

Globular clusters are different in every aspect. They are very close groupings of 20,000 to several million stars. About 120 are known, although based on studies of other galaxies, there are probably about a thousand. They form a more or less spherical halo that extends above and below the center of the Milky Way. It was by mapping the positions of the globular clusters that Harlow Shaply showed that our sun is far from the center of the Milky Way. The stars are old, generally red, and composed of only hydrogen and helium, with very little of the heavier elements. The stars in globular clusters are very tightly packed. The nearest star to our sun is Alpha Centauri, about 4.3 light years distant. If 10,000 stars surrounded the sun within a radius of this distance, the packing would be similar to that in a globular cluster. Such tight packing apparently affects stellar processes, because globular clusters are sources of X-ray radiation.

The age of a cluster is determined by plotting its H-R diagram. By knowing the approximate time required for stars to move off the main sequence, astronomers can estimate the ages of clusters. Open clusters range in age from a few million years to 10,000 million years. The globular clusters average 12,500 ± 3000 million years old and are made of the oldest known stars.

Figure 18-21
The Whirlpool galaxy shows nebulae in the spiral arms. (Photo from Lick Observatory.)

Figure 18-22
Population I and II stars in the Andromeda galaxy are similar to those in the Milky Way. (Photo from Palomar Observatory, California Institute of Technology.)

Apparently the globular clusters formed about the same time as the galaxy from nebulae composed mainly of hydrogen and perhaps helium. The open clusters contain a small percentage of heavy elements and are associated with dust clouds. Apparently these much younger stars formed from dust clouds that originated from nova explosions; the earlier stars that exploded had formed the heavier elements.

Mapping of the locations of the open clusters reveals that they are in the spiral arms of the Milky Way. Other galaxies have spiral arms (Figure 18-21), and dust clouds are prominent in the arms, all of which leads to the idea that the Milky Way is a spiral galaxy similar to many others visible with large telescopes. **F 18–21**

The young, hot stars with heavy elements and associated with dust clouds are called **population I stars.** They are generally found in open clusters in the arms of the Milky Way. **Population II stars** are older, generally cooler stars, composed mainly of hydrogen and helium. Their homes are in the globular clusters that surround the center of the galaxy (Figure 18-22). **F 18–22**

INTERSTELLAR MATERIAL

Spiral galaxies are composed of both stars and *clouds of dust and gas called* **nebulae** (singular **nebula**), the Latin word for *cloud*. The dust clouds clearly show in photographs of spiral galaxies (Figure 18-23), **F 18–23** especially in the arms where stars are forming. Two types of nebula are obvious in the photos, **dark nebulae** and **bright nebulae.** In gen-

Figure 18-23
The dust clouds, or nebulae, in the spiral arms of this galaxy in the constallation Virgo show very clearly. (Photo from Palomar Observatory, California Institute of Technology.)

eral, there is about one dust particle for every 1000 hydrogen atoms in a nebula and only about one hydrogen atom in every cubic centimeter; thus, the density of a nebula is less than that of even an excellent vacuum here on earth. In spite of this low density, it is mainly the few dust particles that obscure the starlight in a dark nebula. Nebulae are generally quite large, and, even with their low densities, their total mass may be that of a star.

Bright nebulae can form in two ways. **Reflection nebulae** are dust clouds that simply reflect light from a nearby star, and the spectrum of a reflection nebula is similar to that of the star whose light is reflected. In the second case, a nebula near a star, especially a hot, bright star, may fluoresce. The gas absorbs some of the star's energy and reradiates it at a different wavelength. The process is similar to what happens in a fluorescent light or a television tube, in that radiant energy hitting the fluorescent material causes it to emit light energy. The spectrum of a **fluorescent nebula** is recognized by its bright-line, or emission, spectrum.

Much smaller amounts of dust and gas in space scatter the light from distant stars, making a star appear redder than it is. This effect is most obvious when we look in the direction of the disk of the Milky Way, and it limits our observation.

MILKY WAY GALAXY

To us, the Milky Way galaxy is huge. The disk is about 100,000 light years (30,000 parsecs) in diameter and between 650 and 1300 light years (200 and 400 parsecs) in thickness; the center is a bulge about 10,000 light years (3000 parsecs) thick (Figure 18-24). The Milky Way F 18-2

Figure 18-24
The Milky Way
galaxy.

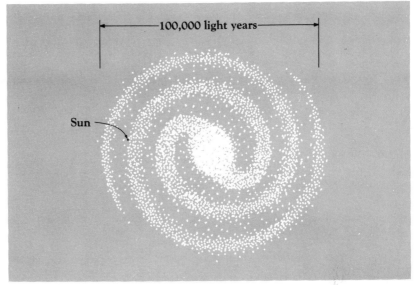

Top view

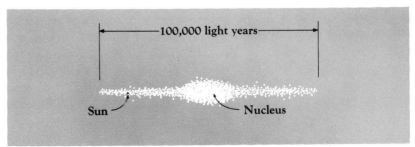

Side view

contains about 100,000 million (100 billion) stars. Our sun is at the edge of one of the spiral arms and is estimated to be between 26,000 and 32,000 light years (8000 and 10,000 parsecs) from the center of the galaxy. These statistics were very hard to obtain, primarily because the dust clouds, or nebulae, in our part of the disk obscure our view, especially toward the center of the galaxy. In many respects it is easier to measure a distant galaxy. The disk shape, the size, the sun's location, and the spiral arms were all determined by mapping star clusters. The Milky Way galaxy rotates around the center with a period of 230 million years, determined by Doppler shifts of spectra.

The *galactic nucleus* is perhaps the least known or understood part of the Milky Way. We cannot view it because of nebulae and dust, but we can observe it at infrared and radio frequencies because it is a very strong energy source of both. We can see that the nuclei of other galaxies, such as Andromeda, are much brighter than the other parts of the galaxies, suggesting that the conditions in the nucleus are different. It is estimated that the very center, a region 32 light years (10

Figure 18-25
The Local Group of
galaxies.

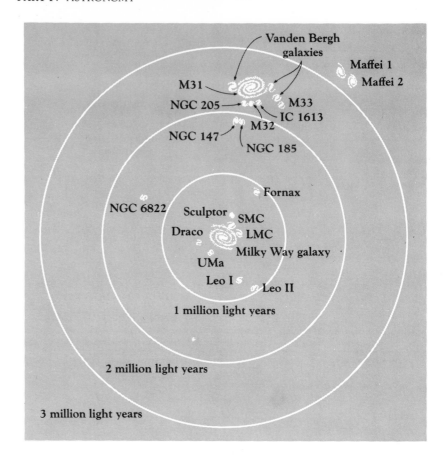

Figure 18-25
The Local Group of galaxies.

parsecs) in diameter, contains about 60 million stars. Perhaps some of the stars there are very large or otherwise unstable, and explosions may be the source of the great energy of galactic nuclei.

GALAXIES

Galaxies occur in groups called **clusters.** The Milky Way is one of a group of 26 nearby galaxies called the **Local Group,** which extends outward from us over 3 million light years (1 million parsecs) (Figure 18-25). F 18-25 Groups of this size are hard to comprehend, but clusters of galaxies also occur in groups called **superclusters.** Why galaxies form in clusters is not known.

Galaxies have been divided into ellipticals, spirals, and irregulars; and each of these types is further subdivided as shown in Figure 18-26. This diagram shows only a classification of galaxies and is not F 18-26 meant to show their evolution. Approximately 70 per cent of all known galaxies are **ellipticals,** which range from almost circular to quite elongate. The flattening may only be apparent, because it may depend on our angle of view. **Spiral galaxies** are about 15 per cent of all galaxies. There are two types, *barred spirals* and *normal spirals,* further subdivided by how tightly the spiral is wound. They have bright central regions like that of our own Milky Way, a spiral. Andromeda, also a member of the Local Group, is a spiral galaxy believed to be very similar to the Milky Way. **Irregular galaxies,** also about 15 per cent of all galaxies, show no regularities at all.

There is no way to tell the ages of galaxies. All have old population II stars, so none of them can be young. Also, there are no masses of gas and dust big enough to form a galaxy, so it appears that none is forming presently.

Differences among galaxies can be observed. Following the types, shown in Figure 18-26, from elliptical, through spiral, to irregular galaxies, the amount of gas increases. The elliptical galaxies have very little gas, so no stars are forming in them. Stars are still forming from the gas in the arms of the spiral galaxies and in the irregular galaxies; thus, stars are younger and larger as we go from elliptical to spiral to irregular galaxies.

RED SHIFT

Most galaxies are too far away for us to record the spectrum of a single star, so we record the spectrum of the whole galaxy. By the late 1920s, enough of these spectra had been recorded to show that all were shifted toward the red **(red shift).** The Doppler effect is such that, if an emitter is moving away from the observer, the spectral lines are shifted toward the red end. Thus, all the galaxies appear to be going away from us; also, the farther away a galaxy is, the faster it is moving from us. Once it was realized that the relationship held for all galaxies whose distance was known, the amount of the red shift was used to determine the distance to a galaxy.

This relationship has some very important consequences. That all of the galaxies are moving away from us does not mean that we are at the center of the universe. Imagine a balloon with many spots on it or a cake batter with many raisins. As the balloon is blown up, or as the cake rises in the oven, each of the spots or raisins is moving away

Figure 18-26
Types of galaxies.
(After Hubble.)

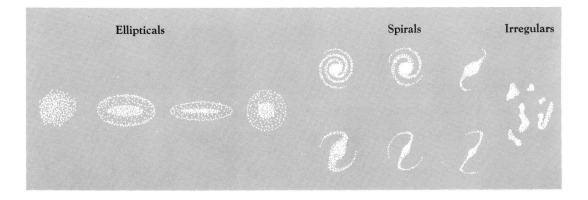

Ellipticals   Spirals   Irregulars

from every other one. Thus, no matter where we are in the universe, all of the other galaxies would appear to be receding from us.

This relationship does imply that the universe is expanding. The fact that the galaxies are accelerating suggests the possibility that at one time they were all together and an explosive event started them all accelerating from that place. This has been called the **big bang** theory and is one of the theories of the formation of the universe. If such an explosion did take place, some of the radiation created by the explosion should still be detectable. Theoretical calculations predict that radiation should be found in all directions in the sky and that it should be similar to radiation from a body at a temperature a few degrees above absolute zero. The predicted radiation has been found—very strong evidence in favor of the big bang theory.

Knowing the velocities and accelerations of galaxies, it is possible to project back in time to when the big bang should have occurred. Such a calculation gives an age of the universe of between 16,000 and 19,000 million (16 and 19 billion) years, certainly a possible age. The oldest rocks on earth are about 4000 million years old, meteorites are about 5000 million years old, and globular clusters are estimated to be 12,500 ± 3000 million years old.

Because of these immense implications, the red shift has not gone unchallenged. If the source of the light is a region of very high gravity, such as might be found near a neutron star or a black hole, Einstein's theory of relativity predicts that the light would be shifted toward the red. That does not seem to be the explanation in this case, because careful study of galaxies shows that the red shift correlates with distance and not at all with mass, size, or type of galaxy. Perhaps a reason for the red shift other than velocity will be discovered, but presently we know of none.

QUASARS

Radio telescopes are also used to study galaxies, and many very energetic radio sources have been found. Big optical telescopes were pointed at some of these radio sources, and at a few of them faint points of light were found. The enigmatic objects are called **quasars** (for quasi-stellar radio source). Over 100 have been found, and more are being discovered constantly.

The spectra from quasars were obtained, and they resemble no other spectra from any astronomical object. In 1963, Maarten Schmidt discovered that the spectrum of the quasar he was studying was that of hydrogen with an extreme red shift (Doppler effect), implying that the quasar is traveling away from us at a very high speed. According to the red shift of galaxies, this high velocity implies that quasars are very distant—the most distant objects so far found by astronomers.

The amount of energy radiated by quasars in optical and radio wavelengths indicates that they must be very massive objects. The amount of radiation varies over periods of a few months to about a year, suggesting that these objects are not very large in diameter because an object whose radiation varies over a period of a year cannot be larger than one light year in diameter. If it were larger, it would take more than a year for the change to move across the object.

If the conventional interpretation of red shift is used, quasars are very distant, very energetic, small objects. To be observed at their great distances, they must be up to 10,000 times brighter than ordinary galaxies and hundreds of times stronger than any other radio galaxy—all this and they must be small, too. Perhaps they are composed of supermassive stars that are unstable and produce supernovae; the resulting dust is recycled into new, massive stars. No process that can produce this much energy in a small volume is known.

Most astronomers believe that quasars are very distant objects. A few argue that they are really much closer and are receding from us at great speed. Evidence supporting the quasars' being at the edge of the observable universe was reported in 1980. According to Einstein's equations, gravity can bend light, just as a prism would. A quasar was found to show multiple images because of the gravity of a massive galaxy lying between the quasar and ourselves. Thus, the quasar is more distant than the galaxy. The light from such remote objects has been traveling toward us for billions of years. We are seeing what conditions were like, perhaps, before the sun formed. Perhaps quasars are the process by which galaxies form. They are the frontier in the study of the universe.

SUMMARY

- The sun is 1,390,000 kilometers (864,000 miles) in diameter and is at an average distance of 150 million kilometers (93 million miles) from the earth. The sun is composed of gas, mainly hydrogen and helium, and its equator rotates slightly faster than its polar regions. The period of rotation is about 27 days. The sun's energy is produced in its core by nuclear reactions.

- The layer of the sun that we see is the *photosphere*. It is a thin layer with a temperature of 5400°C (9800°F). Convection causes *granules* and *supergranules*, patterns that only last a few minutes. *Spicules* are eruptions of incandescent gas associated with supergranules.

- Above the photosphere is the *chromosphere*, a pink-glowing layer visible only during an eclipse.

- Above the chromosphere is the *corona*, a very low-density layer that extends a few million kilometers but is only visible with instruments.

- *Sunspots* are dark, slightly cooler areas on the sun's surface. Sunspots appear in cycles with maxima every 11 years, but the complete cycle requires about 22 years. *Prominences* are large flows of gas associated with sunspots. *Flares* are more violent gas flows and are associated with large sunspots.

- *Solar wind* is the material ejected from sunspots. These high-energy particles interact with the earth's magnetic field to form the *Van Allen radiation belts*. The particles eventually move toward the earth's magnetic poles where they collide with molecules in the atmosphere, causing the *aurorae*

- *Parallax* is the most direct means of measuring the distance to a star. It uses a right triangle with the radius of the earth's orbit as its base. Because very small angles are measured, the method is useful for only the close stars—those within about 100 parsecs. A star's distance in *parsecs* is 1 divided by the angle measured in seconds of arc (parsecs = 1/angle). One parsec is 3.26 light years.

- The *luminosity* of a star is a measure of how much energy it radiates.

- The *apparent magnitude* of a star is its magnitude as seen from earth. The *absolute magnitude* is what the magnitude would be if the star moved to a distance of 10 parsecs (32.6 light years). If the apparent and the absolute magnitudes of a star are known, its distance can be calculated, because the brightness of light varies inversely with the square of the distance.

- Stars are classified according to their *spectral types*.

- The *H-R (Hertzsprung-Russell) diagram* is a plot of spectral type versus luminosity or absolute magnitude. It can only be made for stars close enough for us to know their distance (from parallax) and, therefore, their absolute magnitude. Once the relationship between absolute magnitude and spectral type was established from the H-R diagram, it became possible to find the absolute magnitude of a distant star from its spectral type. In this way, distance to distant stars can be determined.

- Spectral studies can also reveal the composition of the gases surrounding a star, the temperature, the magnetic field, and the rate of rotation of a star.

- Widely spaced optical or radio telescopes can be used to measure the diameter of a few large stars.

- *Binary stars* are two stars that revolve around a mutual center of mass. Their study can reveal their masses.

- *Cepheid variable stars* are stars whose luminosity varies with a regular period.

- On the H-R diagram, most stars plot in a diagonal band that is called the *main sequence*.

- *Giant stars* are those that plot above the main sequence.

- *Dwarf stars* are those that plot below the main sequence.

- The main-sequence stars produce energy by converting hydrogen into helium. The larger, hotter stars use their hydrogen at a much higher rate than the smaller stars.

- Stars form from dust and gas clouds called *nebulae*. A nebula compresses due to gravity, and, when the pressure in the core is great enough to cause nuclear reactions, a main-sequence star is born.

- When a star has used its available hydrogen, a new nuclear process in which helium is converted into heavier elements begins, and the star moves off the main sequence. At first most stars contract, and then, as their cores become hotter, they expand and become giant stars. Giant stars use their helium rapidly, so this phase does not last as long as the main-sequence phase.

- When the helium is used, most stars become *variable stars*.

- The variable phase ends spectacularly. Average-size stars shed much of their masses, forming nebulae, and the core collapses into a *white dwarf*. Binary stars become *novas* as material from one star falls into the other, causing an explosion. A nova is a star that becomes much brighter for a short time. The more massive stars become *supernovas*. These processes are very important because these are the ways in which heavy elements are made available for the formation of new stars or planets.

- White dwarfs are the final stages of many stars. They are planet-sized stars made of atoms from which the electrons have been stripped by the high pressure.

- When the more massive stars collapse, the nuclei of the atoms also collapse, and *neutron stars* are the result. They are extremely dense matter. The collapse of a very large star may produce a *black hole,* a star so dense that not even radiation can escape its gravity field.

- *Pulsars* are rapidly spinning neutron stars that emit pulses of radio energy.

- *Galaxies* are large groups of stars held together by gravity.

- *Star clusters* are much smaller groups than galaxies.

- *Open clusters* are irregular, loose groupings of 100 to 1000 stars. The stars are young, blue, hot stars with some heavy elements. Stars of this type are called *population I stars,* and they are generally associated with dust clouds.

- *Globular clusters* are groups of 20,000 to several million old, red stars composed of hydrogen and helium. These stars are called *population II stars.*

- Interstellar space contains very small amounts of gas and dust.

- Nebulae may be either bright or dark. Bright nebulae may be lighted by reflection or fluorescence from a close star.

- The Milky Way galaxy is about 100,000 light years (30,000 parsecs) in diameter and contains about 100,000 million (100 billion) stars. The galactic nucleus is the central thicker part that is obscured by dust and nebulae.

- Galaxies occur in groups called *clusters.* The Milky Way is part of the *Local Group* of 26 galaxies.

- Most galaxies have elliptical shapes; some, like the Milky Way, are spirals, and some are irregulars.

- New stars are forming from nebulae in the spiral arms.

- Most galaxies are moving away from us, and the farther away they are, the faster they are moving. This is revealed by the shift toward the red of their spectra, the *red shift.*

- The red shift suggests that the universe may have been created in a great explosion called the *big bang.* The radiation that would be created in such an event has been detected.

- *Quasars* are faint stars that emit radio energy. Their red shifts are very large, suggesting that they are moving at very high speeds and so probably are very distant. If so, they emit more energy than any known source.

KEY TERMS

chemical reaction	second of arc	giant star	population I stars
nuclear reaction	parsec	supergiant	population II stars
photosphere	luminosity	dwarf star	dark nebula
granules	magnitude	stable star	bright nebula
supergranules	apparent magnitude	nebula (nebulae)	reflection nebula
spicule	absolute magnitude	cocoon star	fluorescent nebula
chromosphere	inverse square law	variable star	galaxy cluster
corona	spectral types of stars	nova	Local Group
sunspot	Hertzsprung-Russell,	supernova	supercluster of
prominence	or H-R, diagram	neutron star	galaxies
flare	binary stars	pulsar	elliptical galaxy
solar wind	visual binary stars	black hole	spiral galaxy
Van Allen radiation	eclipsing binary stars	Milky Way	irregular galaxy
belts	spectrographic binary	galaxy	red shift
aurora borealis,	stars	star cluster	big bang
aurora australis	cepheid variable star	open star cluster	quasar
parallax	main sequence	globular star cluster	

REVIEW QUESTIONS

1. What are the principal elements in the sun? Which element is the most abundant?

2. What is the source of the sun's energy? What is this kind of reaction called? What is the principal way in which this reaction differs from the type of reaction that occurs when coal is burned? Where in the sun is the energy produced?

3. Describe the three layers of the sun's atmosphere. Under what conditions is each layer seen from earth?

4. Why do sunspots appear dark? Describe an 11-year sunspot cycle. In what way can the cycle be said to run 22 years?

5. (a) What is the source of the solar wind?
   (b) How are the Van Allen radiation belts created?
   (c) What are the aurorae? What causes them?

6. (a) Explain how parallax is used to measure the distance of stars.
   (b) Define parsec.

7. (a) Define luminosity.
   (b) Is a star of magnitude 1 dimmer or brighter than a star of magnitude 5? How much dimmer or brighter?
   (c) What is the difference between the apparent magnitude of a star and its absolute magnitude? Which of these magnitudes measures a star's luminosity?
   (d) What two quantities must be known in order to calculate a star's absolute magnitude? Explain why the inverse square law is used in calculating absolute magnitude.

8. (a) On what is a star's spectral type based?
   (b) What is the relationship between spectral type and temperature?
   (c) Explain how a star's composition, temperature, magnetic field, and rotation are determined from the star's spectrum.

9. What relationship does the H-R diagram show? How is the diagram used to find the luminosity of a star whose distance is too great to be measured by parallax?

10. Explain why a star's diameter can be calculated from its temperature and absolute magnitude.

11. (a) How do astronomers determine the total mass of a pair of binary stars?
    (b) Which star of the binary pair has the greater mass—the one closer to the center of gravity or the one farther away?
    (c) How are astronomers able to estimate the mass of a star that does not belong to a binary pair?

12. Why are cepheid variable stars such handy tools for astronomers?

13. Review the H-R diagram and answer the questions about each of the following stars:
    (a) *Main-sequence stars.* Describe the band that they form on the diagram.
    (b) *Large main-sequence stars.* Where in the band are most of these stars found? How do their rate of energy production and their longevity compare with the smaller main-sequence stars?
    (c) *Giants.* In what general area of the diagram do they lie? Are their diameters larger or smaller than main-sequence stars of the same spectral class? Do they radiate more or less energy than main-sequence stars of the same class?
    (d) *Supergiants.* Compare their position on the diagram, diameter, and amount of radiation with those of giants.
    (e) *Dwarfs.* In what area of the diagram do they lie? Are their diameters larger or smaller than those of main-sequence stars of the same spectral type? Do they radiate more or less energy than main-sequence stars of the same spectral type?

14. (a) How is a star born?
    (b) Define a stable star.

15. What is the source of energy for a main-sequence star? What is the source of energy for a giant star?

16. When a star collapses, it may become a white dwarf, a neutron star, or possibly a black hole.

(a) What happens to the atoms in a white dwarf?

(b) What happens to the atoms in a neutron star?

(c) Compare the size and density of a white dwarf with the size and density of a neutron star.

(d) What is a black hole?

17. Describe the history of each of the following stars from the time they leave the main sequence until their collapse is completed:

(a) G type on main sequence.

(b) G type on main sequence that is part of a binary pair.

(c) O type on main sequence.

18. In what way are we children of the stars?

19. Answer these questions for both a. and b.: Describe the shape and location of the cluster. State the relative age, relative temperature, and composition of the typical star in the cluster. Is dust associated with the cluster? Describe the history of the cluster.

(a) open cluster

(b) globular cluster

20. (a) Describe the composition and density of a nebula. How large might a nebula's mass be? What does this say about the size of a typical nebula?

(b) What makes the difference between a dark and a bright nebula?

(c) What causes a nebula to reflect?

(d) What causes a nebula to fluoresce?

(e) Why is it easier to measure other galaxies than to measure our own?

21. Make a simple sketch of the Milky Way. Place an X approximately where our solar system is. What kind of galaxy is the Milky Way?

22. To obtain a sense of the immensity of the universe, think about the following: How many galaxies, in addition to our Milky Way, belong to the Local Group? How many light years does the Local Group extend out from us? What is the speed of light? Do other clusters of galaxies exist? What is a supercluster?

23. What evidence indicates that no new stars are forming in elliptical galaxies? Are new stars forming in spiral galaxies? If so, where? Are new stars forming in irregular galaxies?

24. What evidence indicates that the universe is expanding? Describe the big bang theory. Calculations based on that theory indicate that the universe is how old? How old is our solar system?

25. In what ways is the term *quasar* descriptive of certain objects? Why are quasars believed to be very distant, very energetic, and small? Why are scientists puzzled by their small size?

SUGGESTED READINGS

Bok, B. J. "The Milky Way Galaxy," *Scientific American* (March 1981), Vol. 244, No. 3, pp. 92–120.

Clark, George. "X-ray Stars in Globular Clusters," *Scientific American* (October 1977), Vol. 237, No. 4, pp. 42–51.

Eddy, J. A. "The Case of the Missing Sunspots," *Scientific American* (May 1977), Vol. 236, No. 5, pp. 80–96.

Gott, J. R., and J. E. Gunn. "Will the Universe Expand Forever?" *Scientific American* (March 1976), Vol. 234, No. 3, pp. 62–79.

Heiles, Carl. "The Structure of the Interstellar Medium," *Scientific American* (January 1978), Vol. 238, No. 1, pp. 74–84.

Hodge, P. W. "The Andromeda Galaxy," *Scientific American* (January 1981), Vol. 244, No. 1, pp. 92–101.

Kaufmann, W. J., III. *Stars and Nebulas*. San Francisco: W. H. Freeman, 1978, 204 pp. (paperback).

Kaufmann, W. J., III. *Galaxies and Quasars*. San Francisco: W. H. Freeman, 1979, 226 pp.

Kirshner, Robert. "Supernovas in other Galaxies," *Scientific American* (December 1976), Vol. 235, No. 6, pp. 88–101.

Stephenson, F. R., and David Clark. "Historical Supernovas," *Scientific American* (June 1976), Vol. 234, No. 6, pp. 100–107.

Williams, R. E. "The Shells of Novas," *Scientific American* (April 1981), Vol. 244, No. 4, pp. 120–131.

Wilson, O. C., A. H. Vaughan, and Dmitri Mihalas. "The Activity Cycles of Stars," *Scientific American* (February 1981), Vol. 244, No. 2, pp. 104–119.

Zeilik, Michael. "The Birth of Massive Stars," *Scientific American* (April 1978), Vol. 238, No. 4, pp. 110–118.

APPENDIXES

A. UNIT SYSTEMS

By international agreement, the *Système Internationale* (SI) was adopted by most scientists in 1960. In most places this text uses the International System and gives the equivalent in the more familiar English system. The International System has seven fundamental units:

Physical Quantity	Unit	Abbreviation
Length	meter	m
Mass	kilogram	kg
Time	second	s (or sec)
Electric current	ampere	A
Thermodynamic temperature	kelvin	K
Amount of substance	mole	mol
Luminous intensity	candela	cd

All of the other units can be derived from these seven fundamental units. For example, using Newton's first law,

$$F = ma$$
$$\text{Force} = \text{mass} \times \text{acceleration}$$
$$\text{Force} = \text{kilograms} \times \text{meters/seconds}^2$$

Thus the unit of force is kilogram meter/second^2. This is an awkward unit for a common physical quantity, so the *newton* was defined as 1 kilogram meter/second^2, and it is the unit commonly used for force in the SI.

Work and energy are force times distance, so they are kilogram meter^2/second^2 in fundamental SI units, or newton-meter in derived units. A *joule* is defined as a newton-meter, and this is the more common unit for energy or work.

Power is the rate at which work is done, and it is the number of joules per second. The *watt* is the common unit, and one watt is defined as one joule/second.

Pressure is force exerted on an area or newtons per square meter. The *pascal* is defined as one newton/meter^2.

The units used in the SI are not convenient for all uses, so they are commonly multiplied or divided by decimal units as follows:

Multiple	Prefix	Fraction	Prefix
10	deka-	$\frac{1}{10}$	deci-
100	hecto-	$\frac{1}{100}$	centi-
1000	kilo-	$\frac{1}{1000}$	milli-
1,000,000	mega-	$\frac{1}{1,000,000}$	micro-
		$\frac{1}{1,000,000,000}$	nano-

Notice that the fundamental unit of mass in the SI is the kilogram, meaning 1000 grams. The basic unit should be 1, not 1000. This is a logical inconsistency in the SI. The kilogram was used as a basic unit because its size is much more convenient than is the gram's.

CONVERSION FACTORS

Length

1 millimeter = 0.0394 inch
1 centimeter = 0.394 inch
1 meter = 39.4 inches
1 kilometer = 0.62 mile
1 inch = 2.54 centimeters
1 foot = 0.305 meter
1 mile = 1.61 kilometers
1 angstrom = 0.1 nanometer

The angstrom is a unit that has been in use a long time. Unfortunately, it is not one of the standard fractions of a meter defined above.

Volume

1 liter = 1000 cubic centimeters = 1/1000 cubic meter
1 liter = 1.06 quarts
1 quart = 946 milliliters

Mass

1 kilogram = 2.2 pounds
1 pound = 454 grams

Work and Energy

1 joule = 0.2390 calorie
1 calorie = 4.184 joules

The calorie used in thermodynamics is the gram-calorie. The calorie used in nutrition is the kilogram-calorie and is 1000 times larger.

Pressure

1 pascal = 1 newton/meter^2
1 atmosphere = 101,325 pascals
1 millibar = 100 pascals

The millibar is not a standard SI unit, but it is commonly used in reporting atmospheric pressure.

Temperature

$$\frac{F - 32}{180} = \frac{C}{100}$$

F = Fahrenheit
C = Celsius
K = Kelvin

$$C = \frac{5}{9}(F - 32)$$

$$F = \frac{9}{5}C + 32$$

$$K = C + 273$$

B. MINERAL IDENTIFICATION TABLE

ROCK-FORMING MINERALS—ARRANGED BY HARDNESS

A. Light-Colored Minerals

Mineral	Composition	Hardness	Luster	Color	Streak	Cleavage	
talc	$Mg_3(Si_4O_{10})OH_2$	1	pearly to dull	green, gray, white	white	1	feels greasy
gypsum	$CaSO_4 \cdot 2H_2O$	2	glassy to pearly	clear to white or gray	white	1 good 2 poor	
kaolinite	$Al_4Si_4O_{10}(OH)_8$	2	earthy	white	white	1 (only seen with microscope)	a clay
muscovite	$KAl_2(AlSi_3O_{10})OH_2$	2–2.5	glassy	colorless to light green	white	1	splits into thin flexible sheets
halite	$NaCl$	2.5	glassy	clear, white	white	3 at right angle	salty taste
calcite	$CaCO_3$	3	glassy to earthy	clear, white	white	3 not at right angle	reacts with dilute hydrochloric acid
anhydrite	$CaSO_4$	3–3.5	glassy to pearly	clear to blue or gray	white	3 at right angle	
dolomite	$CaMg(CO_3)_2$	3.5–4	glassy to pearly	clear, white, pink, or brownish	white	3 not at right angle	reacts with dilute hydrochloric acid when powdered
orthoclase	$KAlSi_3O_8$	6	glassy	white to pink	white	2 at right angle	
plagioclase	$NaAlSi_3O_8$ and $CaAl_2Si_2O_8$	6	glassy	white to light green	white	2 at right angle	a continuous mixture of sodic and calcic end members

Mineral	Composition	Hardness	Luster	Color	Streak	Cleavage	
perthite	——	6	glassy	white, pink, or light gray; irregular intergrowth	white	2 at right angle	an intergrowth of orthoclase and sodic plagioclase
olivine	$(MgFe)_2SiO_4$	6.5–7	glassy	olive green	white to gray	none	
quartz	SiO_2	7	glassy	clear, white, or any color	none	none	conchoidal fracture
chalcedony	SiO_2	7	waxy	may be any color	none	none	very fine-grained quartz

B. Dark-Colored Minerals

Mineral	Composition	Hardness	Luster	Color	Streak	Cleavage	
chlorite	$(Mg,Fe,Al)_6(Al,Si)_4O_{10}(OH)_8$	2.5	glassy to earthy	green	white to light green	1	splits into thin inelastic sheets
biotite	$K(Mg,Fe)_3(AlSi_3O_{10})(OH)_2$	2.5	glassy to pearly	dark green, brown, black	white to gray	1	splits into thin flexible sheets
serpentine	$Mg_6(Si_4O_{10})(OH)_8$	2–6	greasy-waxy	dark green	white	none	
apatite	$Ca_5(PO_4)_3(F,Cl,OH)$	5	glassy	green, brown, or blue	white	1 poor	
augite	$(Ca,Na)(Mg,Fe^{II},Fe^{III},Al)(Si,Al)_2O_6$	5–6	glassy	dark green to black	white to gray	2 at right angle	a pyroxene
hornblende	$Ca_2Na(Mg,Fe^{II})_4(Al,Fe^{III},Ti)(Al,Si)_8O_{22}(O,OH)_2$	5–6	glassy	black or dark green	white to gray	2 not at right angle	an amphibole
olivine	$(MgFe)_2SiO_4$	6.5–7	glassy	olive green	white to gray	none	

ROCK FORMING MINERALS

B. Dark-Colored Minerals (cont'd.)

Mineral	Composition	Hardness	Luster	Color	Streak	Cleavage	
garnet	—	6.5–7.5	glassy	red, red-brown	white or reddish	none	a mineral group; generally form crystals in metamorphic rocks
quartz	SiO_2	7	glassy	clear, white, or any color	none	none	conchoidal fracture
chalcedony	SiO_2	7	waxy	may be any color	none	none	very fine-grained quartz

OTHER COMMON MINERALS—ARRANGED BY HARDNESS

A. Nonmetallic Luster

Mineral	Composition	Hardness	Luster	Color	Streak	Cleavage	
graphite	C	1–2	metallic to earthy	black	black	1	greasy feel
sulfur	S	1.5–2.5	resinous	yellow	yellow	none	
sphalerite	ZnS	3.5–4	resinous	dark brown to yellow-brown	yellow to light brown	6	
limonite	hydrous iron oxide; not a mineral	5–5.5	earthy	yellow-brown to dark brown	yellow to brown	none	
opal	$SiO_2 \cdot nH_2O$ not a mineral	5–6	glassy	any	none	none	conchoidal fracture
hematite	Fe_2O_3	5–6.5	earthy	red to red-brown	red to red-brown	none	apparent hardness may be much lower

B. Minerals with Metallic Luster (or Submetallic Luster)

Mineral	Composition	Hardness	Luster	Color	Streak	Cleavage	
graphite	C	1–2	metallic to earthy	black	black	1	greasy feel
galena	PbS	2.5	metallic	lead gray	lead gray	3 at right angle	has high specific gravity
chalcopyrite	$CuFeS_2$	3.5–4	metallic	brass-yellow; commonly tarnished	greenish black	none	
magnetite	Fe_3O_4	6	metallic	black	black	none	attracted by a magnet
pyrite	FeS_2	6–6.5	metallic	brass-yellow	black	none	commonly in cubic crystals

C. TOPOGRAPHIC MAPS

Topographic maps use various methods to show the third, or vertical, dimension. Layer-tinted maps using various shades of green for low areas and shades of brown for mountains are common examples. The most useful way to show elevation is probably the contour line, and contour maps are in wide use.

A **contour** *is an imaginary line all points of which are at the same elevation.* A simple example is the shoreline of an island. If the island is in the ocean, its shoreline would be the zero-elevation contour line. If sea level rose, say, 10 feet, the new shoreline would be the 10-foot contour on our island. This illustrates an important point: contour lines always close on themselves, forming loops. These generally very irregular loops may not close on any one particular map, but if the map covers a large enough area, they will.

Another characteristic of contours is that they can never cross, because that would imply that the place where they cross has two elevations. At a vertical or overhanging cliff they would cross; but such places are very rare, and generally a different symbol is used to indicate a cliff.

The **contour interval** *is the vertical distance between successive contours.* It is generally stated in the legend or explanation at the bottom of the map and should not be confused with the scale of the map. Generally every fourth or fifth contour is a heavier line and is labelled with the elevation.

One problem with contour maps is that the elevation of a point between two contours is not indicated, except, of course, that it is higher than the lower contour and lower than the upper contour. This limitation is especially noticeable at hills and other peaks where the summit elevations are of interest. The solution is to show spot elevations at such places as peaks, road junctions, and other points of interest. The point is indicated by an **x**, and the elevation is given to the nearest meter or foot. Some of these places have a brass marker permanently fixed on the ground. Such markers are called *bench marks,* and they are indicated on the map by the letters BM preceding the elevation.

Learning to visualize the topography from the contours is not difficult. Figure C-1 shows a sketch of a landscape and the corresponding contour map. Where the contours are close together, the terrain is steep; where they are spread apart, the slope is gentle. The contour lines showing hills and ridges form loops; and at valleys, the contours are V-shaped, with the Vs pointing upstream.

Figure C-1
Topographic map and landscape. (After USGS.)

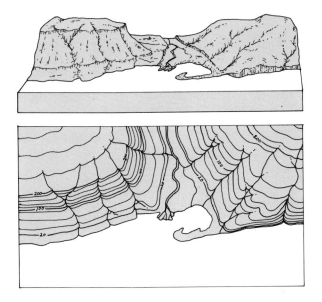

Most of the contour maps in use in the United States are produced by the United States Geological Survey. On these maps, the contour lines are printed in brown; the water features are in blue; roads, human-built features, and survey lines are in red and black; and trees, orchards, and woods are shown by green shading on some maps. USGS topographic maps of most of the United States and its possessions are available.

D. RELATIVE HUMIDITY AND DEW POINT TABLES

RELATIVE HUMIDITY

Dry-Bulb Temperature (°F)	Wet-Bulb Depression (°F) (Dry-bulb − wet-bulb temperature)																
	1	2	3	4	5	6	7	8	9	10	15	20	25	30	35	40	45
10	78	60	34	13													
15	82	67	46	29	11												
20	85	70	55	40	26	12											
25	87	74	62	49	37	25	13										
30	89	78	67	56	46	36	26	16	6								
35	91	81	72	63	54	45	36	27	19	10							
40	92	83	75	68	60	52	45	37	29	22							
45	93	86	78	71	64	57	51	44	38	31							
50	93	87	80	74	67	61	55	49	43	38	10						
55	94	88	82	76	70	65	59	54	49	43	19						
60	94	89	83	78	73	68	63	58	53	48	26	5					
65	95	90	85	80	75	70	66	61	56	52	31	12					
70	95	90	86	81	77	72	68	64	59	55	36	19	3				
75	96	91	86	82	78	74	70	66	62	58	40	24	9				
80	96	91	87	83	79	75	72	68	64	61	44	29	15	3			
85	96	92	88	84	80	76	73	70	66	62	46	32	20	8			
90	96	92	89	85	81	78	74	71	68	65	49	36	24	13	3		
95	96	93	89	86	82	79	76	72	69	66	52	38	28	18	8		
100	96	93	89	86	83	80	77	73	70	68	54	41	30	21	12	4	
105	97	93	90	87	84	80	78	74	72	69	56	44	34	24	15	8	1

WEATHER BUREAU BULLETIN 235, 1941.

DEW POINT TEMPERATURE IN DEGREES FAHRENHEIT

Dry-bulb Temperature (°F)	Wet-bulb Depression (°F) (Dry-bulb − wet-bulb temperature)															
	1	2	3	4	5	6	7	8	9	10	15	20	25	30	35	40
10	5	-2	-10	-27												
15	11	6	0	-9	-26											
20	16	12	8	2	-7	-21										
25	22	19	15	10	5	-3	-15	-51								
30	27	25	21	18	14	8	2	-7	-25							
35	33	30	28	25	21	17	13	7	0	-11						
40	38	35	33	30	28	25	21	18	13	7						
45	43	41	38	36	34	31	28	25	22	18						
50	48	46	44	42	40	37	34	32	29	26	0					
55	53	51	50	48	45	43	41	38	36	33	15					
60	58	57	55	53	51	49	47	45	43	40	25	-8				
65	63	62	60	59	57	55	53	51	49	47	34	14				
70	69	67	65	64	62	61	59	57	55	53	42	26	-11			
75	74	72	71	69	68	66	64	63	61	59	49	36	15			
80	79	77	76	74	73	72	70	68	67	65	56	44	28	-7		
85	84	82	81	80	78	77	75	74	72	71	62	52	39	19		
90	89	87	86	85	83	82	81	79	78	76	69	59	48	32	1	
95	93	93	91	90	89	87	86	85	83	82	74	66	56	43	24	
100	99	98	96	95	94	93	91	90	89	87	86	72	63	52	37	12

WEATHER BUREAU BULLETIN 235, 1941.

E. STAR MAPS

These maps show the sky at 8:30 PM, standard time, on March 22, June 22, September 22, and December 22. The appropriate map can be oriented for other dates and times by turning it so that an easily

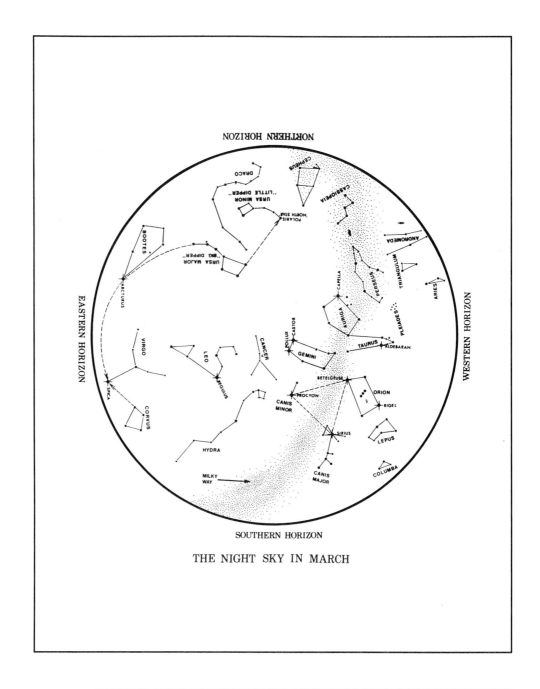

THE NIGHT SKY IN MARCH

These maps are from Griffith Observatory.

recognized constellation, such as the Big Dipper, is in the observed position. The maps were drawn for 34 degrees latitude but are useful between 25 and 50 degrees.

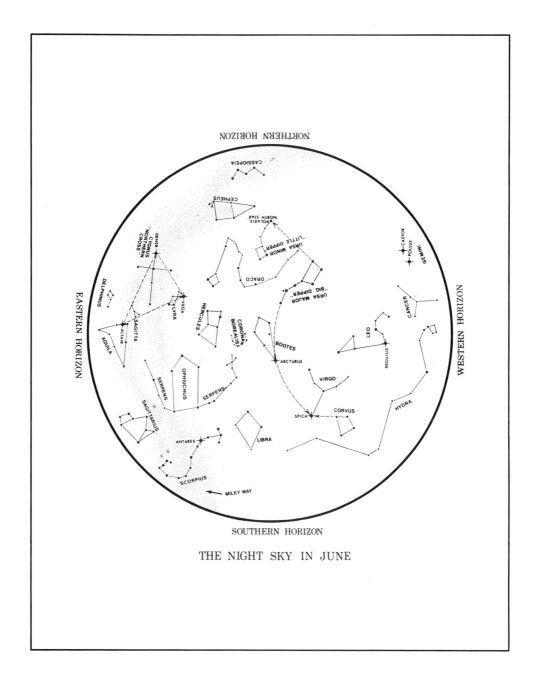

THE NIGHT SKY IN JUNE

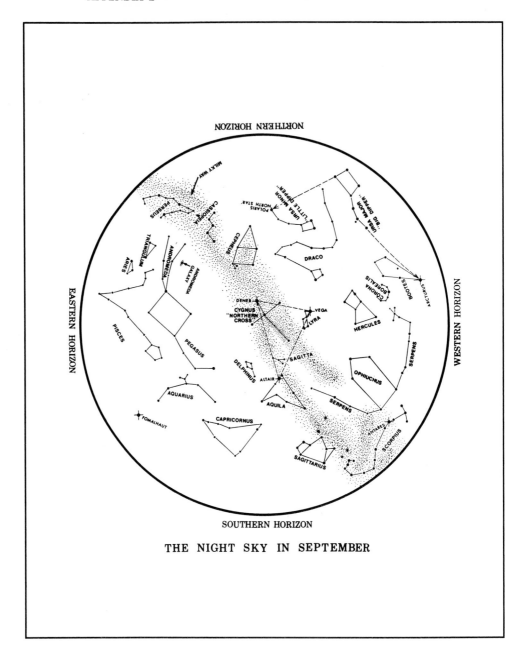

THE NIGHT SKY IN SEPTEMBER

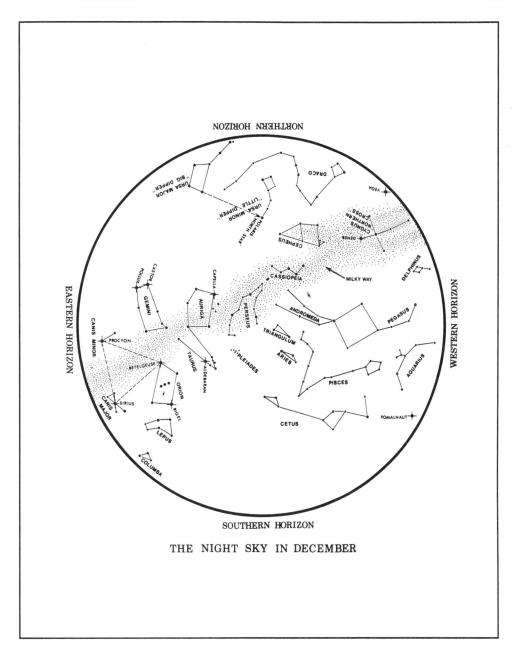

THE NIGHT SKY IN DECEMBER

GLOSSARY

abrasion. The wearing or grinding of bedrock by movement of the bed load of a river.

absolute humidity. The weight of water vapor in a given volume of air, in grams per cubic meter.

absolute magnitude. What the magnitude of a star would be if it were moved to a standard distance of 10 parsecs (32.6 light years) from earth.

absorption, or dark-line, spectrum. A continuous spectrum with some dark lines; the spectrum of a hot, solid object with the radiation passed through and partially absorbed by a cooler gas.

abyssal hill. A hill rising less than 300 meters (1000 feet) from the ocean basin floor.

abyssal plain. Smooth, flat surface on the ocean basin floor between the continental rise and the mid-ocean ridge.

adiabatic. Describing a process in which no heat is gained or lost by an air parcel.

adiabatic lapse rate. The temperature change in an air parcel that is rising or falling without gaining or losing heat.

advection fog. Fog that forms when moist air moves across a cooler surface.

air mass. A large volume of air that has uniform characteristics at any given altitude.

albedo. The amount of energy reflected by a surface.

alluvial fan. The deposit of river-carried sediment where a river flows out of mountains onto relatively flat plains.

alpine, or mountain, glacier. A glacier that forms in existing stream valleys in mountains.

alto-. A prefix used to designate members of the medium-height cloud family.

altocumulus cloud. A medium-height, puffy, cottonlike cloud.

altostratus cloud. A medium-height, generally gray, layered cloud, through which the sun can be seen.

amphibian. Vertebrate that has gills and lives in water while young but develops lungs as an adult. It is dependent on environment for its body temperature.

amphibole. One of a family of complex ferromagnesian silicate minerals. The common amphibole is *hornblende.*

analog forecast. A weather forecast based on similar situations in the past.

anemometer. An instrument that measures wind speed.

aneroid barometer. Without-liquid barometer. A small, sealed, partially evacuated chamber that changes size as the atmospheric pressure increases or decreases.

angiosperm. A plant that has true flowers and enclosed seeds.

angular momentum. Mass times spinning speed times radius.

angular unconformity. A break in the sedimentary record in which the beds below the unconformity are not parallel to those above.

anorthosite. Rock composed mainly of plagioclase, a feldspar mineral.

Antarctic circle. Latitude 66½ degrees south. The latitude marking the boundary of all-day or all-darkness at the solstices in the southern hemisphere.

anthracite. Hard coal. Relatively rare type of coal.

anticline. Hill-like fold. The oldest rocks are in the core of the fold.

anticyclone. A high-pressure system. In the northern hemisphere, the wind is clockwise; in the southern, counterclockwise.

aphelion. The point on earth's (or any planet's) orbit farthest from the sun.

Apollo asteroid. One of a group of about 25 asteroids whose orbits cross earth's.

apparent magnitude. The magnitude of a star as seen on earth.

aquifer, or **reservoir rock.** A rock that has enough well-interconnected pore space that it can store adequate water and permit it to flow into a well.

Arctic circle. Latitude 66½ degrees north. The latitude marking the boundary of all-day or all-darkness at the solstices in the northern hemisphere.

artesian well. A well in which the water rises in or flows from the well without pumping because the water is under pressure in the aquifer.

ash. Pyroclastic material up to the size of sand.

asteroid. One of a large number of small, planet-like bodies, whose orbits about the sun are mainly between Mars and Jupiter.

asthenosphere. The layer of the upper mantle below the lithosphere in which the rocks are not rigid.

Astronomical Unit (A.U.). The average distance between earth and the sun.

atmospheric pressure. Pressure that is caused by the weight of the air; the gravitational attraction between the air molecules and the earth.

atoll. A ring-shaped coral island or group of islands surrounding a central lagoon.

atom. The smallest particle of an element that retains the chemical properties of the element.

augite. The common member of the pyroxene mineral family, recognized by its black or dark green color and two cleavages at right angles.

aurora borealis and **aurora australis; northern** and **southern lights.** Colored lights in the skies, generally seen at high latitudes, that result from the particles from the solar wind passing through the atmosphere.

autumnal equinox. First day of autumn. September 21 or 22 in the northern hemisphere, and March 21 or 22 in the southern hemisphere.

bar. A unit of pressure; 100,000 newtons per square meter.

barrier island. Long, low, wave-built island off a coastline.

basalt. A common volcanic igneous rock, composed mainly of plagioclase feldspar and pyroxene.

base level. The elevation of the mouth of a river; thus the limit to which it can erode.

batholith. A large discordant intrusive igneous body that crystallized slowly so that it has a coarse texture.

bauxite. Aluminum ore formed by extreme weathering of soil.

baymouth bar. A sandbar that extends across a bay.

bedding. The layers in which sedimentary rocks form.

bed load. Large particles pushed or rolled by a river along its bed.

bedrock. The solid rock that underlies the unconsolidated surface layer of soil and weathered material.

bench or **terrace, wave-cut** or **surf-cut.** A surface cut by wave abrasion at a coastline.

big bang. A theory of the formation of the universe in which all galaxies were together and an explosive event started them all accelerating from that place.

binary stars. Two stars that revolve around a mutual center of mass. Double stars.

biogenous sediment. Oceanic sediment that consists of the remains of marine organisms.

biological oceanography. The study of the life of the oceans.

biologic sedimentary rock. A sedimentary rock that was formed or precipitated by organisms.

biotite. Black mica. A common mineral identified by its dark color and perfect cleavage.

bituminous coal. Soft coal. The most common coal.

black hole. A star so dense that not even radiation could escape its gravity field; the theoretical result of the collapse of a very massive supernova.

block, volcanic. Pyroclastic fragment larger than sand size.

bomb. A pyroclastic fragment that was still viscous when it was ejected.

brackish water. Water whose salinity is between salt and fresh.

breccia. Clastic sedimentary rock formed of angular fragments larger than 2 millimeters.

bright-line, or **emission, spectrum.** Spectrum of an energized gas, which emits a series of bright lines.

bright nebula. A nebula that shows light.

Buys Ballot's law. Law stating that if you are standing with your back to the wind, low pressure is to your left in the northern hemisphere, and to your right in the southern hemisphere.

calcite. A common mineral composed of calcium, carbon, and oxygen, recognized by three cleavages not at right angles, and by effervescence with dilute hydrochloric acid. The mineral of which limestone and marble are composed.

calorie. The amount of energy necessary to increase the temperature of 1 gram of water 1°C.

carbon dioxide (CO_2). A gaseous product of combustion of fossil fuel. It is a minor constituent of natural air, but increasing amounts of it may cause climatic change.

carbon monoxide (CO). The gaseous air pollutant introduced into the atmosphere in the largest quantities by human activity; the product of incomplete combustion.

centrifugal force. The force that tends to make a rotating body move away from the center of rotation.

cephalopod. One of the group of swimming predators with tentacles, whose shells evolved so rapidly as to make them index fossils in the Mesozoic Era.

cepheid variable star. A star whose luminosity varies with a regular period.

chalcedony. Very fine-grained quartz.

chalk. A type of limestone composed of shells of very tiny organisms.

change of state. A change in the molecular structure of a substance, as from a solid to a liquid, or a liquid to a gas.

chemical oceanography. The study of the chemistry of sea water.

chemical reaction. A reaction in which only the electrons that form a cloud around the nucleus of an atom are involved.

chemical sedimentary rock. A sedimentary rock that was precipitated directly from sea water.

chemical weathering. The rearrangement of the elements in a rock to form new minerals.

chert. A nonclastic sedimentary rock composed of very fine-grained silica.

chinook (western North America) or **foehn** (Alps). Warm, dry wind that flows downslope. Produced by a low-pressure system on the lee side of a mountain causing air to rush down to the low and to heat by compression.

chlorite. "Green mica," recognized by its color and perfect cleavage.

chromatic aberration. An inherent defect of lenses in which the light of different colors comes to focus at different distances from the lens.

chromosphere. Pink-glowing layer above the sun's photosphere, visible only during an eclipse.

chronometer. A clock that can measure time accurately; specifically used in navigation to determine longitude.

cinder cone. Relatively small, steep-sided volcanic cone formed from viscous pyroclastic material ejected from a single vent.

cirque. The amphitheater-shaped head of a glacially eroded valley.

cirrocumulus clouds. High clouds that resemble a layer of cotton balls. If similar in appearance to fish scales, known as "mackerel sky."

cirrostratus clouds. High, thin layers through which the sun can penetrate easily, although a halo may form.

cirrus cloud. Thin, wispy cloud. A member of the high-cloud family (mares' tails), or used as a descriptive term in the names of other high clouds.

clastic rock. Sedimentary rock composed of fragments broken from preexisting rocks.

clastic texture. Texture of a sedimentary rock composed of broken fragments.

clay. One of a family of complex minerals that result from weathering.

cleavage. The direction of weak bonding in a mineral along which the mineral tends to break, forming smooth, plane surfaces that reflect light.

climate. The typical combination of weather elements that has characterized an area over a long term, taking into account seasonal changes as well as ranges and extremes of weather elements.

cloud seeding. The treating of clouds with silver iodide or dry ice to make snow that melts, forming raindrops.

coal. Sedimentary deposit formed by alteration of buried plant material.

coalescence. The formation of raindrops by falling cloud droplets colliding and merging with other smaller cloud droplets.

cocoon star. Just-formed star, still surrounded by dust.

cold front. The boundary that develops between the air masses when a cold air mass moves into a region previously occupied by warm air.

cold-type occluded front. The boundary that develops between two air masses when a more rapidly moving cold front overtakes a warm front, and the invading cold air is colder than the cool air ahead of the original warm front.

coma of comet. A bright, diffuse glow surrounding the nucleus of the head of a comet.

comet. An object in the solar system with a very elliptical orbit; a small chunk of frozen gas and dust, consisting of a head and a very tenuous tail.

composite volcano, or **stratovolcano.** Large, cone-shaped volcano composed of both lava and pyroclastic material. Eruptions may be explosive.

conchoidal fracture. The rounded, concave fracture that forms when glass is broken.

concordant intrusive body. An igneous rock body that is more or less parallel to the bedding of the rocks into which it intrudes.

condensation. The process of changing into a denser state, as from gas into a liquid.

condensation nucleus. A surface on which water vapor can condense in the atmosphere.

conduction. Heat transfer in which the heat energy travels through a material by molecular collisions.

cone of depression. The cone-shaped depression of the water table around a well that has been pumped faster than water can flow into it.

conglomerate. Clastic rock formed of rounded fragments larger than 2 millimeters.

contact-metamorphic deposit. Ore deposit formed by reaction of magma and its fluids with intruded rocks.

contact, or **thermal, metamorphism.** Metamorphism that occurs at the boundary of intrusive igneous bodies.

continental drift. The movement of continents on the earth's surface.

continental glacier. A glacier that results from the buildup of an ice cap over a large area.

continental rise. The transitional area between the steep continental slope and the relatively flat ocean basin floor.

continental shelf. The geologically quiet shallow-ocean area at the margins of many continents.

continental slope. The relatively steep transition between the continental shelf and the ocean floor.

continuous spectrum. One in which all colors or wavelengths are present in a continuous display. Spectrum of a hot, solid object.

convection, or **convective flow.** Heat transfer by the rising of a warm, therefore less-dense, fluid, as cooler, denser fluid falls to take its place.

convergence. Moving of air inward into an area.

convergent boundary. The boundary between two crustal plates that are coming together. The collision of two plates.

coquina. A type of limestone composed of fossil shells.

core. The inner layer of the earth, consisting of a liquid outer core and a solid inner core, believed to be composed of iron and nickel.

Coriolis effect. The apparent deflection of a moving object because of earth's rotation beneath it as it moves.

corona. The layer of the sun above the chromosphere, visible without instruments only during an eclipse.

correlation. The determination that a rock layer exposed in one area is part of a similar layer exposed elsewhere, or that the two rock layers are of the same age.

cosmogenous sediment. Oceanic sediment consisting of the few particles, such as tiny meteorites, that are of extraterrestrial origin.

creep. Very slow downslope movement of surface material.

creep on fault. Very small movements of a fault that occur relatively frequently.

crescent moon. Shape of moon between new moon and first quarter, and between third quarter and new moon; less than a half-circle.

crevasse. A big crack in a glacier, opened by different rates of movement of the ice.

crust. The thin, outermost layer of the earth, above the Moho, composed of continental granitic rocks and oceanic basalt.

crystal. The form, with smooth faces, taken by a crystalline substance that grows in unrestricted space. The shape is the result of the organized internal structure of crystalline materials.

crystalline. Having an organized internal structure.

crystal settling. The formation of ore deposits by the sinking of dense, early-formed crystals to the bottom of still-liquid magma.

crystal systems. The classification of crystals into six systems based on the degree of symmetry of the crystals.

cumulonimbus cloud. Thunderhead. The biggest vertical cloud, typically with an anvil-shaped top.

cumulus cloud. Cotton-like cloud, generally with a flattish bottom and rounded top. A member of the vertical cloud family. Also used as a descriptive term in the names of other clouds.

Curie point. The temperature of a substance above which it loses its magnetism.

cycle of erosion. The stages of erosional development.

cyclone. A low-pressure system. In the northern hemisphere, the wind is counterclockwise; in the southern, clockwise.

daily, or **diurnal, tide.** One high and one low tide at the same place in one tidal day.

dark-line, or **absorption, spectrum.** A continuous spectrum with some dark lines; the spectrum of a hot, solid object with the radiation passed through and partially absorbed by a cooler gas.

dark nebula. A nebula in which the dust particles obscure the starlight.

daughter product. The element that results from the radioactive decay of an element.

debris slide. Rapid downslope movement of overburden that is broken off on a shear surface.

deep zone. The bottom layer of the water of the open ocean, in which the temperature and salinity of the water are quite uniform.

deflation. The lifting and carrying away of particles of sand or dust by wind.

delta. The deposit of river-carried sediment where a river meets the ocean.

desert armor, or **desert pavement.** A surface covered by the pebbles that were too large for the wind to remove.

dew. Condensation of water vapor directly from the air onto surfaces on or near the ground.

dew point. The temperature at which the air becomes saturated—that is, the temperature at which the relative humidity is 100 per cent.

diatomite. A nonclastic sedimentary rock composed of the silica fossils of very tiny organisms.

dike. A relatively thin discordant intrusive igneous body.

dinosaur. One of the group of dominant reptiles of the Mesozoic Era.

direct lifting. The lifting and carrying of small particles a distance downstream by turbulent river water.

discordant intrusive body. An igneous rock body that cuts across the bedding of the rocks into which it intrudes.

disseminated magmatic deposits. Ore deposits of magmatic origin in which the valuable mineral is scattered throughout the igneous body.

dissolved load. The material that a river carries in solution.

diurnal tide or **daily tide.** One high and one low tide at the same place in one tidal day.

divergence. Air moving outward away from an area.

divergent boundary. The boundary between two plates that are moving away from each other.

doldrums. Zone of light surface winds near the equator.

dolomite. A mineral composed of calcium, magnesium, carbon, and oxygen, recognized by effervescence with dilute hydrochloric acid only after powdering. *Also,* a nonclastic sedimentary rock composed of the mineral dolomite.

Doppler effect. Change in wavelength that occurs because of motion of either the emitter or the observer.

downslope movement, or **mass movement.** Erosional downhill movement, commonly of massive amounts of material.

drift. General term for all types of glacial deposits.

drizzle. Water drops less than 0.5 millimeter in diameter.

dry adiabatic lapse rate. The rate of cooling of a rising air parcel as long as no condensation occurs in that parcel; 1°C per 100 meters (5.4°F per 1000 feet).

dune. An accumulation of wind-deposited sand.

dwarf stars. Very dense stars that plot below the main sequence on an H-R diagram and that radiate less energy than main-sequence stars of the same spectral type.

dynamic metamorphism. Metamorphism that occurs when rocks are broken, sheared, or ground by movement.

earthflow. Downslope movement of wet surface material, characterized by a definite scarp at the head and a tongue at the foot.

earthquake. The vibrations caused by movements along a fault.

earthshine. Illumination of the moon by sunlight reflected from earth.

eclipse of the moon. The earth's shadow falling on the moon.

eclipse of the sun. The moon's shadow falling on the earth when the moon is directly between the sun and the earth.

eclipsing binary stars. Binary stars that pass in front of each other, causing the overall amount of light reaching us to change abruptly.

ecliptic. The apparent path of the sun in front of the stars.

Ekman spiral. The increased deflection of successive layers in the ocean by the Coriolis effect until at some depth the water will be moving in the opposite direction to the surface water.

electromagnetic radiation. Radiation that has electric and magnetic fields that vary rapidly, such as light, X rays, and radio waves.

element. The most fundamental subdivision of matter that can be made by ordinary chemical methods.

elliptical galaxy. Galaxy whose shape ranges from almost circular to a quite elongate ellipse; approximately 70 per cent of all galaxies are elliptical.

emergent coast. A coast where the continent was uplifted or sea level went down.

emission, or **bright-line, spectrum.** Spectrum of an energized gas, which emits a series of bright lines.

environmental lapse rate. The average rate at which temperature drops with elevation.

epicenter. The point on the surface directly above where an earthquake occurs.

epicycles. Smaller circles inside the orbits of planets in the Ptolemaic system.

equinox. The day that the earth's axis is at a right angle to a line between the centers of earth and sun. March 21 or 22, and September 21 or 22.

erosion. The wearing down of the crust.

erratic boulder. Boulder that is different from the local rock type that was transported to its location by a glacier.

estuary. A place where fresh water mixes with salt water, as at a river mouth and in a bay into which rivers flow.

evaporation. The process of changing state into a gas.

evaporation fog, or **sea smoke,** or **steam fog.** Fog that forms when cold, dry air moves slowly over a warm, moist surface.

evaporite. A sedimentary rock precipitated by the evaporation of water containing dissolved material.

extratropical cyclone, or **wave cyclone.** Cyclonic storm of the middle latitude; a low-pressure system containing both a warm front and a cold front. Opposed to *cyclone,* which only means "low-pressure system," and *tropical cyclone,* which only means "hurricane, or tropical storm without fronts."

extrusive igneous rock. An igneous rock that crystallized at the earth's surface.

eyepiece. The small telescope lens that enables the viewer to see the image gathered by the objective lens.

fall. Rapid downslope erosion by free fall, bouncing, and rolling.

fault. A break in rocks along which movement has occurred.

feldspar. A member of the most important and abundant family of silicate minerals.

ferromagnesian mineral. A silicate mineral containing iron and magnesium.

fetch. The distance over which the wind blows in creating ocean waves.

first-quarter moon. Moon seen as half-circle, rising at noon.

fissure eruption. Quiet eruption of very fluid lava from fissures. The resulting flows cover large areas.

flare. A large, violent cloud of gas, shot from the sun's surface, originating from a large, active sunspot.

flat-topped seamount, or **guyot.** A mountain rising more than 300 meters (1000 feet) from the ocean basin floor that has a flat top.

flood plain. The flat area on each side of a river in a broad valley; the area that is normally flooded during times of large river flow.

flow. Downslope erosion, generally of overburden, that behaves like a viscous fluid. The speed is influenced by the amount of water.

fluorescent nebula. A bright nebula whose gas absorbs some of a nearby hot, bright star's energy and reradiates it at a different wavelength.

focal length. Distance between the lens and the place where parallel light rays entering the lens form an image.

focus. The point in the earth where an earthquake occurs.

foehn (Alps) or **chinook** (western North America). Warm, dry winds that flow downslope, produced by a low-pressure system on the lee side of a mountain causing air to rush down to the low and to heat by compression.

fog. A cloud layer at or nearly at ground level.

fog drip, or **mist.** Moisture deposited on surfaces by fog or clouds moving past the surface.

folding. The bending of rock layers.

foliated metamorphic rock. Rock that has a layered structure because it crystallized under stress or directed pressure.

fossil. Evidence of life at the time the enclosing rocks formed.

fossil, or **remanent, magnetism.** The magnetism frozen in a rock.

fracture. The way that a mineral breaks in a noncleavage direction.

fracture zone. The area of irregular topography where a transform fault extends onto the deep ocean floor.

freezing rain and **freezing drizzle.** Rain and drizzle that freeze on reaching the ground.

frequency. Rate of variation of electric and magnetic fields in electromagnetic radiation.

front. The boundary separating two air masses of different temperatures.

frost. Condensation of water vapor on a surface at or near the ground if the temperature of condensation is below freezing.

frost point temperature. Temperature of condensation if it is below freezing.

full moon. Phase when moon is seen as a full circle, rising at sunset.

funnel cloud, tornado, or **twister.** The most destructive storm; a small, rapidly spinning body of rising air, associated with thunderstorms on active cold fronts.

galaxy. A very large grouping of stars held together by mutual gravitational attraction among the stars, usually also containing gas and dust clouds.

galaxy cluster. A group of galaxies.

Galilean moons. The four largest satellites of Jupiter, discovered by Galileo.

geocentric or **Ptolemaic system.** Belief that the earth is at the center of the universe.

geographic axis. The imaginary line connecting the north and south poles; the axis about which the earth spins each day.

geological oceanography. The study of the shape of the ocean bottoms and their rocks.

geologic time scale. A scale of relative time based on fossils.

geology. Science of the study of the solid rock earth.

geostrophic current. A current formed by water flowing around the low hill of water formed by the Ekman spiral, in which the force of gravity is balanced by the Coriolis effect.

geostrophic wind. A wind that moves parallel to the isobars.

geosyncline. A huge, elongate basin in which sediments are deposited; later deformed into a mountain range.

geothermal energy. Heat from within the earth.

geyser. A hot spring that erupts steam and water periodically.

giant stars. Stars that plot above the main sequence on an H-R diagram and that are very much larger than main-sequence stars with similar spectral types.

gibbous moon. Shape of moon between first quarter and full moon, and between full moon and third quarter; between a half-circle and a full circle.

glacier. A mass of moving ice.

globular star clusters. Very close groupings of 20,000 to several million old, generally red stars, composed of hydrogen and helium.

Glossopteris. A genus of fossil plants found on the southern continents, whose presence on all southern continents is interpreted as evidence for continental drift.

gneiss. A high-grade foliated metamorphic rock, containing alternating bands of feldspar and darker minerals.

Gondwanaland. The large continent that broke up to form the present southern continents.

grade, metamorphic. Relative intensity of metamorphic conditions such as temperature and pressure.

graded bedding. Layering of sediments in which the coarsest and densest particles settled to the bottom before the finer material did.

graded river. A river whose flow is balanced between its load and the factors, such as slope, amount of water, and velocity, that determine its ability to transport that load.

granite. A common coarse-grained igneous rock, containing mainly quartz and feldspar.

granules. The upper surfaces of convection cells on the surface of the sun's photosphere, lasting only a few minutes each.

great, outer, or **Jovian, planets.** Larger planets composed mainly of hydrogen and helium, with lesser amounts of ammonia, methane, and water ice. Jupiter, Saturn, Neptune, and Uranus.

greenhouse effect. The heating of a planet's atmosphere, certain gases of which are transparent to short-wavelength energy from the sun but opaque to longer-wavelength radiation from the planet, to the point that the planet's radiation can penetrate the atmosphere.

guyot or **flat-topped seamount.** A mountain rising more than 300 meters (1000 feet) from the ocean basin floor that has a flat top.

gypsum. A mineral that forms from the evaporation of saline waters.

gyre. Nearly closed loop in the map pattern of the ocean's surface currents.

hail. Ice, generally with a layered structure and a spherical shape, commonly 5 to 50 millimeters (0.1 to 1 inch) in diameter.

half-life. The unit of measurement of radioactive decay; the time it takes for one-half of the atoms of the parent element to decay into the daughter product.

halite. Rock salt. A mineral, recognized by its salty taste, that forms from the evaporation of saline waters.

halocline. The zone in the waters of the open ocean in which the salinity changes rapidly with depth.

hanging valley. A tributary valley that is higher than the main valley at their junction because of deeper glacial erosion of the main valley.

hardness. The measure of a mineral's resistance to scratching or abrasion.

haze. Condensation in the atmosphere when the relative humidity is less than 100 per cent, caused by the presence of water-seeking (hygroscopic) condensation nuclei.

heat. A form of energy. The heat of an object is proportional to the total motion, or kinetic energy, of all of its molecules.

heat lightning. Lightning too far away for the observer to hear the thunder.

heliocentric system. Belief that the sun is at the center of the universe.

hematite. An iron oxide mineral, recognized by its distinctive red to red-brown streak.

Hertzsprung-Russell diagram, H-R diagram. A plot of the relationship between absolute magnitude and spectral type (or temperature) of stars.

highland. Light-colored, relatively rough area of the moon's surface, higher in elevation than the maria. Terra.

horn. The pyramid-shaped peak formed by cirques on different sides of a glacially eroded mountain.

hornblende. The common member of the amphibole mineral family, recognized by its black or dark green color and two cleavages not at right angles.

hornfels. A thermal- or contact-metamorphic rock.

horse latitudes. Zones of light surface winds near 30° latitude.

hot spot. An area in the earth's mantle where conditions are right for the formation of magma, so that a line of volcanoes forms as the sea floor moves over the area.

hot spring. A spring formed of water that has circulated deep enough in the ground to be heated.

H-R diagram, Hertzsprung-Russell diagram. A plot of the relationship between absolute magnitude and spectral type (or temperature) of stars.

humidity. A measure of the amount of water vapor in the atmosphere.

humus. Plant material that has been partly decomposed by bacteria and molds in the soil.

hurricane; tropical cyclone; cyclone (Indian Ocean); **typhoon** (Pacific Ocean); **willy-willy** (Australia). A tropical storm without fronts that has wind speeds exceeding 120 kilometers per hour (75 miles per hour).

hydrocarbon. A compound containing only hydrogen and carbon; hydrocarbons are involved in production of photochemical smog.

hydroelectric energy. Energy from water power.

hydrogenous sediment. Oceanic sediment that is precipitated directly from sea water by chemical reactions.

hydrothermal deposit. Ore deposit, usually in quartz veins, formed by deposition of minerals by hot-water solutions.

hygrometer. The simplest and least accurate instrument used to measure relative humidity, using hair or a fiber that changes length.

hygroscopic nucleus. A surface on which water vapor can condense in the atmosphere, composed of water-seeking material such as salt.

ice crystal process. The formation of raindrops by the evaporation of the water droplets in the cloud and the deposition of the cloud's water vapor on the ice crystals in the cloud.

igneous rock. A rock that was once melted.

igneous texture. The size of mineral grains, generally controlled by the rate of cooling.

impact crater. Crater caused by the impact of materials such as meteorites.

index fossil. A fossil that by itself can establish the age of the containing rocks because it is sufficiently abundant, widespread, and of a short-enough time span.

industrial smog. The most serious type of air pollution, caused mainly by smoke and oxides of sulfur released by burning fossil fuels.

infrared radiation. Heat wavelengths. Radiation whose wavelength is somewhat longer than the wavelength of the red end of visible light.

inner or **terrestrial planets.** Those more or less similar to earth in size and composition. Mercury, Venus, Earth, and Mars.

insulator. A material that is a poor conductor.

intensity. A measure of how much damage an earthquake caused.

intrusive igneous rock. An igneous rock that crystallized below the earth's surface.

inverse square law. Law stating that the brightness of light varies inversely with the square of the distance of the source from the observer.

invertebrate. An animal without a backbone.

ion. Atom that has gained or lost one or more electrons.

irregular galaxy. A galaxy with no regularities. About 15 per cent of all galaxies are irregular.

isobar. A line drawn on a weather map through points of equal pressure.

isostasy. The concept that the lithosphere floats in the asthenosphere.

jet stream. Narrow river of high-speed wind aloft, moving from west to east.

Jovian, great, or **outer, planets.** Larger planets composed mainly of hydrogen and helium, with lesser amounts of ammonia, methane, and water ice. Jupiter, Saturn, Neptune, and Uranus.

kaolinite. A clay mineral.

karst topography. The landscape in an area with many caves, in which the surface is pockmarked with sinkholes.

kinetic theory. The assumption that all matter is composed of molecules that are in motion.

laccolith. A concordant intrusive igneous body that domes up the overlying rocks to some extent.

land breeze. A nighttime wind caused by the convective flow of warm air rising over the ocean and cooler air from land taking its place.

landslide. Rapid downslope movement of bedrock and overburden that are broken off on a shear surface.

lapse rate. The rate of decrease of temperature with height in the atmosphere.

late magmatic deposit. Ore deposit formed by crystallization of the fluid containing the valuable mineral late in the crystallizing of the magma.

latent heat. Heat required to cause a change in state of a substance; also called hidden heat because it does not cause the temperature of the substance to change.

lateral moraine. A deposit of material carried by a glacier along the glacier's side.

law of gravity. Law stating that every piece of mass in the universe attracts every other mass with a force that is proportional to the product of the masses and inversely proportional to the square of the distance between them.

levee. River-carried sediment that was deposited along the banks of a river when the flow slowed abruptly, as during flooding.

libration. Slight wobble of the moon.

lightning, or **lightning flash.** The spark that flows between the static electrical positive and negative charges that have been separated by the processes in a thunderhead when the electrical field between them becomes great enough.

light year. Distance light travels in a year.

lignite. Altered peat; a step in the development of coal.

limestone. A nonclastic sedimentary rock composed of calcite.

liquid metallic hydrogen. Liquid hydrogen that is tightly compressed so that the electrons can move freely through the liquid, which can thus conduct electricity like a metal.

liquid molecular hydrogen. Liquid hydrogen in which the electrons are held tightly by the molecules.

lithogenous sediment. Oceanic sediment that comes primarily from weathering and erosion of continental rocks and also from volcanic eruptions.

lithosphere. The layer of the earth above the low-seismic-velocity zone in which the rocks are rigid.

Local Group. The cluster of galaxies of which the Milky Way is a member.

locked fault. A fault on which forces have built up over a long period of time.

loess. Wind-deposited dust.

longshore current. The movement of water and the material it carries along a shore. Caused by the oblique approach of breaking waves and the return flow straight down the slope of the beach.

low-seismic-velocity zone. A zone in the earth's upper mantle in which seismic waves are slowed.

luminosity. A measure of how much energy a star radiates.

luster. The way that a mineral reflects light.

magma. Melted mineral material that may include suspended crystals and generally includes dissolved gases, mainly steam.

magnetite. An iron oxide mineral, recognized by being attracted by a magnet.

magnification of telescope. The focal length of the objective lens divided by the focal length of the eyepiece.

magnitude, earthquake. A measure of how much energy an earthquake released.

magnitude, stellar. A scale used to measure the brightness of stars.

main asteroid belt. The main orbital belt of asteroids between Mars and Jupiter.

main sequence. The diagonal band on the H-R diagram in which most stars plot.

mammal. A vertebrate that bears live young, produces milk to feed them, and controls its own body temperature.

manganese nodule. A potentially economic oceanic sediment, containing significant manganese and iron, precipitated directly from sea water.

mantle. The middle layer of the earth, from the Moho to the core, believed to be composed of peridotite.

marble. Metamorphosed limestone.

mare (pl., **maria**). Dark-colored, relatively smooth area of the moon's surface.

mass movement or **downslope movement.** Erosional downhill movement, commonly of massive amounts of material.

maturity. The second stage cycle of erosion, with no flat interstream divides and maximum relief.

M-discontinuity or **Mohorovičić discontinuity** or **Moho.** The change in seismic velocity that separates the crust from the mantle of the earth.

meanders. The more or less regular patterns of curves developed by a river as it widens its valley.

mechanical weathering. The breaking up of an existing rock into smaller fragments.

mercury barometer. The classical device for measuring atmospheric pressure, using mercury in a long tube sealed at one end inverted in a container of mercury. The atmospheric pressure on the mercury in the container balances the height of the column of mercury in the tube.

mesopause. The top of the mesosphere.

mesosphere. The third layer of the atmosphere, extending from the stratopause to about 85 kilometers, a zone of falling temperature.

metamorphic grade. Relative intensity of metamorphic conditions such as temperature and pressure.

metamorphic rock. Rock changed in mineralogy or texture or both by processes that occur without melting.

meteor. A streak of light in the sky, or "shooting star," caused by interplanetary material burning by friction in the earth's atmosphere.

meteorite. A fragment of interplanetary material that is not burned up by friction in a planet's atmosphere and so falls to the surface.

meteorology. The study of the atmosphere and its changes and movements.

meteor shower. A time when the number of meteors exceeds the average.

methane. A hydrocarbon (CH_4) produced naturally by decaying plant material.

mica. One of a group of complex silicate minerals characterized by one perfect cleavage.

micrometer. One-millionth of a meter.

mid-ocean ridge, or **rise.** A geologically active, continuous ridge extending throughout the central portion of the ocean floors, in some places high enough to form islands.

Milky Way. The galaxy in which our sun is a star.

millibar (mb). One one-thousandth of a bar; 100 newtons per square meter.

mineral. Naturally occurring, crystalline, inorganic substance, with a definite chemical composition.

mixed tide. Two high tides and two low tides, all of different heights, at the same place in one tidal day.

mixing zone or **surface zone.** The upper, least dense layer of the water of the open ocean; the warmest part of the ocean.

Moho, or **Mohorovičić discontinuity,** or **M-discontinuity.** The change in seismic velocity that separates the crust from the mantle.

Mohs hardness scale. The standard scale of mineral hardness.

monsoon. In summer, a moist wind that comes from the Indian Ocean over the warmer land, rising as it approaches the Himalaya Mountains; the resulting cooling produces heavy rainfall.

moraine. The deposit formed by the dropping of material carried by a glacier.

mountain, or **alpine, glacier.** A glacier that forms in existing stream valleys in mountains.

mountain wind. Wind produced by the nighttime cooling of highlands, causing the cool air to flow down the slopes.

mud crack. A sedimentary feature formed by periodic wetting and drying.

mudflow. Rapidly moving downslope movement of very wet surface material.

muscovite. White mica, recognized by its light or clear color and perfect cleavage.

neap tide. The high tide at first-quarter and third-quarter moon when the tidal forces of the sun and moon are at right angles; the lowest high tides.

nebula. An interstellar cloud of dust, gas, or both.

neutron star. Very dense star formed by the collapse of a supernova, in which the atomic nuclei collapse and the protons and electrons combine to form neutrons.

new moon. Phase when the moon is between the sun and the earth, illuminated only by earthshine and rising at sunrise.

nimbo-, -nimbus. Terms, used in describing clouds, meaning rainstorm.

nimbostratus cloud. A low, layered cloud from which rain or other precipitation is falling.

nonclastic sedimentary rock. Sedimentary rock formed from the products of chemical weathering by biologic or chemical precipitation, rather than from fragments broken from preexisting rock.

nonfoliated metamorphic rock. Metamorphic rock that has no obvious directional character.

normal fault. A fault caused by tension. The lateral area occupied by the rocks is enlarged.

normal remanent magnetism. Fossil magnetism in which the field is the same as the earth's present field.

northern and **southern lights, aurora borealis** and **australis.** Colored lights in the skies, generally seen at high latitudes, that result from the particles from the solar wind passing through the atmosphere.

nova. A star that becomes very much brighter for a short time as the result of an explosion.

nuclear reaction. A reaction in which the atoms are close enough together that interactions among the nuclei occur.

nucleus of comet. The bright point in the head of a comet.

numerical weather prediction. Prediction based on analysis of atmospheric data by computers programmed to apply physical laws to the behavior of the atmosphere.

objective lens. The large telescope lens that gathers light.

obsidian. Volcanic glass, formed by cooling so quickly that no crystals formed.

occluded front. The boundary that develops between the air masses when a more rapidly moving cold front overtakes a warm front.

oceanography. The study of the oceans.

old age. The stage in the cycle of erosion in which the interstream areas are low and rounded.

olivine. A silicate mineral containing iron, magnesium, silicon, and oxygen.

ooze. Oceanic sediment that consists of at least 30 per cent remains of marine organisms, generally named for the type of organism.

open star clusters. Irregular, loose groupings, generally of 100 to 1000 young, hot, blue stars, with some heavy elements and dust.

ore. Material that can be mined at a profit.

organized thunderstorms. Thunderstorms associated with fronts, forming in squall lines.

orographic lifting. Lifting of air parcels by wind that forces the air to rise over mountains.

orthoclase. A feldspar composed of potassium, aluminum, silicon, and oxygen.

outer, great, or **Jovian, planets.** Larger planets composed mainly of hydrogen and helium, with lesser amounts of ammonia, methane, and water ice. Jupiter, Saturn, Neptune, and Uranus.

outwash plain. Morainal material redeposited by meltwater from a glacier.

overburden, or **regolith.** The unconsolidated surface layer, generally consisting of soil and weathered bedrock.

oxides of nitrogen. Gaseous air pollutants, nitrogen oxide (NO), produced in high-temperature combustion, which reacts with oxygen to produce red-brown nitrogen dioxide (NO_2), a cause of photochemical smog.

oxides of sulfur. The most toxic and dangerous of gaseous air pollutants, sulfur dioxide (SO_2) and sulfur trioxide (SO_3), which react with water to form sulfuric acid (H_2SO_4).

ozone. An oxygen molecule composed of three oxygen atoms (O_3) instead of the two atoms (O_2) in a molecule of ordinary oxygen.

ozone layer, or **ozonosphere.** A part of the stratosphere, extending from about 10 to 50 kilometers (6 to 30 miles), in which the maximum concentration of ozone is about 5 parts per million at about 30 kilometers (18 miles).

parallax. The angle that is a measure of the apparent shift of the position of a close star relative to distant stars as the earth travels its orbit. Used to measure stellar distances.

parsec. The distance that has a PARallax of one SECond of arc. 3.26 light years.

peat. Partially decomposed plant material; the first stage in the formation of coal.

pedalfer. Acidic soil formed in relatively humid areas, in which iron and aluminum are deposited in the subsoil, and calcite and dolomite are removed.

pediment. The bedrock surface that has been cut during the erosion of desert mountains.

pedocal. Soil formed in relatively arid areas, in which calcite is deposited in the subsoil.

pegmatite. An extremely coarse-grained rock body.

peneplain. The rare ultimate stage of the cycle of erosion, in which the interstream areas are almost flat.

penumbra. Light part of moon's shadow during an eclipse of the sun, causing an area of partial eclipse.

perched water table. The saturation of ground with water whose downward percolation was stopped by a relatively impervious layer.

peridotite. A rock composed of olivine and pyroxene, believed to be the composition of the mantle.

perihelion. Closest approach of the earth (or any planet) to the sun during its orbital trip around the sun.

period, geologic. A division of the geologic time scale.

period of a wave. The time between successive crests.

permeability. A measure of how well the pore spaces of a rock are interconnected.

persistence forecast. A forecast used locally for short-range weather prediction.

phases of the moon. Regular changes in both the visible shape of the moon and the time of day it is visible.

photochemical reaction. A chemical reaction produced by the sun's energy.

photochemical smog. Air pollution caused by automobile exhaust gases, oxides of nitrogen and unburned hydrocarbons, reacting by means of the sun's energy to form irritating and damaging compounds.

photosphere. The layer of the sun that we see, a cloud-like layer that is the base of the sun's atmosphere.

physical oceanography. The study of the motions of sea water.

placer. Deposit of dense, resistant minerals, concentrated where the river or ocean-beach currents that carried them were slowed.

plagioclase. A feldspar composed of calcium, sodium, aluminum, silicon, and oxygen.

planetary circulation. The largest possible scale of movements of air.

plate tectonics. The process in which large, relatively thin units of the earth's crust move, causing the structures of the crust.

plucking or **quarrying.** The removal of generally big fragments of bedrock by a glacier.

plunging wave. A breaking wave in which the curling top of the wave traps air in a pocket, causing foam and splashing as it escapes.

polar front. The place where the westerly winds of the temperate areas meet cold polar air, generally between 40° and 60° latitude.

polar front jet stream. Narrow stream of wind aloft, flowing from west to east in the area of the polar front.

Population I stars. Young, hot stars with heavy elements and associated with dust clouds, generally in open clusters.

Population II stars. Older, generally cooler stars, composed mainly of hydrogen and helium, in globular clusters.

porosity. The amount of pore space (open space) between the grains of a rock.

precession. A slow wobble of the earth's axis.

pressure gradient. The measure of how much air pressure changes with distance.

prominence. Large cloud of gas shot up hundreds of thousands of kilometers above the sun's surface.

Ptolemaic system, or **geocentric system.** Belief that the earth is at the center of the universe; named for Ptolemy (second century A.D.).

pulsar. A radio source that radiates strong pulses of raio waves at regular intervals; believed to be a neutron star.

pumice. Very light volcanic rock formed by the froth caused by gas escaping from the magma.

P wave. Primary wave. The earthquake wave that arrives first at a point. A push wave, causing compression and rarefaction as it moves.

P-wave shadow zone. A zone on the earth's surface in which no P wave is received; caused by refraction at the boundary between the mantle and the core.

pycnocline zone or **transition zone.** The layer of water of the open ocean between the surface zone and the deep zone, in which the density of water changes rapidly with depth.

pyrite. A mineral, known as "fool's gold," recognized by its brass-yellow color and black streak.

pyroclastic rock. The explosive material ejected from a volcano.

pyroxene. One of a family of complex ferromagnesian silicate minerals. The common pyroxene is *augite.*

quarrying, or **plucking.** The removal of generally big fragments of bedrock by a glacier.

quartz. An abundant silicate mineral composed of only silicon and oxygen.

quartzite. Metamorphosed quartz sandstone.

quasar. Quasi-stellar radio source. A radio source that is a very distant, very energetic, very massive small object.

radial velocity. Velocity directly toward or away from the observer.

radiation. Transfer of energy by electromagnetic waves.

radiation fog. Fog caused when the ground surface loses its heat by radiation on a clear, calm night, and cool, heavy, moist air drains downslope to the low areas.

radioactivity. The process whereby certain elements change into other elements by the spontaneous emission of particles from their nuclei.

radio telescope. A large antenna that collects weak radio energy from space.

rain. Water drops over 0.5 millimeter in diameter, typically 1 or 2 millimeters in diameter.

rain gauge. An instrument used to measure precipitation in centimeters or inches.

red clay. Oceanic sediment consisting of clay-sized particles that sank so slowly that they were oxidized before they settled, giving the iron in them a rust color.

red shift. The shifting of the lines of a spectrum toward the red end, caused by the Doppler effect of an emitter moving away from the observer.

reflecting telescope. A telescope that uses a curved mirror to focus the light.

reflection nebula. A bright nebula in which the dust cloud reflects light from a nearby star.

refracting telescope. A telescope that gathers and focuses light through lenses.

refraction. The change in the direction of travel of a wave. The bending of light rays by a lens or prism.

regional metamorphism. Metamorphism that occurs over a large area.

regolith or **overburden.** The unconsolidated surface layer, generally consisting of soil and weathered bedrock.

relative humidity. Ratio of the amount of water vapor actually in the air to the maximum amount of water vapor that the air can hold, expressed as a percentage.

remanent or **fossil magnetism.** The magnetism frozen in a rock.

reptile. Vertebrate that breathes with lungs, lays shelled eggs, has dry, usually scaled skin, and is dependent on environment for its body temperature.

reservoir rock or **aquifer.** A rock that has

enough well-interconnected pore space that it can store adequate water and permit it to flow into a well.

retrograde motion or **spin.** Motion or spin in the opposite or backward direction. In the case of the solar system, the clockwise direction.

return strokes. The visible part of a lightning flash, consisting of a series of strokes that return to the original cloud after the stepped leader has ionized the path.

reverse fault. A fault caused by compression. The lateral area occupied by the rocks is reduced.

reverse remanent magnetism. Fossil magnetism in which the field is the reverse of the earth's present field; that is, fossil north is toward the present south pole.

revolution. Orbital trip of the earth (or any planet) about the sun.

rift valley. A valley formed near the crest of a mid-ocean ridge.

rille. Sinuous valley on the moon's surface.

rip current. The concentration of the return flow of water from a beach to the ocean in a narrow channel.

ripple mark. A sedimentary feature formed by current action.

Roche's limit. The distance from a planet within which a large moon cannot exist.

rock. Naturally occurring, firm material that forms part of the earth's crust.

rockfall. The fall of bedrock through the air from a cliff.

rockslide. Rapid downslope movement of bedrock that is broken off on a shear surface.

salinity. The amount by weight of dissolved material in sea water, usually expressed in parts per thousand.

sandstone. Clastic sedimentary rock composed of fragments between 2 millimeters and 1/16 millimeter.

Santa Ana wind. In southern California, a strong wind produced by warm, dry air from the high deserts moving west toward low pressure.

saturated or **wet adiabatic lapse rate.** The rate of cooling of a rising air parcel if condensation is occurring in that parcel; 0.6°C per 100 meters (3°F per 1000 feet).

scattering. The changing of the direction of a light ray by interaction with gas molecules and dust particles.

schist. A medium-grade foliated metamorphic rock, containing easily visible mica.

sea breeze. A daytime wind caused by convective flow of warm air rising over land and cooler air from the ocean taking its place.

sea-floor spreading. The process of formation of new oceanic crust at the mid-ocean ridge, pushing the older crust aside away from the ridge to make room.

seamount, or **submarine mountain.** A mountain rising more than 300 meters (1000 feet) from the ocean basin floor.

sea smoke or **evaporation fog** or **steam fog.** Fog that forms when cold, dry air moves slowly over a warm, moist surface.

second. Basic unit of time, originally based on a solar day: $1/60 \times 60 \times 24$, or $1/86{,}400$ of an average solar day. Now defined using vibration frequency of cesium atoms.

second of arc. An angular measure. A circle has 360 degrees, each degree is divided into 60 minutes, and each minute is divided into 60 seconds.

sedimentary rock. Rock that formed at the earth's surface, generally in layers or beds.

seismology. The study of vibrations in the earth caused by earthquakes or explosions.

semidaily, or **semidiurnal, tide.** Two high tides of about the same height, and two low tides of about the same height, at the same place in one tidal day.

shale. Clastic sedimentary rock composed of clay particles (less than 1/256 millimeter).

shield volcano. A volcano formed by quiet flows of fluid lava from a single vent.

sidereal day. The period between successive times that a given star is at the same point in the sky.

sidereal month. Period of moon's rotation around earth measured relative to the stars.

silicate mineral. A mineral whose structure is based on the silicon-oxygen unit of one silicon atom surrounded by four oxygen atoms.

sill. A relatively thin concordant intrusive igneous body.

siltstone. Clastic sedimentary rock composed of fragments between 1/16 and 1/256 millimeter (silt).

sinkhole. A depression formed by the collapse of the roof of a cave.

slate. A foliated metamorphic rock formed by low-grade metamorphism of shale.

sleet. In American usage, rain or drizzle that was frozen before reaching the ground; also, partially melted snow or a mixture of rain and snow.

slide. Downslope erosion by sliding, generally on a shear surface.

sling psychrometer. An instrument to measure relative humidity, using wet-bulb and dry-bulb thermometers to compare the temperatures after evaporation from the wet bulb.

slump. Rapid slide in which the bedrock remains intact and is only rotated and moved downslope.

smog. A contraction of smoke and fog.

snowflake. Aggregate of ice crystals that can be up to several centimeters in diameter.

snowgrains. Small ice crystals; the solid equivalent of drizzle.

soil. The material at the earth's surface that supports plant life.

solar day. The period from (for example) one noon until the next noon.

solar energy. Energy directly from the sun.

solar wind. An extension of the corona, consisting of high-energy particles of material ejected from sunspots.

solifluction. Downslope flow of wet surface material above frozen ground.

solstice. The day that either pole is most tilted (23½°) toward the sun. June 21 or 22, and December 21 or 22.

specific gravity. The ratio of the weight of a material to the weight of the same volume of water.

specific heat. A measure of how much heat is required to raise 1 gram of a material 1°C; thus a measure of how much heat a material can absorb.

spectral types of stars. A classification of stars based on differences in their spectra, which correspond to their temperatures. From hottest to coolest: O B A F G K M.

spectrographic binary stars. Binary stars whose rotation is detected by either a change in spectral type or from Doppler shift.

spectrum. A display of electromagnetic radiation spread out by frequency or wavelength.

spicule. Incandescent gas shot above the surface of the sun.

spilling wave. A breaking wave in which the oversteepened front spills forward, reducing its height, as the wave continues to advance.

spiral galaxy. Galaxy with arms and a bright central region; about 15 per cent of all galaxies are spirals, including the Milky Way.

spit. A deposit, generally of sand, extending from a headland.

spring tide. The high tide at new and full moon when the tidal forces of the sun and moon are additive; the highest high tides.

squall line. A line of thunderstorms that may precede an active, fast-moving cold front in a wave or extratropical cyclone. It may have associated showers, hail, or tornadoes.

stable air. A rising air parcel that tends to return to its original position.

stable star. A star in which the pressure of the outward-moving energy balances the gravitational contraction.

stalactite. A deposit formed on the roof of a cave by evaporation of water containing dissolved calcite.

stalagmite. A deposit formed on the floor of a cave by evaporation of water containing dissolved calcite.

star clusters. Groupings of stars within galaxies.

stationary front. A boundary between warm and cold air masses that does not move appreciably.

statistical forecast. A weather forecast based on the most probable weather for the situation.

steam fog, or **evaporation fog,** or **sea smoke.** Fog that forms when cold, dry air moves slowly over a warm, moist surface.

stepped leader. The almost invisible beginning of a lightning flash, in which the negative charge moves downward in a series of advances.

storm surge. A very high tide caused by the combination of strong onshore winds and reduced pressure at a hurricane center.

stratocumulus cloud. A low cloud with a flat bottom and a rounded, puffy top; may appear as a long ridge.

stratopause. The upper boundary of the stratosphere.

stratosphere. The second layer of the atmosphere, from the tropopause to about 50 kilometers (30 miles).

stratovolcano, or **composite volcano.** Large, cone-shaped volcano composed of both lava and pyroclastic material. Eruptions may be explosive.

stratus cloud. Layered cloud; a member of the low-cloud family. Also used as a descriptive term in the names of other clouds.

streak. Color of a powdered mineral.

subduction zone. The place where one plate moves downward under another.

submarine canyon. A deep, commonly V-shaped valley cut into the continental slope.

submarine mountain, or **seamount.** A mountain rising more than 300 meters (1000 feet) from the ocean basin floor.

submergent coast. A coast where the continent was depressed or sea level rose.

summer solstice. First day of summer. June 21 or 22 in the northern hemisphere, and December 21 or 22 in the southern hemisphere.

sunspot. A cooler and therefore darker area on the sun, occurring in cycles, that is associated with magnetic activity on the sun.

supercluster of galaxies. A group of clusters of galaxies.

supercooled water. Undisturbed pure water that is at a temperature well below its freezing point yet not frozen.

supergiants. Giant stars brighter than type O.

supergranules. Larger patterns in the granules on the surface of the sun's photosphere.

supernova. The result of an explosion when a star with a mass more than 30 times that of the sun collapses; the most energetic stellar explosion known.

superposition. The principle that the oldest rocks are at the bottom and the youngest at the top of an undisturbed section of sedimentary rocks.

supersaturated. A condition in which the relative humidity is more than 100 per cent and the temperature is below the dew point; caused by the lack of surfaces on which the water vapor can condense.

surface wave. The earthquake wave that travels along the surface. The slowest of the waves but the one with the most amplitude; the most damaging wave.

surface zone, or **mixing zone.** The upper, least dense layer of the water of the open ocean; the warmest part of the ocean.

surf-cut, or **wave-cut bench,** or **terrace.** A surface cut by wave abrasion at a coastline.

suspended load. Fine particles carried by a river that, if abundant, cause the water to look muddy.

S wave. Secondary wave. The earthquake wave that arrives second at a point. A shear wave.

S-wave shadow zone. A zone on the earth's surface in which no S wave is received because a shear wave cannot pass through the liquid core.

swell. Large, long-period wave after it has left its source area.

syncline. Basin-like fold. The youngest rocks are in the core of the fold.

synodic month. The period of the moon's rotation around the earth, measured, for example, as full moon to full moon.

synoptic chart. A chart that is a snapshot of the weather at a particular moment.

system. The rocks deposited during a geologic period.

talus. The surface formed by the accumulation of fallen fragments below a cliff.

temperature. A measure of the intensity of heat. The temperature of an object is proportional to the average energy of each of its molecules.

temperature inversion. A situation in which cold air is near the surface and the temperature increases in the inversion layer; above it, the temperatures are normal and fall with increasing elevation.

terminal moraine. A deposit of material carried by a glacier at the glacier's snout.

terra (pl., **terrae**). Light-colored, relatively rough highland area of the moon's surface.

terrace, surf-cut or **wave-cut.** A surface cut by wave abrasion at a coastline.

terrestrial, or **inner, planets.** Those more or less similar to earth in size and composition. Mercury, Venus, Earth, and Mars.

thermal, or **contact metamorphism.** Metamorphism that occurs at the boundary of intrusive igneous bodies.

thermocline. The zone in the waters of the open ocean in which the temperature changes rapidly with depth.

thermohaline circulation. Circulation of the deep water of the ocean, principally controlled by temperature and salinity, which control the density of the water.

thermosphere. The upper zone of the atmosphere, extending for several hundred kilometers, in which temperature increases but the atmosphere is very thin.

third-quarter moon. Moon seen as a half-circle, rising at midnight.

thrust fault. A reverse fault in which the fault plane is nearly horizontal. The movement on thrust faults is often large.

thunder. The sound of the explosive expansion of air caused by the very high temperatures created by lightning.

thunderstorm. The most widespread severe storm, formed typically when differences in surface heating are greatest, and when warm, moist air is forced to rise by mountains or a cold front.

tidal day. The interval between two overhead passages of the moon at the same place, about 24 hours 50 minutes.

till. Ice-deposited sediment.

time scale, geologic. A scale of relative time based on fossils.

tombolo. A sandbar extending between an island and the shore.

tornadic waterspout. A tornado that has moved from land to water; it can be quite destructive.

tornado, twister, or **funnel cloud.** The most destructive storm; a small, rapidly spinning body of rising air, associated with thunderstorms on active cold fronts.

tornado warning. The alert given when a tornado is actually reported.

tornado watch. The alert given when the severe thunderstorms that can cause tornadoes are noted on radar screens.

trade wind. Surface wind blowing generally toward the west from 30° latitude toward the equator.

transform fault. The boundary between two plates moving past each other.

transition zone, or **pycnocline zone.** The layer of water of the open ocean between the surface zone and the deep zone, in which the density of water changes rapidly with depth.

trench. A deep, long depression of the ocean floor, associated with a volcanic arc. Trenches are the deepest places in the oceans.

tropical cyclone, hurricane, cyclone (Indian Ocean); **typhoon** (Pacific Ocean); **willy-willy** (Australia). A tropical storm without fronts that has wind speeds exceeding 120 kilometers per hour (75 miles per hour).

Tropic of Cancer. Latitude 23½° north. The latitude at which the sun's rays are perpendicular to earth at the northern hemisphere's summer solstice, June 21 or 22.

Tropic of Capricorn. Latitude $23\frac{1}{2}°$ south. The latitude at which the sun's rays are perpendicular to earth at the southern hemisphere's summer solstice, December 21 or 22.

tropopause. The top of the troposphere.

troposphere. The lowest part of the atmosphere, in which the temperature falls on average about 0.65°C per 100 meters (3.5°F per 1000 feet) as one rises.

tsunami. Wave generated by an earthquake.

tuff. Rock formed from volcanic ash.

turbidity current. A mixture of sediment and water that, because it is denser than sea water, moves downslope along the bottom.

twister, tornado, or **funnel cloud.** The most destructive storm; a small, rapidly spinning body of rising air, associated with thunderstorms on active cold fronts.

typhoon (Pacific Ocean); **hurricane; tropical cyclone; cyclone** (Indian Ocean); **willy-willy** (Australia). A tropical storm without fronts that has wind speeds exceeding 120 kilometers per hour (75 miles per hour).

ultraviolet radiation. Radiation whose wavelength is somewhat shorter than the wavelength of the violet end of visible light.

umbra. Dark part of the moon's shadow during an eclipse of the sun, causing area of total eclipse.

unconformity. A break in the sedimentary record.

uniformitarianism. The principle that processes now active on earth are the same processes that have always operated.

unstable air. A rising air parcel that tends to continue to rise.

urban heat island. The prevailing higher temperatures in a city than in the surrounding countryside.

valley breeze. Wind produced by the sun's heating of a valley's sides, causing the warmed air to rise.

Van Allen radiation belts. Two belts around the earth that result from the interaction of the high-energy particles of the solar wind with the earth's magnetic field.

variable star. A star whose energy radiation varies.

ventifact. A pebble with smooth faces cut by wind abrasion.

vernal equinox. First day of spring. March 21 or 22 in the northern hemisphere, and September 21 or 22 in the southern hemisphere.

vertebrate. An animal with a backbone.

visual binary stars. Double stars that can be distinguished with the naked eye or with telescopes.

volcanic arc. A line of active volcanoes along a continental margin.

volcanic island arc. A line of active volcanoes forming islands that have a curved map pattern.

warm front. The boundary that develops between the air masses when a warm air mass moves into an area previously occupied by cold air.

warm-type occluded front. The boundary that develops between two air masses when a more rapidly moving cold front overtakes a warm front, and the invading cold air is warmer than the cold air ahead of the original warm front.

waterspout. A small, rapidly spinning body of rising air that tends to occur over warm, shallow water during fair weather; not so destructive as a tornado.

water table. That level in the ground below which the ground is saturated with water.

wave-cut, or **surf-cut bench** or **terrace.** A surface cut by wave abrasion at a coastline.

wave cyclone, or **extratropical cyclone.** Cyclonic storm of the middle latitude; a low-pressure system containing both a warm front and a cold front, as opposed to *cyclone*, which only means "low pressure system," and *tropical cyclone*, which only means "hurricane, or tropical storm without fronts."

wavelength. The distance between successive crests of a wave.

weather. Short-term changes in the atmosphere.

weathering. The changes that take place in a rock as the result of the rock's exposure to conditions at the earth's surface.

welded tuff. A pyroclastic rock formed from hot ash that welded together when it landed.

wet, or **saturated, adiabatic lapse rate.** The rate of cooling of a rising air parcel if condensation is occurring in that parcel; 0.6°C per 100 meters (3°F per 1000 feet).

willy-willy (Australia); **hurricane; tropical cyclone; cyclone** (Indian Ocean); **typhoon** (Pacific Ocean). A tropical storm without fronts that has wind speeds exceeding 120 kilometers per hour (75 miles per hour).

wind. The movement of air parallel to the surface. Winds are named for the direction from which they come.

wind vane. An instrument that measures wind direction.

winter solstice. First day of winter. December 21 or 22 in the northern hemisphere, and June 21 or 22 in the southern hemisphere.

year. The time for earth to complete one revolution around the sun.

youth. The first stage in the cycle of erosion, characterized by flat interstream divides and intrenched rivers.

zodiac. A zone about 9° on each side of the ecliptic that is the apparent path of the planets and the moon across the sky.

INDEX